Science and Society

The History of Modern Physical Science in the Twentieth Century

Edited with introductions by

Peter Galison
Harvard University

Michael Gordin
Harvard University

David Kaiser
Massachusetts Institute of Technology

Routledge
Taylor & Francis Group
LONDON AND NEW YORK

Series Content

Volume 1
MAKING SPECIAL RELATIVITY

Volume 2
MAKING GENERAL RELATIVITY

Volume 3
PHYSICAL SCIENCE AND THE LANGUAGE OF WAR

Volume 4
QUANTUM HISTORIES

Science and Society

The History of Modern Physical Science in the Twentieth Century

Volume 1

Making Special Relativity

Edited with introductions by

Peter Galison
Harvard University

Michael Gordin
Harvard University

David Kaiser
Massachusetts Institute of Technology

Routledge
Taylor & Francis Group
LONDON AND NEW YORK

Published in 2001 by
Routledge
Taylor & Francis Group
711 Third Avenue,
New York, NY 10017

Published in Great Britain by
Routledge
Taylor & Francis Group
2 Park Square, Milton Park,
Abingdon, Oxon OX14 4RN

Routledge is an imprint of the Taylor & Francis Group, an Informa business

Library of Congress Cataloging-in-Publication Data

Science and society: the history of modem physical science in the twentieth century / edited with introduction by Peter Galison, Michael Gordin, and David Kaiser.
p.cm.
Includes bibliographical references.
Contents: v. 1 Making special relativity-- v. 2 Making general relativity-- v. 3 Physical science and the language of war --v. 4 Quantum histories.

1. Physics--History--20th century. I. Galison, Peter Louis. II. Gordin, Michael. III. Kaiser, David.

QC7.5 .S39 2001
530'.09'04-dc21

2001038718

9780415937146 SET ISBN
9780415532181 POD ISBN
9780415937153 V1
9780415937160 V2
9780415937177 V3
9780415937184 V4

Contents

Volume Introduction

For almost the entire twentieth century Albert Einstein's paper "On the Electrodynamics of Moving Bodies" has riveted physicists, historians, mathematicians, philosophers, artists, and the public. Published in 1905, Einstein's *anno mirabilis*, the paper remains a magnet for commentary. To find a scientific work of comparable cultural fascination, one would have to turn to Charles Darwin's *Origin of Species* or Sir Isaac Newton's *Principia Mathematica.* Written in a lucid, almost discursive style, Einstein's relativity paper began with his apparently elementary reflections on a coil, a magnet, and their mutual interactions. He went on to reflect on the nature of space and time, how they ought to be measured, and how, from the behavior of clocks, rulers, and light signals, one could fix ideas about simultaneity. From these simple beginnings, Einstein launched a new kind of physics, upending attempts to find an electrodynamic basis for inertial mass and kinematics more generally. He started his mechanics and electrodynamics from a new kinematics—from the force-free behavior of moving bodies. Only then did he proceed to reinterpret the meaning of James Maxwell's equations and, in the follow-on paper, introduce that most famous of equations: $E=mc^2$.

Because special relativity has so powerfully seized the imagination of many generations, and because its ideas are so tightly woven into the subsequent development of so many branches of the physical sciences, several questions arise simultaneously. What cultural, technical, and scientific resources did Einstein bring to bear on this work? How has the work itself been interpreted and understood across different technical, national, and subnational cultures over the decades following 1905? The papers assembled in this volume touch on these and other questions, building, necessarily, on many years of Einstein scholarship. Although it is impossible to reproduce the full gamut of first-rate scholarship on special relativity, the papers chosen here provide an excellent entry into this work. In addition to the many works referred to in the individual papers, the article by David Cassidy offers a fine bibliography to many other works, including foundational work by Russell McCormmach, Tetu Hirosige, Martin Klein, and, of course, the indispensable early volumes of Einstein's complete works edited by John Stachel and his collaborators.

On first reading, Einstein's "On the Electrodynamics of Moving Bodies" appears to issue, like Athena from the head of Zeus, full grown. It contains no substantial reference to other works; it seems isolated from the world around it. Einstein, however, was anything but an isolated sport of culture. Deeply interested in philosophy, for example, he and his coterie of friends read intently if eclectically across a wide range of literary, philosophical, and scientific works. Gerald Holton, in the third article of this volume, explores some of those cultural resources that "resonated with and conditioned [Einstein's] science." Einstein's interest in classical and romantic authors began long before university as he made his way through a *Gymnasium* curriculum that included Friedrich von Schiller, Johann Wolfgang von Goethe, and Johann Gottfried von Herder, as well as Cicero, Caesar, Ovid, and Virgil. One does well to bear in mind, for example, that Einstein's lifelong interest in Immanuel Kant began with his immersion in the *Critique of Pure Reason* at the age of thirteen, an engagement that he maintained throughout his career.

What Holton makes of these resources is this: Einstein's triad of papers in 1905 (on Brownian motion, the light quantum, and special relativity) all have a similar form. Each begins with a perceived asymmetry, a gap, that Einstein renders troubling before he attempts to resolve it. For example, the relativity paper opens by highlighting an asymmetry that bothered Einstein in the explanations usually invoked between what happens when a moving magnet approaches a coil at rest (in the ether) and when a coil approaches a magnet at rest (in the ether). By the end of the paper, Einstein, having dispensed with the ether and presented new quasi-operational definitions of space and time, concludes that there is only one movement (magnet and coil approach each other) and to that single relative movement, a single, unified explanation. In the cultural roots of Einstein's striving for unity, Holton sees his commitment to certain literary and philosophical texts of Goethe. The Goethean-Faustian search for the Sign of the Macrocosmos became for many physicists, including Ludwig Boltzmann, a Sign of the Mathematical Physics in the form of Maxwell's equations. For Einstein that unity led to a lifelong search for new equations and interpretations that would render physics whole. Closer to the physics, Holton, in his "Search for Reality" essay, tracks Einstein's gradual drift away from Ernst Mach. Although Einstein began his career attracted by a Machian program of stamping out that which could not be experienced, through rethinking his early work, and then through general relativity, he came to vest the physical universe with a Platonic mathematical structure.

If Einstein appropriated various pieces of literary and philosophical culture, artists, authors, and philosophers seized on his scientific work with a matched enthusiasm, some to attack, others to applaud. In passing, Thomas Glick notes some of the many uses to which Einstein's relativity was put: from Catalan nationalism and Mussolinian fascism to Marxist dialectics. To the Argentinean generals of the 1980s Einstein's theory had undermined space and time as surely as Sigmund Freud and Karl Marx had wormed their way through the sturdy foundations of family and society. German Nazis lambasted the theory as Jewish physics in the 1920s, 1930s, and 1940s. While some American physicists deplored relativity as undemocratic, certain French writers recoiled from its Germanness. Enabling this "slide" from the scientific into the broadly cultural and back was, Glick argues, a partially shared world of concepts—including time, space, and energy.

As immersed as Einstein was in the wider world of late-nineteenth-century German *Kultur*, he was also standing in the midst of a vast turn-of-the-century electrodynamic upheaval. Telegraphs, electric streetcars, electric lighting, generator plants, and transmission wires cut through the coal and steam landscape of the industrial revolution. Einstein was in the thick of it—his father and uncle ran an electro-technical company, and he himself was working at the Bern patent office, where he had been hired to evaluate electromagnetic inventions. Within the study of electrodynamics no question of mathematical physics loomed larger at the beginning of the twentieth century than the problematic relation between Maxwell's equations and how those equations ought to be modified for matter flowing through the world ether.

We are aided here by a profusion of studies on the state of turn-of-the-century electrodynamics. The best of these have captured the self-appointed tasks of the various approaches rather than depict the work of anyone other than Einstein as simply an obstacle to the effective promulgation of special relativity. Reckoning with the scientific heterogeneity in these technical subcultures allows one to see more precisely how Einstein appropriated, used, and partly rejected the electrodynamics of Hendrik Lorentz and Henri Poincaré, as Arthur Miller, Olivier Darrigol, and Michel Janssen argue. Conversely, Stanley Goldberg, Lewis Pyenson, Glick, and Andrew Warwick analyze how others, within their own trajectories of work, seized, modified, and selectively ignored the work of Einstein.

No physicist was more important to Einstein than Lorentz. Father figure, adviser, confessor, Lorentz offered Einstein a model of being a physicist as well as a specific physics that widened his conception of the world. Casting aside the reductionist picture that saw the ether as the fundamental entity of which all else was to be constructed, Lorentz populated the world with material, charged electrons, and the fields with which they interacted. Out of these building blocks he and his followers aimed to build a theory of matter, one able to cope with such difficult problems as optical, spectral, and electromagnetic effects. Faced with the striking difficulty of establishing motion with respect to the ether, Lorentz postulated, first approximately and then precisely, a compensatory contraction of matter that would balance the slowing down of light in the ether wind. In this way, beginning in 1895, he hoped to explain the surprising null result of experiments, like Albert Michelson and Edward Morley's, that aspired to measure our absolute velocity through a universal ether. Bit by bit Lorentz improved on his initial gambit, altering the theory to account for more accurate and more diverse experiments.

For many years historians, philosophers, and physicists used the contrast between Einstein and Lorentz as a modern instantiation of a somewhat parodically rendered transition from Ptolemy to Copernicus. Janssen contends here that such an analogy fails on both legs: just as late-twentieth-century scholarship has completely debunked the picture of a radical Copernicus overthrowing the sterile Ptolemaic program, it is futile to paint the contrast between Lorentz and Einstein as a single revolutionary shift in scientific program, conceptual scheme, or paradigm. Lorentz's efforts were directed toward an explanation of the nature of matter. In this respect, his theories were "constructive," in Einstein's sense. Lorentz's program was eventually superseded not by special relativity but by quantum mechanics and condensed matter physics. Einstein's 1905 paper, by contrast, as Miller stresses, was not so much a theory of matter per se as a theory of principle that would guide and constrain future theories of matter. And it was not any predictive advantage of special relativity that gave it an edge on the final Lorentz theory. Instead, as Janssen argues, it was the bringing together of diverse phenomena under a single explanatory framework that gave relativity its persuasive power: common cause rather than simplicity or heuristic power.

Unlike Einstein, Poincaré worked on Lorentz's scheme. But Poincaré transformed the theory as he used it—synthesizing and re-expressing the work of Oliver Heaviside, Hermann Helmholtz, Maxwell, Sir Joseph Larmor, and Hertz as well as Lorentz himself. Whereas Einstein withheld all reference to other physicists in his "On the Electrodynamics of Moving Bodies," Poincaré produced whole treatises adumbrating, reconciling, and reinterpreting the works of others.

An important example: as Miller and Darrigol show in their contributions to this volume, for Lorentz the idea of "local time" was not at all physical. It was a calculational device, a fictional new variable t', useful for transforming the effects in an artificial moving frame back into the real ether-rest reference frame. But when Poincaré took up Lorentz's local time, he did so explicitly and re-expressed it as a physical, "apparent" time; Darrigol makes it clear that by 1900 when Poincaré spoke of local time, he had in mind the *measurable* passage of time in a moving frame. Added to the Lorentz contraction and supplemented by various assumptions about the dynamics of electrons and the forces that held matter together, Lorentz and Poincaré assembled an empirically powerful if awkward theory that developed from month to month. Poincaré was not satisfied with this state of affairs: "Of hypotheses there is never a lack," he famously quipped. Principles were needed, principles like that of action/reaction and, at least for matter alone, the relativity principle itself (Darrigol). Einstein's analytical separation of constructive and principled theories does not quite capture Poincaré's ambition, which might be thought of as both at the same time: building a constructive theory from an amalgam of special assumptions and principled starting points.

Peter Galison's contribution to this volume, "Einstein's Clocks," pursues one particularly critical component of relativity: the stipulation of simultaneity as the coordination of clocks by the exchange of an electromagnetic signal. This coordinative process has tended to be viewed as a result, in Einstein's case, of a kind of Machian operationalism—and in Poincaré's, as an instance of his conventionalism. True enough, but there is clearly more to the story. In the 1890s Poincaré was deeply engaged with (indeed in 1899 president of) the Bureau des Longitudes, the agency charged with establishing longitude and mapmaking inside France—coordinating clocks by the exchange of electromagnetic signals was a very real concern. Einstein, by contrast, was not in the longitude business at all, but he was evaluating electromechanic patents in Bern. And through Bern came a steady stream of such patent applications concerned with electric clock signals that typically ran along train lines. Here was no obscure technology but rather one of the proud exemplars of modernity—setting electrically coordinated clocks on street corners, steeples, and railroad stations. In these concerns the contextual-technological joins with the scientific-philosophical: a form of abstract concreteness.

Relativity stands so large in our view of the past century that it takes an enormous effort to imagine ourselves back to a world in which it was one view among many. One aim of scholarship has been to analyze turn-of-the-century physics projects in their own right, not as mere stepping-stones or obstacles to the establishment of relativity. This effort has paid off for Poincaré and Lorentz—and also in the studies (see the bibliography in the Cassidy chapter) of Max Abraham and other contemporaries. But another goal of studies done in the last decades of the twentieth century was to understand the disciplining of the physicists, the training processes that set particular value on the inculcation of specific sets of skills and practices.

In some of the most striking arguments presented by Pyenson, Goldberg, and Warwick, it becomes clear how the various educational systems of Germany and Britain created the conditions under which Einstein's paper was read and used. Pyenson, for example, contrasts the elite, literary, and mathematical *Gymnasia* with the more popular, technical, and engineering-oriented *Realschulen*. Intriguingly, his point is not that the graduates of *Realschulen* were either for or against Einstein. Instead, and more subtly, he argues that the real correlation is between the *Gymnasia* graduates, who tended to be supporters and followers of Hermann Minkowski, who geometrized relativity theory and espoused a preestablished harmony between mathematics and physics, and the *Realschulen* graduates, who tended to stick with the much less mathematically oriented presentation of Einstein himself.

Goldberg, in one of the older papers reproduced here, asked, inter alia, why relativity seemed to tap so lightly on the doors of turn-of-the-century British physicists. The answer, he contends, lay in the prominent role that the ether played in British thought since at least the time of Thomas Young. Ether made possible the kind of local action and causal sequencing that British physicists prized so highly, and the disciplined drilling for the tripos examination stressed the manipulation of the ether. As an example, Goldberg takes up the case of Ebenezer Cunningham, the physicist who introduced relativity into Britain. For Goldberg, Cunningham's immersion in the tripos vision of the world presented an almost insuperable hurdle toward an understanding of relativity and in particular obfuscated Einstein's slaying of the ether.

Warwick's article takes up the education problem precisely where Goldberg's ends. For Warwick, as for other authors in this volume, the goal is to locate relativity without making the world orbit around it. So instead of depicting British electrodynamics as an ether-preserving obstacle, Warwick wants to describe, positively, what it was that British electrodynamics sought in its own right. From within the tripos culture, he pinpoints which pieces of Einstein's relativity were taken up by whom and to what end. This leads, for example, to a parallel study of Norman Campbell, who came to relativity from the rather anti-tripos Cavendish and consequently read the theory as having nothing to do with the ether and everything to do with experiment. Contemplating these various regimes of schooling (*Gymnasia*, *Realschulen*, tripos, Cavendish) suggests how a new and very effective twist on reception theory might be put into play, one that would be more sensitive to the skills and practices imparted in training and that would put correspondingly less weight on abstract, and more mutable, commitments to ontology.

No doubt, every age will have its Einstein. Ours is an Einstein whose physics was deeply enmeshed in the turn-of-the-century German-speaking world: a world longing for worldviews, absolutes, and neo-Kantian philosophy. We find in Einstein's world a culture torn between its contradictory affinity for granite-columned classical stability and its hankering for revolutionary crisis; a world, it now seems, looking to escape from material reality into abstraction and yet proud beyond measure of its technological prowess; a world that took up Einstein as a symbol of an age and mounted his image on the banners of warring ambitions.

Suggested Further Reading

Darrigol, Olivier. *Electrodynamics from Ampère to Einstein.* Oxford, U.K.: Oxford University Press, in press.

Goldberg, Stanley. *Understanding Relativity: Origin and Impact of a Scientific Revolution.* Boston: Birkhauser, 1984.

Hirosige, Tetu. "The Ether Problem, the Mechanistic World-View, and the Origins of the Theory of Relativity." *Historical Study of the Physical Sciences* 7 (1976): 3–82.

Holton, Gerald. *Thematic Origins of Scientific Thought; Kepler to Einstein.* 2nd ed. Cambridge, Mass.: Harvard University Press, 1988.

Howard, Don, and John Stachel, eds. *Einstein: The Formative Years, 1879–1909.* Boston: Birkhauser, 2000.

Janssen, Michel Heinrich Paul. "A Comparison between Lorentz's Ether Theory and Special Relativity in the Light of the Experiments of Trouton and Noble." Ph.D. diss., University of Pittsburgh, 1995.

Klein, Martin J. "Thermodynamics in Einstein's Thought." *Science* 157 (1967): 509–16.

McCormmach, Russell. "Einstein, Lorentz, and the Electron Theory." *Historical Study of the Physical Sciences* 2 (1970): 41–87.

Miller, Arthur I. *Albert Einstein's Special Theory of Relativity: Emergence (1905) and Early Interpretation (1905–1911).* Reading, Mass.: Addison-Wesley, 1981.

Pais, Abraham. *"Subtle Is the Lord . . .": The Science and the Life of Albert Einstein.* New York: Oxford University Press, 1982.

Stachel, John. "History of Relativity." In *Twentieth Century Physics.* Edited by Laurie Brown, Abraham Pais, and Brian Pippard, pp. 249–356. Philadelphia: Institute of Physics Publishing, 1995.

Stachel, John, Robert Schulmann, et al., eds. *The Collected Papers of Albert Einstein.* Princeton, N.J.: Princeton University Press, 1987–.

Science and Society

The History of Modern Physical Science in the Twentieth Century

Edited with introductions byr

Peter Galison
Harvard University

Michael Gordon
Harvard University

David Kaiser
Massachusetts Insitute of Tecnology

ROUTLEDGE
New York/London

ON THE ELECTRODYNAMICS OF MOVING BODIES

By A. Einstein

That Maxwell's electrodynamics – the way in which it is usually understood – when applied to moving bodies, leads to asymmetries which do not appear to be inherent in the phenomena is well known. Consider, for example, the reciprocal electrodynamic interaction of a magnet and a conductor. The observable phenomenon here depends only on the relative motion of the conductor and the magnet, whereas the customary conception draws a sharp distinction between the two cases in which either the one or the other of these bodies is in motion. For if the magnet is in motion and the conductor at rest, there arises in the neighborhood of the magnet an electric field with a certain definite energy, producing a current at the places where parts of the conductor are situated. But if the magnet is at rest and the conductor in motion, no electric field arises in the neighborhood of the magnet. In the conductor, however, we find an electromotive force, to which in itself there is no corresponding energy, but which gives rise – assuming equality of relative motion in the two cases discussed – to electric currents of the same path and intensity as those produced by the electric forces in the former case.

Examples of this sort, together with the unsuccessful attempts to discover any motion of the earth relatively to the "light medium," lead to the conjecture that to the concept of absolute rest there correspond no properties of the phenomena, neither in mechanics, nor in electrodynamics, but rather that as has already been shown to quantities of the first order, for every reference system in which the laws of mechanics are valid*, the laws of electrodynamics and optics are also valid.

We will raise this conjecture (whose intent will from now on be referred to as the "Principle of Relativity") to a postulate, and moreover introduce another postulate, which is only apparently irreconcilable with the former: light is always propagated in empty space with a definite velocity c which is independent of the state of motion of the emitting body. These two postulates suffice in order to obtain a simple and consistent theory of the electrodynamics of moving bodies taking as a basis Maxwell's theory for bodies at rest. The introduction of a "luminiferous ether" will prove to be superfluous because the view here to be developed will introduce neither an "absolutely resting space" provided with special properties, nor associate a velocity-vector with a point of empty space in which electromagnetic processes occur.

The theory to be developed is based – like all electrodynamics – on the kinematics of the rigid body, since the assertions of any such theory concern

[* The preceeding memoir by Lorentz was not at this time known to the author. (A.S.)]

the relationships between rigid bodies (coordinate systems), clocks, and electromagnetic processes. Insufficient consideration of this circumstance is the root of the difficulties with which the electrodynamics of moving bodies presently has to contend.

I. KINEMATICAL PART

§1. Definition of Simultaneity

Let us consider a coordinate system in which the equations of Newtonian mechanics hold.* For precision of demonstration and to distinguish this coordinate system verbally from others which will be introduced later, we call it the "resting system."

If a material point is at rest relatively to this coordinate system, its position can be defined relative to it by rigid measuring rods employing the methods of Euclidean geometry, and can be expressed in Cartesian coordinates.

If we wish to describe the *motion* of a material point, we give the values of its coordinates as functions of the time. Now we must bear carefully in mind that a mathematical description of this kind has no physical meaning unless we are quite clear as to what we will understand by "time". We have to take into account that all our judgments in which time plays a role are always judgments of *simultaneous events*. If, for instance, I say, "That train arrives here at 7 o'clock," I mean something like this: "The pointing of the small hand of my watch to 7 and the arrival of the train are simultaneous events."**

It might appear possible that all the difficulties concerning the definition of "time" can be overcome by substituting "the position of the small hand of my watch" for "time." In fact such a definition is satisfactory when we are concerned with defining a time exclusively for the place where the watch is located; but it is no longer satisfactory when we have to connect in time series of events occurring at different places, or – what comes to the same thing – to evaluate the times of events occurring at places remote from the watch. We could in principle content ourselves to time events by using an observer located at the origin of the coordinate system, and equipped with a clock, who coordinates the arrival of the light signal originating from the event to be timed and traveling to his position through empty space, to be timed with the hands of his clock. Yet as we know from experience, this coordination has the disadvantage that it is not independent of the standpoint of the observer with the clock. We arrive at a much more practical arrangement by means of the following considerations.

[* i.e. to the first approximation. (A.S.)]

** We shall not here discuss the inexactitude which lurks in the concept of simultaneity of two events at (approximately) the same place, which must be removed through introducing an abstract concept.

If at the point A of space there is a clock, an observer at A can time the events in the immediate vicinity of A by coordinating the positions of the hands which are simultaneous with these events. If there is at the space point B another clock – and we wish to add, "a clock being of exactly the same characteristics as the one at A" – then it is possible for an observer at B to time the events in the immediate neighborhood of B. But, without further definitions it is not possible to compare, in respect with time, an event at A with an event at B. Thus far we have defined only an "A time" and a "B time", but no common "time" for A and B. The latter time can now be defined by requiring that by definition the "time" necessary for light to travel from A to B be identical to the "time" necessary to travel from B to A. Let a ray of light start at the "A time" t_A from A toward B, let it at the "B time" t_B be reflected at B in the direction of A, and arrive again at A at the "A time" t'_A. The two clocks run in synchronization by definition if

$$t_B - t_A = t'_A - t_B. \qquad [\S1.1]$$

We assume this definition of synchronization to be free of any possible contradictions, applicable to arbitrarily many points, and that the following relations are universally valid: –

1. If the clock at B synchronizes with the clock at A, the clock at A synchronizes with the clock at B.

2. If the clock at A synchronizes with the clock at B and also with the clock at C, the clocks at B and C also synchronize with each other.

Thus with the help of certain (imaginary) physical experiments we have defined what is to be understood by synchronous stationary clocks located at different places, and have clearly obtained a definition of "simultaneous," or "synchronous," and of "time." The "time" of an event is the reading simultaneous with the event of a clock at rest and located at the position of the event, this clock being synchronous, and indeed synchronous for all time determinations, with a specified clock at rest.

In addition, in agreement with experience we further require that the quantity

$$\frac{2AB}{t'_A - t_A} = c, \qquad [\S1.2]$$

be a universal constant (the velocity of light in empty space).

It is essential to have time defined by means of clocks at rest in a resting system, and the time now defined being appropriate to the resting system we call "the time of the resting system."

§2. *On the Relativity of Lengths and Times*

The following considerations are based on the principle of relativity and on the principle of the constancy of the velocity of light. We define these two principles thus –

1. The laws by which the states of physical systems undergo changes are independent of whether these changes of state are referred to one or the other of two coordinate systems moving relatively to each other in uniform translational motion.

2. Any ray of light moves in the "resting" coordinate system with the definite velocity c, which is independent of whether the ray was emitted by a resting or by a moving body. Consequently,

$$\text{velocity} = \frac{\text{light path}}{\text{time interval}}$$

where time interval is to be understood in the sense of the definition in §1.

Consider a rigid rod at rest whose length is l when measured by a measuring-rod which is also at rest. We now imagine the axis of the rod lying along the x-axis of the resting coordinate system, and that a uniform motion of parallel translation with velocity v along the x-axis in the direction of increasing x is then imparted to the rod. We now inquire as to the length of the *moving* rod, which we imagine to be determined by means of the following two operations:—

(a) The observer moves together with the given measuring-rod and the rod to be measured, and measures the length of the rod directly by superposing the measuring-rod, in just the same way as if the rod to be measured, observer and measuring rod were at rest.

(b) By means of clocks at rest set up in the resting system and synchronized in accordance with §1, the observer ascertains at what points of the resting system the two ends of the rod to be measured are located at a definite time t. The distance between these two points, measured by the measuring-rod already employed, which in this case is at rest, is also a length which may be designated "the length of the rod."

According to the principle of relativity the length to be discovered by the operation (a)—we will call it "the length of the rod in the moving system"—must be equal to the length l of the rod at rest.

The length to be discovered by the operation (b) we will call "the length of the (moving) rod in the resting system," which we shall determine on the basis of our two principles. We shall find that it differs from l.

Current kinematics assumes tacitly that the lengths determined by these two operations are precisely equal, or in other words, that a moving rigid body at the instant of time t may in geometrical respects be perfectly represented by *the same* body *at rest* in a definite position.

We imagine further that at the two ends A and B of the rod, clocks are placed which synchronize with the clocks of the resting system—that is, that their indications correspond at any instant to the "time of the resting system" at the places where they happen to be. Consequently these clocks are "synchronous in the resting system."

We imagine further that with each clock there is a moving observer, and that these observers apply to both clocks the criterion established in §1 for the synchronization of two clocks. Let a ray of light depart from A at the time* t_A, let it be reflected at B at the time t_B, and reach A again at the time t'_A. Taking into consideration the principle of the constancy of the velocity of light we find that

$$t_B - t_A = \frac{r_{AB}}{c - v} \quad \text{and} \quad t'_A - t_B = \frac{r_{AB}}{c + v}$$

where r_{AB} denotes the length of the moving rod—measured in the resting system. Observers moving with the moving rod would thus find that the two clocks were not synchronous, while observers in the resting system would declare the clocks to be synchronous.

Thus we see that we can attribute no *absolute* meaning to the concept of simultaneity, but that two events which, examined from a coordinate system, are simultaneous, can no longer be interpreted as simultaneous events when examined from a system which is in motion relatively to that system.

§3. Theory of the Transformation of Coordinates and Times from a Resting System to another System in Uniform Motion of Translation Relatively to the Former

Let us in "resting" space take two systems of coordinates, i.e., two systems, each of three rigid material lines, perpendicular to one another, and originating from a point. Let the X-axes of the two systems coincide, and their Y- and Z-axes, respectively, be parallel. Let each system be provided with a rigid measuring-rod and a number of clocks, and let the two measuring-rods, and likewise all the clocks of the two systems, be in all respects identical.

Now to the origin of one of the two systems (k) let a (constant) velocity v be imparted in the direction of the increasing x of the other resting system (K), and let this velocity be communicated to the coordinate axes, the relevant measuring-rod, and the clocks. To any time t of the resting system K there then will correspond a definite position of the axes of the moving system, and from reasons of symmetry we are entitled to assume that the motion of k may be such that the axes of the moving system are at the time t (this "t" always denotes a time of the resting system) parallel to the axes of the resting system.

We now imagine space to be measured from the resting system K by means of the measuring-rod at rest, and also from the moving system k by means of the measuring-rod moving with it; thus we determine the coordinates x, y, z, and ξ, η, ζ, respectively. Further, let the time t of the resting system be determined for all points thereof at which there are clocks by means of light

* "Time" here denotes "time of the resting system" and also "position of hands of the moving clock located at the place under discussion."

signals in the manner indicated in §1. Similarly let the time τ of the moving system be determined for all points of the moving system at which there are clocks at rest relatively to that system by applying the method, given in §1, of exchanging light signals between the points at which the latter clocks are located.

To any system of values x, y, z, t, which completely defines the place and time of an event in the resting system, there belongs a system of values ξ, η, ζ, τ, determining that event relatively to the system k, and our task is now to find the system of equations connecting these quantities.

In the first place it is clear that the equations must be *linear* on account of the properties of homogeneity which we attribute to space and time.

If we set $x' = x - vt$, it is clear that a point at rest in the system k belongs to a system of values x', y, z, independent of time. We first define τ as a function of x', y, z, and t. To do this we have to express in equations that τ is nothing else than the collection of the data of clocks at rest in system k, which have been synchronized according to the rule given in §1.

From the origin of system k let a ray be emitted at the time τ_0 along the X-axis to x', and at the time τ_1 be reflected back to the coordinate origin, arriving there at the time τ_2; thus we must have $\frac{1}{2}(\tau_0 + \tau_2) = \tau_1$, or, by inserting the arguments of the function τ and applying the principle of the constancy of the velocity of light in the resting system:—

$$\frac{1}{2}\left[\tau(0, 0, 0, t) + \tau\left(0, 0, 0, t + \frac{x'}{c - v} + \frac{x'}{c + v}\right)\right] = \tau\left(x', 0, 0, t + \frac{x'}{c - v}\right). \quad [\S 3.1]$$

Hence, if x' be chosen infinitesimally small,

$$\frac{1}{2}\left(\frac{1}{c - v} + \frac{1}{c + v}\right)\frac{\partial \tau}{\partial t} = \frac{\partial \tau}{\partial x'} + \frac{1}{c - v}\frac{\partial \tau}{\partial t}, \quad [\S 3.2]$$

or

$$\frac{\partial \tau}{\partial x'} + \frac{v}{c^2 - v^2}\frac{\partial \tau}{\partial t} = 0. \quad [\S 3.3]$$

It is to be noted that instead of the origin of the coordinates we might have chosen any other point for the point of origin of the ray, and the equation just obtained is therefore valid for all values of x', y, z.

A similar analysis — applied to the H- and Z-axes* — taking into account that when considered from the resting system light propagates along these axes with the velocity $\sqrt{(c^2 - v^2)}$, yields the results

[* Einstein's letters H and Z are the Greek upper case symbols for eta and zeta, and refer to axes in the system k. For consistency in notation, in line 39 Einstein should have employed Ξ instead of X. Since in line 90 below appears Ξ, then his first use of X was an oversight. (A.I.M.)]

$$\frac{\partial \tau}{\partial y} = 0, \qquad \frac{\partial \tau}{\partial z} = 0.$$

Since τ is a *linear* function, it follows from these equations that

$$\tau = a\left(t - \frac{v}{c^2 - v^2}x'\right) \qquad [§3.4]$$

where a is a function $\phi(v)$ at present unknown, and where for brevity it is assumed that at the origin of k, $\tau = 0$, when $t = 0$.

With the help of this result we easily determine the quantities ξ, η, ζ by expressing in equations that light (as required by the principle of the constancy of the velocity of light, in combination with the principle of relativity) is also propagated with velocity c when measured in the moving system. For a ray of light emitted at the time $\tau = 0$ in the direction of the increasing ξ

$$\xi = c\tau \quad \text{or} \quad \xi = ac\left(t - \frac{v}{c^2 - v^2}x'\right). \qquad [§3.5]$$

But the ray moves relatively to the initial point of k, when measured in the resting system, with the velocity $c - v$, so that

$$\frac{x'}{c - v} = t. \qquad [§3.6]$$

If we insert this value of t in the equation for ξ, we obtain

$$\xi = a\frac{c^2}{c^2 - v^2}x'. \qquad [§3.7]$$

In an analogous manner we find, by considering rays moving along the two other axes, that

$$\eta = c\tau = ac\left(t - \frac{v}{c^2 - v^2}x'\right) \qquad [§3.8]$$

when

$$\frac{y}{\sqrt{(c^2 - v^2)}} = t, \qquad x' = 0. \qquad [§3.9]$$

Thus

$$\eta = a\frac{c}{\sqrt{(c^2 - v^2)}}y \quad \text{and} \quad \zeta = a\frac{c}{\sqrt{(c^2 - v^2)}}z. \qquad [§3.10]$$

Substituting for x' its value, we obtain

$$\tau = \phi(v)\beta(t - vx/c^2), \qquad [§3.11]$$

$$\xi = \phi(v)\beta(x - vt), \qquad [\S 3.12]$$

$$\eta = \phi(v)y, \qquad [\S 3.13]$$

$$\zeta = \phi(v)z, \qquad [\S 3.14]$$

where

$$\beta = \frac{1}{\sqrt{(1 - v^2/c^2)}}, \qquad [\S 3.15]$$

and ϕ is an as yet unknown function of v. If no assumption whatever be made as to the initial position of the moving system and as to the zero point of τ, an additive constant is to be placed on the right side of each of these equations.

We now have to prove that any ray of light, measured in the moving system, is propagated with the velocity c, if, as we have assumed, this is the case in the resting system; for we have not as yet furnished the proof that the principle of the constancy of the velocity of light is compatible with the principle of relativity.

At the time $t = \tau = 0$, when the coordinate origins of the two systems coincide, let a spherical wave be emitted from a source at the origin of both systems, and be propagated with the velocity c in system K. If (x, y, z) is a point just reached by this wave, then

$$x^2 + y^2 + z^2 = c^2t^2. \qquad [\S 3.16]$$

Transforming this equation with the aid of our equations of transformation we obtain after a simple calculation

$$\xi^2 + \eta^2 + \zeta^2 = c^2\tau^2. \qquad [\S 3.17]$$

The wave under consideration is therefore no less a spherical wave with velocity of propagation c when considered in the moving system. Consequently our two fundamental principles are compatible with each other.*

In the equations of transformation which have been developed there enters an unknown function ϕ of v, which we will now determine.

For this purpose we introduce a third system of coordinates K′, which relatively to the system k is in a state of parallel translational motion parallel to the Ξ-axis such that its coordinate origin moves with velocity $-v$ on the Ξ-axis relative to k. At the time $t = 0$ let all three origins coincide, and when $t = x = y = z = 0$ let the time t' of the system K′ be zero. We call the coordinates, measured in the system K′, x', y', z', and by a twofold application of our transformation equations, we obtain

[* The equations of the Lorentz transformation may be more simply deduced directly from the condition that in virtue of those equations the relation $x^2 + y^2 + z^2 = c^2t^2$ shall have as its consequence the second relation $\xi^2 + \eta^2 + \zeta^2 = c^2\tau^2$. (A.S.)]

$$t' = \phi(-v)\beta(-v)(\tau + v\xi/c^2) = \phi(v)\phi(-v)t, \qquad [\S 3.18]$$

$$x' = \phi(-v)\beta(-v)(\xi + v\tau) \quad = \phi(v)\phi(-v)x, \qquad [\S 3.19]$$

$$y' = \phi(-v)\eta \quad = \phi(v)\phi(-v)y, \qquad [\S 3.20]$$

$$z' = \phi(-v)\zeta \quad = \phi(v)\phi(-v)z. \qquad [\S 3.21]$$

Since the relations between x', y', z' and x, y, z do not contain the time t, the systems K and K′ are at rest with respect to one another, and it is clear that the transformation from K to K′ must be the identity transformation. Thus

$$\phi(v)\phi(-v) = 1. \qquad [\S 3.22]$$

We now inquire into the meaning of $\phi(v)$. We fix our attention on that part of the H-axis of system k which lies between $\xi = 0$, $\eta = 0$, $\zeta = 0$ and $\xi = 0$, $\eta = l$, $\zeta = 0$. This part of the H-axis is a rod moving perpendicularly to its axis with velocity v relatively to system K, whose ends have in K the coordinates

$$x_1 = vt, \qquad y_1 = \frac{l}{\phi(v)}, \qquad z_1 = 0 \qquad [\S 3.23]$$

and

$$x_2 = vt, \qquad y_2 = 0, \qquad z_2 = 0. \qquad [\S 3.24]$$

The length of the rod measured in K is therefore $l/\phi(v)$; and this gives us the meaning of the function $\phi(v)$. From reasons of symmetry it is now evident that the length of a given rod moving perpendicularly to its axis, measured in the resting system, can depend on only the velocity and not on the direction and the sense of the motion. Thus, the length of the moving rod measured in the resting system does not change when v and $-v$ are interchanged. Consequently $l/\phi(v) = l/\phi(-v)$, or

$$\phi(v) = \phi(-v). \qquad [\S 3.25]$$

It follows from this relation and the previous one that $\phi(v) = 1$, so that the transformation equations which have been found become

$$\tau = \beta(t - vx/c^2), \qquad [\S 3.26]$$

$$\xi = \beta(x - vt), \qquad [\S 3.27]$$

$$\eta = y, \qquad [\S 3.28]$$

$$\zeta = z, \qquad [\S 3.29]$$

where

$$\beta = 1/\sqrt{(1 - v^2/c^2)}. \qquad [\S 3.30]$$

§4. Physical Meaning of the Equations Obtained Concerning Moving Rigid Bodies and Moving Clocks

We consider a rigid sphere* of radius R, at rest relatively to the moving system k, and whose center is at the coordinate origin of k. The equation of the surface of this sphere moving relatively to the system K with velocity v is

$$\xi^2 + \eta^2 + \zeta^2 = R^2. \qquad [\S 4.1]$$

The equation of this surface is expressed in x, y, z at the time $t = 0$ as

$$\frac{x^2}{(\sqrt{(1 - v^2/c^2)})^2} + y^2 + z^2 = R^2. \qquad [\S 4.2]$$

A rigid body which, measured in a state of rest, has the form of a sphere, therefore has in a state of motion – viewed from the resting system – the form of an ellipsoid of revolution with the axes

$$R\sqrt{(1 - v^2/c^2)},\ R,\ R. \qquad [\S 4.3]$$

Thus, whereas the Y- and Z-dimensions of the sphere (and therefore of every rigid body of arbitrary form) do not appear modified by the motion, the X dimension appears shortened in the ratio $1 : \sqrt{(1 - v^2/c^2)}$, i.e., the greater the value of v, the greater the shortening. For $v = c$ all moving objects – viewed from the "resting" system – shrivel up into plane figures. For velocities greater than that of light our deliberations become meaningless; we shall, however, find in what follows that the velocity of light in our theory plays the role, physically, of an infinitely great velocity.

It is clear that the same results hold good of bodies at rest in the "resting" system, viewed from a system in uniform motion.

Further, we imagine one of the clocks which are qualified to mark the time t when at rest relatively to the resting system, and the time τ when at rest relatively to the moving system, to be located at the origin of the coordinates of k, and so adjusted that it marks the time τ. What is the rate, of this clock, when viewed from the resting system?

Between the quantities, x, t and τ, which refer to the position of the clock, the equations clearly hold – and

$$\tau = \frac{1}{\sqrt{(1 - v^2/c^2)}}(t - vx/c^2)$$

and

$$x = vt.$$

Therefore,

* That is, a body possessing spherical form when examined at rest.

$$\tau = t\sqrt{(1 - v^2/c^2)} = t - (1 - \sqrt{(1 - v^2/c^2)})t$$

whence it follows that the time marked by the clock (viewed in the resting system) is slow by $1 - \sqrt{(1 - v^2/c^2)}$ seconds per second, or—neglecting magnitudes of fourth and higher order—by $\frac{1}{2}v^2/c^2$ seconds.

From this there ensues the following peculiar consequence. If at the points A and B of K there are stationary clocks which, viewed in the resting system, are synchronous; and if the clock at A is moved with the velocity v along the line AB to B, then on its arrival at B the two clocks no longer synchronize, but the clock moved from A to B lags behind the other which has remained at B by $\frac{1}{2}tv^2/c^2$ seconds (up to magnitudes of fourth and higher order), t being the time required to move the clock from A to B.

It is at once apparent that this result still holds good if the clock moves from A to B in any polygonal line, and also when the points A and B coincide.

If we assume that the result proved for a polygonal line is also valid for a continuously curved line, we obtain the theorem: If one of two synchronous clocks at A is moved in a closed curve with constant velocity until it returns to A, the journey lasting t seconds, then the clock that moved runs $\frac{1}{2}tv^2/c^2$ seconds slower than the one that remained at rest. Thus we conclude that a balance-clock* at the equator must go more slowly, by a very small amount, than a precisely similar clock situated at one of the poles under otherwise identical conditions.

§5. *The Theorem of Addition of Velocities*

In the system k moving along the X-axis of the system K with velocity v, let a point move in accordance with the equations

$$\xi = w_\xi \tau, \qquad \eta = w_\eta \tau, \qquad \zeta = 0,$$

where w_ξ and w_η denote constants.

Required: the motion of the point relatively to the system K. If with the help of the transformation equations developed in §3 we introduce the quantities x, y, z, t into the quations of motion of the point, we obtain

$$x = \frac{w_\xi + v}{1 + vw_\xi/c^2}t,$$

$$y = \frac{\sqrt{(1 - v^2/c^2)}}{1 + vw_\xi/c^2}w_\eta t,$$

$$z = 0.$$

Thus the law of the parallelogram of velocities is valid according to our theory

[* Not a pendulum-clock, which is physically a system to which the earth belongs. This case had to be excluded. (A.S.)]

only to a first approximation. We set*

$$V^2 = \left(\frac{dx}{dt}\right)^2 + \left(\frac{dy}{dt}\right)^2,$$

$$w^2 = w_\xi^2 + w_\eta^2,$$

$$\alpha = \tan^{-1} w_\eta/w_\xi,$$

α is then to be looked upon as the angle between the velocities v and w. After a simple calculation we obtain

$$V = \frac{\sqrt{[(v^2 + w^2 + 2vw\cos\alpha) - (vw\sin\alpha)^2/c^2]}}{1 + vw\cos\alpha/c^2}$$

It is worthy of remark that v and w enter into the expression for the resultant velocity in a symmetrical manner. If w also has the direction of the (Ξ-axis) we get

$$V = \frac{v + w}{1 + vw/c^2}.$$

It follows from this equation that from a composition of two velocities which are less than c, there always results a velocity less than c. For if we set $v = c - \kappa$, $w = c - \lambda$, κ and λ being positive and less than c, then

$$V = c\frac{2c - \kappa - \lambda}{2c - \kappa - \lambda + \kappa\lambda/c} < c.$$

It follows, further, that the velocity of light c cannot be altered by composition with a velocity less than that of light. For this case we obtain

$$V = \frac{c + w}{1 + w/c} = c.$$

We might also have obtained the formula for V, for the case when v and w have the same direction, by compounding two transformations in accordance with §3. If in addition to the systems K and k figuring in §3 we introduce still another system of coordinates k′ moving parallel to k, its initial point moving on the Ξ-axis with the velocity w, we obtain equations between the quantities x, y, z, t and the corresponding quantities of k′, which differ from the equations found in §3 only in that the place of "v" is taken by the quantity

$$\frac{v + w}{1 + vw/c^2};$$

[* Here the *Annalen* version contains the following typographical error:

$$\alpha = \tan^{-1} w_y/w_x$$

(A.I.M.).]

from which we see that such parallel transformations – necessarily – form a group.

We now have deduced the essential theorems of the kinematics corresponding to our two principles, and we proceed to exhibit their application to electrodynamics.

II. ELECTRODYNAMICAL PART

§6. *Transformation of the Maxwell-Hertz Equations for Empty Space. On the Nature of the Electromotive Forces Occurring in a Magnetic Field During Motion*

Let the Maxwell-Hertz equations for empty space be valid for the resting system K, so that we have

$$\begin{aligned}
\frac{1}{c}\frac{\partial X}{\partial t} &= \frac{\partial N}{\partial y} - \frac{\partial M}{\partial z}, & \frac{1}{c}\frac{\partial L}{\partial t} &= \frac{\partial Y}{\partial z} - \frac{\partial Z}{\partial y}, \\
\frac{1}{c}\frac{\partial Y}{\partial z} &= \frac{\partial L}{\partial z} - \frac{\partial N}{\partial x}, & \frac{1}{c}\frac{\partial M}{\partial t} &= \frac{\partial Z}{\partial x} - \frac{\partial X}{\partial z}, \\
\frac{1}{c}\frac{\partial Z}{\partial t} &= \frac{\partial M}{\partial x} - \frac{\partial L}{\partial y}, & \frac{1}{c}\frac{\partial N}{\partial t} &= \frac{\partial X}{\partial y} - \frac{\partial Y}{\partial x},
\end{aligned} \qquad [§6.1]$$

where (X, Y, Z) denotes the electric-force vector and (L, M, N) the magnetic-force vector.

If we apply to these equations the transformation developed in §3, by referring the electromagnetic processes to the coordinate system there introduced, moving with the velocity v, we obtain the equations

$$\frac{1}{c}\frac{\partial X}{\partial \tau} = \frac{\partial}{\partial \eta}\left\{\beta\left(N - \frac{v}{c}Y\right)\right\} - \frac{\partial}{\partial \zeta}\left\{\beta\left(M + \frac{v}{c}Z\right)\right\}, \qquad [§6.2]$$

$$\frac{1}{c}\frac{\partial}{\partial \tau}\left\{\beta\left(Y - \frac{v}{c}N\right)\right\} = \frac{\partial L}{\partial \zeta} - \frac{\partial}{\partial \xi}\left\{\beta\left(N - \frac{v}{c}Y\right)\right\}. \qquad [§6.3]$$

$$\frac{1}{c}\frac{\partial}{\partial \tau}\left\{\beta\left(Z + \frac{v}{c}M\right)\right\} = \frac{\partial}{\partial \xi}\left\{\beta\left(M + \frac{v}{c}Z\right)\right\} - \frac{\partial L}{\partial \eta}, \qquad [§6.4]$$

$$\frac{1}{c}\frac{\partial L}{\partial \tau} = \frac{\partial}{\partial \zeta}\left\{\beta\left(Y - \frac{v}{c}N\right)\right\} - \frac{\partial}{\partial \eta}\left\{\beta\left(Z + \frac{v}{c}M\right)\right\}, \qquad [§6.5]$$

$$\frac{1}{c}\frac{\partial}{\partial \tau}\left\{\beta\left(M + \frac{v}{c}Z\right)\right\} = \frac{\partial}{\partial \xi}\left\{\beta\left(Z + \frac{v}{c}M\right)\right\} - \frac{\partial X}{\partial \zeta}, \qquad [§6.6]$$

$$\frac{1}{c}\frac{\partial}{\partial \tau}\left\{\beta\left(N - \frac{v}{c}Y\right)\right\} = \frac{\partial X}{\partial \eta} - \frac{\partial}{\partial \xi}\left\{\beta\left(Y - \frac{v}{c}N\right)\right\}, \qquad [§6.7]$$

where

$$\beta = 1/\sqrt{(1 - v^2/c^2)}.$$

Now the principle of relativity requires that if the Maxwell-Hertz equations for empty space hold in system K, they also hold in system k. In other words, the vectors of the electric and the magnetic force (X', Y', Z') and (L', M', N') of the moving system k, which are defined by their ponderomotive effects on electric or magnetic substances, respectively, satisfy the following equations:—

$$\frac{1}{c}\frac{\partial X'}{\partial \tau} = \frac{\partial N'}{\partial \eta} - \frac{\partial M'}{\partial \zeta}, \qquad \frac{1}{c}\frac{\partial L'}{\partial \tau} = \frac{\partial Y'}{\partial \zeta} - \frac{\partial Z'}{\partial \eta},$$

$$\frac{1}{c}\frac{\partial Y'}{\partial \tau} = \frac{\partial L'}{\partial \zeta} - \frac{\partial N'}{\partial \xi}, \qquad \frac{1}{c}\frac{\partial M'}{\partial \tau} = \frac{\partial Z'}{\partial \xi} - \frac{\partial X'}{\partial \zeta},$$

$$\frac{1}{c}\frac{\partial Z'}{\partial \tau} = \frac{\partial M'}{\partial \xi} - \frac{\partial L'}{\partial \eta}, \qquad \frac{1}{c}\frac{\partial N'}{\partial \tau} = \frac{\partial X'}{\partial \eta} - \frac{\partial Y'}{\partial \xi}.$$

Evidently the two systems of equations found for system k must express exactly the same thing, since both systems of equations are equivalent to the Maxwell-Hertz equations for system K. Since, further, the equations of the two systems agree, with the exception of the symbols for the vectors, it follows that the functions occurring in the systems of equations at corresponding places must agree, with the exception of a factor $\psi(v)$, which is common for all functions of the one system of equations, and is independent of ξ, η, ζ and τ but depends upon v. Thus we have the relations

$$X' = \psi(v)X, \qquad L' = \psi(v)L,$$

$$Y' = \psi(v)\beta\left(Y - \frac{v}{c}N\right), \qquad M' = \psi(v)\beta\left(M + \frac{v}{c}Z\right),$$

$$Z' = \psi(v)\beta\left(Z + \frac{v}{c}M\right), \qquad N' = \psi(v)\beta\left(N - \frac{v}{c}Y\right).$$

If we now form the reciprocal of this system of equations, firstly by solving the equations just obtained, and secondly by applying the equations to the inverse transformation (from k to K), which is characterized by the velocity $-v$, it follows, when we consider that the two systems of equations thus obtained must be identical, that $\psi(v)\psi(-v) = 1$. Further, from reasons of symmetry* $\psi(v) = \psi(-v)$, and therefore

$$\psi(v) = 1,$$

* If, for example, $X = Y = Z = L = M = 0$, and $N \neq 0$, then from reasons of symmetry it is clear that when v changes sign without changing its numerical value, Y' must also change sign without changing its numerical value.

and our equations assume the form

$$X' = X, \qquad L' = L,$$

$$Y' = \beta\left(Y - \frac{v}{c}N\right), \quad M' = \beta\left(M + \frac{v}{c}Z\right),$$

$$Z' = \beta\left(Z + \frac{v}{c}M\right), \quad N' = \beta\left(N - \frac{v}{c}Y\right).$$

As to the interpretation of these equations we make the following remarks: Let a point charge of electricity have the magnitude "one" when measured in the resting system K, i.e., let it when at rest in the resting system exert a force of one dyne upon an equal quantity of electricity at a distance of one cm. By the principle of relativity this electric charge is also of the magnitude "one" when measured in the moving system. If this quantity of electricity is at rest relatively to the resting system, then by definition the vector (X, Y, Z) is equal to the force acting upon it. If the quantity of electricity is at rest relatively to the moving system (at least at the relevant instant), then the force acting upon it, measured in the moving system, is equal to the vector (X', Y', Z'). Consequently the first three equations above allow themselves to be expressed in words in the following two ways:—

1. If a unit electric point charge is moving in an electromagnetic field, there acts upon it, in addition to the electric force, an "electromotive force" which, if we neglect the terms multiplied by the second and higher powers of v/c, is equal to the vector-product of the velocity of the charge and the magnetic force, divided by the velocity of light. (Old manner of expression.)

2. If a unit electric point charge is moving in an electromagnetic field, the force acting upon it is equal to the electric force which is present at the position of this unit charge, and which we determine by transformation of the field to a coordinate system at rest relatively to the electrical unit charge. (New manner of expression.)

The analogy holds with "magnetomotive forces." We see that in the developed theory the electromotive force plays only the part of an auxiliary concept, which owes its introduction to the circumstance that electric and magnetic forces do not exist independently of the state of motion of the coordinate system.

Furthermore it is clear that the asymmetry mentioned in the introduction as arising when we consider the currents produced by the relative motion of a magnet and a conductor, now disappears. Likewise, questions as to the "seat" of electrodynamic electromotive forces (unipolar machines) become meaningless.

§7. *Theory of Doppler's Principle and of Aberration*

In the system K, very far from the coordinate origin, let there be a source of electrodynamic waves, which in a part of space containing the coordinate origin may be represented to a sufficient degree of approximation by the equations

$$X = X_0 \sin \Phi, \qquad L = L_0 \sin \Phi,$$
$$Y = Y_0 \sin \Phi, \qquad M = M_0 \sin \Phi,$$
$$Z = Z_0 \sin \Phi, \qquad N = N_0 \sin \Phi,$$

where

$$\Phi = \omega\left\{t - \frac{1}{c}(lx + my + nz)\right\}.$$

Here (X_0, Y_0, Z_0) and (L_0, M_0, N_0) are the vectors defining the amplitude of the wave-train, and l, m, n the direction-cosines of the wave-normals. We seek the characteristics of these waves, when they are examined by an observer at rest in the moving system k.

Applying the transformation equations found in §6 for electric and magnetic forces, and those found in §3 for the coordinates and the time, we obtain directly

$$X' = X_0 \sin \Phi', \qquad L' = L_0 \sin \Phi',$$
$$Y' = \beta(Y_0 - vN_0/c) \sin \Phi', \qquad M' = \beta(M_0 + vZ_0/c) \sin \Phi',$$
$$Z' = \beta(Z_0 + vM_0/c) \sin \Phi', \qquad N' = \beta(N_0 - vY_0/c) \sin \Phi',$$
$$\Phi' = \omega'\left\{\tau - \frac{1}{c}(l'\xi + m'\eta + n'\zeta)\right\}$$

where

$$\omega' = \omega\beta(1 - lv/c),$$
$$l' = \frac{l - v/c}{1 - lv/c},$$
$$m' = \frac{m}{\beta(1 - lv/c)},$$
$$n' = \frac{n}{\beta(1 - lv/c)}.$$

From the equation for ω' it follows that if an observer is moving with velocity v relatively to an infinitely distant source of light of frequency ν, in such a way that the connecting line "light source-observer" makes the angle ϕ with the velocity of the observer referred to a coordinate system which is at rest relatively to the source of light, the frequency ν' of the light perceived by the observer is given by the equation

$$\nu' = \nu \frac{1 - \cos\phi \cdot v/c}{\sqrt{(1 - v^2/c^2)}}.$$

This is Doppler's principle for arbitrary velocities. When $\phi = 0$ the equation assumes the simple form

$$\nu' = \nu \sqrt{\frac{1 - v/c}{1 + v/c}}.$$

We see that, in contrast with the customary view, when $v = -c$, $\nu' = \infty$.*

If we call the angle between the wave-normal (direction of the ray) in the moving system and the connecting line "light source-observer" ϕ', the equation for l' becomes

$$\cos\phi' = \frac{\cos\phi - v/c}{1 - \cos\phi \cdot v/c}.$$

This equation expresses the law of aberration in its most general form. If $\phi = \frac{1}{2}\pi$, the equation becomes simply

$$\cos\phi' = -v/c.$$

We still have to find the amplitude of the waves, as it appears in the moving system. If we call the amplitude of the electric or magnetic force A or A' respectively, accordingly as it is measured in the resting system or in the moving system, we obtain

$$A'^2 = A^2 \frac{(1 - \cos\phi \cdot v/c)^2}{1 - v^2/c^2},$$

for $\phi = 0$ this equation becomes

$$A'^2 = A^2 \frac{1 - v/c}{1 + v/c}.$$

From the equations developed here, it follows that to an observer approaching a source of light with the velocity c, this source of light must appear of infinite intensity.

[* Here the *Annalen* version has the following typographical error: "for $v = -\infty$, $\nu = \infty$." In the desk copy of Einstein's own set of reprints, which was presented to Professor Gerald Holton at the completion of his work on a *catalogue raisonné* of the Einstein papers in August 1964, there are some additions and corrections in Einstein's own hand. On this page the corrections consist of the deletion of the $-\infty$ (in $v = -\infty$) and its replacement by V which is the symbol that Einstein used for the velocity of light in vacuum c; and the replacement in line 17 of the phrase "connecting line 'light source-observer'" by the phrase "direction of motion." (A.I.M.)]

§8. Transformation of the Energy of Light Rays. Theory of the Pressure of Radiation Exerted on Perfect Reflectors

Since $A^2/8\pi$ equals the energy of light per unit volume, according to the principle of relativity, we have to regard $A'^2/8\pi$ as the energy of light per unit volume in the moving system. Thus A'^2/A^2 would be the ratio of the "measured in motion" to the "measured at rest" energy of a given light complex, if the volume of a light complex were the same, whether measured in K or in k. This is, however, not the case. If l, m, n are the direction-cosines of the wave-normals of the light in the resting system, no energy passes through the surface elements of a spherical surface moving with the velocity of light:—

$$(x - lct)^2 + (y - mct)^2 + (z - nct)^2 = R^2.$$

We may say, therefore, that this surface permanently encloses the same light complex. We inquire into the amount of energy enclosed by this surface, from the viewpoint of the system k—that is, the energy of the light complex relatively to the system k.

The spherical surface—viewed in the moving system—is an ellipsoidal surface, whose equation at the time $\tau = 0$, is

$$(\beta\xi - l\beta\xi v/c)^2 + (\eta - m\beta\xi v/c)^2 + (\zeta - n\beta\xi v/c)^2 = R^2. \qquad [§8.1]$$

If S is the volume of the sphere, and S' the volume of the ellipsoid, then by a simple calculation

$$\frac{S'}{S} = \frac{\sqrt{1 - v^2/c^2}}{1 - \cos\phi \cdot v/c}. \qquad [§8.2]$$

Thus, if we call the light energy enclosed by this surface E when it is measured in the resting system, and E' when measured in the moving system, we obtain

$$\frac{E'}{E} = \frac{A'^2 S'}{A^2 S} = \frac{1 - \cos\phi \cdot v/c}{\sqrt{(1 - v^2/c^2)}}, \qquad [§8.3]$$

and, when $\phi = 0$, this formula simplifies into

$$\frac{E'}{E} = \sqrt{\frac{1 - v/c}{1 + v/c}}. \qquad [§8.4]$$

It is remarkable that the energy and the frequency of a light complex vary with the observer's state of motion in accordance with the same law.

Now let the coordinate plane $\xi = 0$ be a perfectly reflecting surface, at which the plane waves considered in the previous section are reflected. We seek the pressure of light exerted on the reflecting surface, and the direction, frequency, and intensity of the light after reflection.

Let the incident light be defined by the quantities A, $\cos\phi$, ν (referred to system K). Viewed from k the corresponding quantities are

$$A' = A \frac{1 - \cos\phi \cdot v/c}{\sqrt{(1 - v^2/c^2)}}, \qquad [§8.5]$$

$$\cos\phi' = \frac{\cos\phi - v/c}{1 - \cos\phi \cdot v/c}, \qquad [§8.6]$$

$$\nu' = \nu \frac{1 - \cos\phi \cdot v/c}{\sqrt{(1 - v^2/c^2)}}. \qquad [§8.7]$$

Referring the process to system k, we obtain for the reflected light

$$A'' = A', \qquad [§8.8]$$

$$\cos\phi'' = -\cos\phi', \qquad [§8.9]$$

$$\nu'' = \nu'. \qquad [§8.10]$$

Finally, by transforming back to the resting system K, we obtain for the reflected light*

$$A''' = A'' \frac{1 + \cos\phi'' \cdot v/c}{\sqrt{(1 - v^2/c^2)}} = A \frac{1 - 2\cos\phi \cdot v/c + v^2/c^2}{1 - v^2/c^2}, \qquad [§8.11]$$

$$\cos\phi''' = \frac{\cos\phi'' + v/c}{1 + \cos\phi'' \cdot v/c} = -\frac{(1 + v^2/c^2)\cos\phi - 2v/c}{1 - 2\cos\phi \cdot v/c + v^2/c^2}, \qquad [§8.12]$$

$$\nu''' = \nu'' \frac{1 + \cos\phi'' \cdot v/c}{\sqrt{(1 - v^2/c^2)}} = \nu \frac{1 - 2\cos\phi \cdot v/c + v^2/c^2}{1 - v^2/c^2}. \qquad [§8.13]$$

The energy (measured in the resting system) incident per unit time upon the unit of surface of the mirror is evidently $A^2(c\cos\phi - v)/8\pi$. The energy leaving per unit time the unit of surface of the mirror is $A'''^2(-c\cos\phi''' + v)/8\pi$. According to the principle of energy, the difference of these two expressions is the work done per unit time by the pressure of light. If we express this work as the product Pv, where P is the pressure of light, we obtain

$$P = 2 \cdot \frac{A^2}{8\pi} \frac{(\cos\phi - v/c)^2}{1 - v^2/c^2}. \qquad [§8.14]$$

In agreement with experiment and with other theories, to a first approximation we obtain

$$P = 2 \cdot \frac{A^2}{8\pi} \cos^2\phi. \qquad [§8.15]$$

---

* In the *Annalen* version the Eq. [§8.13] is incorrect because the part containing the frequency and angle in K has a denominator of $(1 - v/c)^2$.

All problems in the optics of moving bodies can be solved by the method here employed. The essential point is that the electric and magnetic force of the light, which is influenced by a moving body, be transformed to a coordinate system at rest relative to the body. By this means all problems in the optics of moving bodies are reduced to a series of problems in the optics of bodies at rest.

§9. *Transformation of the Maxwell-Hertz Equations when Convection-Currents are Taken into Account*

We start from the equations

$$\frac{1}{c}\left\{\frac{\partial X}{\partial t}+u_x\rho\right\}=\frac{\partial N}{\partial y}-\frac{\partial M}{\partial z}, \qquad \frac{1}{c}\frac{\partial L}{\partial t}=\frac{\partial Y}{\partial z}-\frac{\partial Z}{\partial y},$$

$$\frac{1}{c}\left\{\frac{\partial Y}{\partial t}+u_y\rho\right\}=\frac{\partial L}{\partial z}-\frac{\partial N}{\partial x}, \qquad \frac{1}{c}\frac{\partial M}{\partial t}=\frac{\partial Z}{\partial x}-\frac{\partial X}{\partial z},$$

$$\frac{1}{c}\left\{\frac{\partial Z}{\partial t}+u_z\rho\right\}=\frac{\partial M}{\partial x}-\frac{\partial L}{\partial y}, \qquad \frac{1}{c}\frac{\partial N}{\partial t}=\frac{\partial X}{\partial y}-\frac{\partial Y}{\partial x},$$

where

$$\rho=\frac{\partial X}{\partial x}+\frac{\partial Y}{\partial y}+\frac{\partial Z}{\partial z}$$

denotes 4π times the charge density, and (u_x, u_y, u_z) the velocity-vector of the charge. If we imagine the electrical substances to be coupled in an unchanging manner to small rigid bodies (ions, electrons), these equations are the electromagnetic basis of Lorentz's electrodynamics and optics of moving bodies.

Let these equations be valid in the system K, and using the transformation equations given in §§3 and 6, transform them to the system k. We then obtain the equations

$$\frac{1}{c}\left\{\frac{\partial X'}{\partial \tau}+u_\xi\rho'\right\}=\frac{\partial N'}{\partial \eta}-\frac{\partial M'}{\partial \zeta}, \qquad \frac{1}{c}\frac{\partial L'}{\partial \tau}=\frac{\partial Y'}{\partial \zeta}-\frac{\partial Z'}{\partial \eta},$$

$$\frac{1}{c}\left\{\frac{\partial Y'}{\partial \tau}+u_\eta\rho'\right\}=\frac{\partial L'}{\partial \zeta}-\frac{\partial N'}{\partial \xi}, \qquad \frac{1}{c}\frac{\partial M'}{\partial \tau}=\frac{\partial Z'}{\partial \xi}-\frac{\partial X'}{\partial \zeta},$$

$$\frac{1}{c}\left\{\frac{\partial Z'}{\partial \tau}+u_\zeta\rho'\right\}=\frac{\partial M'}{\partial \xi}-\frac{\partial L'}{\partial \eta}, \qquad \frac{1}{c}\frac{\partial N'}{\partial \tau}=\frac{\partial X'}{\partial \eta}-\frac{\partial Y'}{\partial \xi},$$

where

$$u_\xi=\frac{u_x-v}{1-u_xv/c^2},$$

$$u_\eta = \frac{u_y}{\beta(1 - u_x v/c^2)},$$

$$u_\zeta = \frac{u_z}{\beta(1 - u_x v/c^2)},$$

and

$$\rho' = \frac{\partial X'}{\partial \xi} + \frac{\partial Y'}{\partial \eta} + \frac{\partial Z'}{\partial \zeta}$$

$$= \beta(1 - u_x v/c^2)\rho.$$

Since – as follows from the addition theorem of velocities (§5) – the vector (u_ξ, u_η, u_ζ) is nothing else than the velocity of the electric substances, measured in the system k. Consequently we have proved that, on the basis of our kinematical principles, the electrodynamic foundation of Lorentz's theory of the electrodynamics of moving bodies is in agreement with the principle of relativity.

In passing, it may be remarked that the following important theorem may easily be deduced from the developed equations: If an electrically charged body is in motion anywhere in space without altering its charge when regarded from a system of coordinates moving with the body, its charge also remains constant – when viewed from the "resting" system K.

§*10. Dynamics of the (Slowly Accelerated) Electron*

Let there be in motion in an electromagnetic field a point particle possessing an electric charge ε (in the sequel this electrically charged particle is called an "electron"). We assume for its law of motion the following: If at a certain instant of time the electron is at rest, then in the next instant of time the electron's motion is described by the equations

$$\mu \frac{d^2 x}{dt^2} = \varepsilon X,$$

$$\mu \frac{d^2 y}{dt^2} = \varepsilon Y,$$

$$\mu \frac{d^2 z}{dt^2} = \varepsilon Z,$$

where x, y, z denote the electron's coordinates, and μ the mass of the electron, as long as its motion is slow.

Now, secondly, let the velocity of the electron at a given instant of time be v. We seek the electron's law of motion in the immediately following instant of time.

Without loss of generality, we may and will assume that the electron, at the moment when we focus our attention on it, is at the coordinate origin, and moves with the velocity v along the X-axis of the system K. It is then clear that at the designated instant ($t = 0$) the electron is at rest relatively to a coordinate system k which is in motion with a velocity v parallel to the X-axis.

From the above assumption, in combination with the principle of relativity, it is clear that in the immediately following time (for small values of t) the electron, considered from the system k, moves in accordance with the equations

$$\mu \frac{d^2\xi}{d\tau^2} = \varepsilon X',$$

$$\mu \frac{d^2\eta}{d\tau^2} = \varepsilon Y',$$

$$\mu \frac{d^2\zeta}{d\tau^2} = \varepsilon Z',$$

in which the symbols ξ, η, ζ, τ, X', Y', Z' refer to the system k. If, further, we decide that when $t = x = y = z = 0$ then $\tau = \xi = \eta = \zeta = 0$, the transformation equations of §§3 and 6 hold, so that we have

$$\xi = \beta(x - vt), \quad \eta = y, \quad \zeta = z, \quad \tau = \beta(t - vx/c^2),$$

$$X' = X, \quad Y' = \beta(Y - vN/c), \quad Z' = \beta(Z + vM/c).$$

With the help of these equations we transform the above equations of motion from system k to system K, and obtain

$$\left.\begin{aligned} \frac{d^2x}{dt^2} &= \frac{\varepsilon}{\mu\beta^3} X, \\ \frac{d^2y}{dt^2} &= \frac{\varepsilon}{\mu\beta}\left(Y - \frac{v}{c}N\right), \\ \frac{d^2z}{dt^2} &= \frac{\varepsilon}{\mu\beta}\left(Z + \frac{v}{c}M\right) \end{aligned}\right\} \quad \cdots \text{(A)}$$

Taking the customary point of view we now inquire as to the "longitudinal" and the "transverse" mass of the moving electron. We write the equations (A) in the form

$$\mu\beta^3 \frac{d^2x}{dt^2} = \varepsilon X = \varepsilon X',$$

$$\mu\beta^2 \frac{d^2y}{dt^2} = \varepsilon\beta\left(Y - \frac{v}{c}N\right) = \varepsilon Y',$$

$$\mu\beta^2 \frac{d^2 z}{dt^2} = \varepsilon\beta\left(Z + \frac{v}{c}M\right) = \varepsilon Z',$$

and remark firstly that $\varepsilon X'$, $\varepsilon Y'$, $\varepsilon Z'$ are the components of the ponderomotive force acting upon the electron, as viewed in a system moving at this moment with the same velocity as the electron. (This force might be measured, for example, by a spring balance at rest in the last-mentioned system.) Now if we call this force simply "the force acting upon the electron,"* and maintain the equation—mass × acceleration = force—and if we also decide that the accelerations are to be measured in the resting system K, we obtain from the above equations

$$\text{Longitudinal mass} = \frac{\mu}{(\sqrt{1 - v^2/c^2})^3},$$

$$\text{Transverse mass} = \frac{\mu}{1 - v^2/c^2}.$$

Naturally, with a different definition of force and acceleration we would obtain other values for the masses. This shows us that in comparing different theories of the motion of the electron we must proceed very cautiously.

We remark that these results as to the mass are also valid for ponderable material points, because a ponderable material point can be made into an electron (in our sense of the word) by the addition of an *arbitrarily small* electric charge.

Next we determine the electron's kinetic energy. If an electron that was initially at rest at the coordinate origin of the system K moves along the X-axis under the influence of an electrostatic force X, it is clear that the energy obtained from the electrostatic field is $\int \varepsilon X\,dx$. As the electron is to be slowly accelerated, and consequently may not emit any energy in the form of radiation, the energy obtained from the electrostatic field must be equal to the electron's energy of motion W. Bearing in mind that during the whole process of motion which we are considering, the first of the equations (A) applies, we obtain

$$W = \int \varepsilon X\,dx = \mu \int_0^v \beta^3 v\,dv$$
$$= \mu c^2 \left\{ \frac{1}{\sqrt{1 - v^2/c^2}} - 1 \right\}.$$

[* The definition of force here given is not advantageous, as was first shown by M. Planck. It is more to the point to define force in such a way that the laws of momentum and energy assume the simplest form. (A.S.)]

Thus, when $v = c$, W becomes infinite. Velocities greater than that of light have – as in our previous results – no possibility of existence.

This expression for the kinetic energy must also, by virtue of the argument stated above, apply to ponderable masses as well.

We will now enumerate the properties of the electron's motion which result from the system of equations (A), and are accessible to experiment.

1. From the second equation of the system (A) it follows that when $Y = Nv/c$, an electric force Y and a magnetic force N have an equally strong deflective action on an electron moving with the velocity v. Thus we see that it is possible by our theory to determine the velocity of the electron from the ratio of the magnetic deflection A_m to the electric deflection A_e, for any velocity, by applying the law

$$\frac{A_m}{A_e} = \frac{v}{c}.$$

This relationship may be tested experimentally, since the velocity of the electron can be directly measured, e.g. by means of rapidly oscillating electric and magnetic fields.

2. From the derivation for the electron's kinetic energy, it follows that between the potential difference traversed and the acquired velocity v of the electron there must be the relationship

$$P = \int X\,dx = \frac{\mu}{\varepsilon}c^2\left\{\frac{1}{\sqrt{1 - v^2/c^2}} - 1\right\}.$$

3. We calculate the radius of curvature R of the electron's path when a magnetic force N is present (the only deflective force) acting perpendicularly to the electron's velocity. From the second of the equations (A) we obtain

$$-\frac{d^2y}{dt^2} = \frac{v^2}{R} = \frac{\varepsilon}{\mu}\frac{v}{c}N\sqrt{1 - \frac{v^2}{c^2}}$$

or

$$R = \frac{\mu c^2}{\varepsilon}\cdot\frac{v/c}{\sqrt{(1 - v^2/c^2)}}\cdot\frac{1}{N}.$$

These three relationships are a complete expression for the laws according to which, by the theory here advanced, the electron must move.

In conclusion I wish to say that in working at the problem here dealt with I have had the loyal assistance of my friend and colleague M. Besso, and that I am indebted to him for several valuable suggestions.

MACH, EINSTEIN, AND THE SEARCH FOR REALITY

IN THE history of ideas of our century, there is a chapter that might be entitled "The Philosophical Pilgrimage of Albert Einstein," a pilgrimage from a philosophy of science in which sensationism and empiricism were at the center, to one in which the basis was a rational realism. This essay, a portion of a more extensive study,[1] is concerned with Einstein's gradual philosophical reorientation, particularly as it has become discernible during the work on his largely unpublished scientific correspondence.[2]

The earliest known letter by Einstein takes us right into the middle of the case. It is dated 19 March 1901 and addressed to Wilhelm Ostwald.[3] The immediate cause for Einstein's letter was his failure to receive an assistantship at the school where he had recently finished his formal studies, the Polytechnic Institute in Zürich; he now turned to Ostwald to ask for a position at his laboratory, partly in the hope of receiving "the opportunity for further education." Einstein included a copy of his first publication, *Folgerungen aus den Capillaritätserscheinungen*[4] (1901), which he said had been inspired (*angeregt*) by Ostwald's work; indeed, Ostwald's ALLGEMEINE CHEMIE is the first book mentioned in all of Einstein's published work.

Not having received an answer, Einstein wrote again to Ostwald on 3 April 1901. On 13 April 1901 his father, Hermann Einstein, sent Ostwald a moving appeal, evidently without his son's knowledge. Hermann Einstein reported that his son esteems Ostwald "most highly among all scholars currently active in physics."[5]

The choice of Ostwald was significant. He was, of course, not only one of the foremost chemists, but also an active "philosopher-scientist" during the 1890's and 1900's, a time of turmoil in the physical sciences as well as in the philosophy of science. The opponents of kinetic, mechanical, or materialistic views of natural phenomena were vociferous. They objected to atomic theory and gained great strength from the victories of thermodynamics, a field in which no knowledge or assumption was needed concerning the detailed nature of material substances (for example, for an understanding of heat engines).

Ostwald was a major critic of the mechanical interpretation of physical phenomena, as were Helm, Stallo, and Mach. Their form of positivism—as against the sophisticated logical positivism developed later in Carnap and Ayer's work—provided an epistemology for the new phenomenologically based science of correlated observations, linking energetics and sensationism. In the second (1893) edition of his influential textbook on chemistry, Ostwald had given up the mechanical treatment of his first edition for Helm's "energetic" one. "Hypothetical" quantities such as atomic entities were to be omitted; instead, these authors claimed they were satisfied, as Merz wrote around 1904, with "measuring such quantities as are presented in observation, such as energy, mass, pressure, volume, temperature, heat, electrical potential, etc., without reducing them to imaginary mechanisms or kinetic quantities." They condemned such conceptions as the ether, with properties not accessible to direct observation, and they issued a call "to consider anew the ultimate principles of all physical reasoning, notably the scope and validity of the Newtonian laws of motion and of the conceptions of force and action, of absolute and relative motion."[6]

All these iconoclastic demands—except anti-atomism—must have been congenial to the young Einstein who, according to his colleague Joseph Sauter, was fond of calling himself "a heretic."[7] Thus, we may well suspect that Einstein felt sympathetic to Ostwald who denied in the ALLGEMEINE CHEMIE that "the assumption of that medium, the ether, is unavoidable. To me it does not seem to be so. . . . There is no need to inquire for a carrier of it when we find it anywhere. This enables us to look upon radiant energy as independently existing in space."[8] It is a position quite consistent with that shown later in Einstein's papers of 1905 on photon theory and relativity theory.

In addition, it is worth noting that Einstein, in applying to Ostwald's laboratory, seemed to conceive of himself as an experimentalist. We know from many sources that in his student years in Zürich, Einstein's earlier childhood interest in mathematics had slackened considerably. In the *Autobiographical Notes*, Einstein reported: "I really could have gotten a sound mathematical education. However, I worked most of the time in the physical laboratory, fascinated by the direct contact with experience."[9] To this, one of his few reliable biographers adds: "No one could stir him to visit the mathematical seminars. . . . He did not yet see the possibility of seizing that formative power resident in mathematics, which later became the guide of his work. . . . He wanted to proceed quite empirically, to suit his scientific feeling of the time. . . . As a natural scientist, he was a pure empiricist."[10]

Ostwald's main philosophical ally was the prolific and versatile Austrian physicist and philosopher Ernst Mach (1838–1916), whose main work Einstein had read avidly in his student years and with whom he was destined to have later the encounters that form a main concern of this paper. Mach's major book, THE SCIENCE OF MECHANICS,[11] first published in 1883, is perhaps most widely known for its discussion of Newton's PRINCIPIA, in particular for its devastating critique of what Mach called the "conceptual monstrosity of absolute space"—a conceptual monstrosity because it is "purely a thought-thing which cannot be pointed to in experience."[12] Starting from his analysis of Newtonian presuppositions, Mach proceeded in his announced program of eliminating all metaphysical ideas from science. As Mach said quite bluntly in the preface to the first edition of THE SCIENCE OF MECHANICS: "This work is not a text to drill theorems of mechanics. Rather, its intention is an enlightening one—or to put it still more plainly, an anti-metaphysical one."

It will be useful to review briefly the essential points of Mach's philosophy. Here we can benefit from a good, although virtually unknown, summary presented by his sympathetic follower, Moritz Schlick, in the essay *Ernst Mach, der Philosoph.*

Mach was a physicist, physiologist, and also psychologist, and his philosophy . . . arose from the wish to find a principal point of view to which he could hew in any research, one which he would not have to change when going from the field of physics to that of physiology or psychology. Such a firm point of view he reached by going back to that which is given before all scientific research: namely, the world of sensations. . . . Since all our testimony concerning the so-called external world relies only on sensations, Mach held that we can and must take these sensations and complexes of sensations to be the sole contents [*Gegenstände*] of those testimonies, and, therefore, that there is no need to assume in addition an unknown reality hidden behind the sensations. With that, the existence *der Dinge an sich* is removed as an unjustified and unnecessary assumption. A body, a physical object, is nothing else than a complex, a more or less firm [we would say, invariant] pattern of sensations, i.e., of colors, sounds, sensations of heat and pressure, etc.

There exists in this world nothing whatever other than sensations and their connections. In place of the word "sensations," Mach liked to use rather the more neutral word "elements." . . . [As is particularly clear in Mach's book ERKENNTNIS UND IRRTUM] scientific knowledge of the world consists, according to Mach, in nothing else than the simplest possible description of the connections between the elements, and it has as its only aim the intellectual mastery of those facts by means of the least possible effort of thought. This aim is reached by means of a more and more complete "accommodation of the thoughts to one another." This is the formulation by Mach of his famous "principle of the economy of thought."[13]

The influence of Mach's point of view, particularly in the German-speaking countries, was enormous—on physics, on physiology, on psychology, and on the fields of the history and the philosophy of science[14] (not to mention Mach's profound effect on the young Lenin, Hofmannsthal, Musil, among many others outside the sciences). Strangely neglected by recent scholarship—there is not even a major biography—Mach has in the last two or three years again become the subject of a number of promising studies. To be sure, Mach himself always liked to insist that he was beleaguered and neglected, and that he did not have, or wish to have, a philosophical system; yet his philosophical ideas and attitudes had become so widely a part of the intellectual equipment of the period from the 1880's on that Einstein was quite right in saying later that "even those who think of themselves as Mach's opponents hardly know how much of Mach's views they have, as it were, imbibed with their mother's milk."[15]

The problems of physics themselves at that time helped to reinforce the appeal of the new philosophical attitude urged by Mach. The great program of nineteenth-century physics, the reconciliation of the notions of ether, matter, and electricity by means of mechanistic pictures and hypotheses, had led to enormities—for example, Larmor's proposal that the electron is a permanent but movable state of twist or strain in the ether, forming discontinuous particles of electricity and possibly of all ponderable matter. To many of the younger physicists of the time, attacking the problems of physics with conceptions inherited from classical nineteenth-century physics did not seem to lead anywhere. And here Mach's iconoclasm and incisive critical courage, if not the details of his philosophy, made a strong impression on his readers.

Mach's Early Influence on Einstein

As the correspondence at the Einstein Archives at Princeton reveals, one of the young scientists deeply caught up in Mach's point of view was Michelange (Michele) Besso—Einstein's oldest and closest friend, fellow student, and colleague at the Patent Office in Bern, the only person to whom Einstein gave public credit for help (*manche wertvolle Anregung*) when he published his basic paper on relativity in 1905. It was Besso who introduced Einstein to Mach's work. In a letter of 8 April 1952 to Carl Seelig, Einstein wrote: "My attention was drawn to Ernst Mach's SCIENCE OF MECHANICS by my friend Besso while a student, around the year 1897. The book exerted a deep and persisting impression upon me . . . , owing to its physical orientation toward fundamental concepts and fundamental laws." As Einstein noted in his *Autobiographical Notes* written in 1946, Ernst Mach's THE SCIENCE OF MECHANICS "shook this dogmatic faith" in "mechanics as the final basis of all physical thinking. . . . This book exercised a profound influence upon me in this regard while I was a student. I see Mach's greatness in his incorruptible skepticism and independence; in my younger years, however, Mach's epistemological position also influenced me very greatly."[10]

As the long correspondence between those old friends shows, Besso remained a loyal Machist to the end. Thus, writing to Einstein on 8 December 1947, he still said: "As far as the history of science is concerned, it appears to me that Mach stands at the center of the development of the last 50 or 70 years." Is it not true, Besso also asked, "that this introduction [to Mach] fell into a phase of development of the young physicist [Einstein] when the Machist style of thinking pointed decisively at observables—perhaps even, indirectly, to clocks and meter sticks?"

Turning now to Einstein's crucial first paper on relativity in 1905, we can discern in it influences of many, partly contradictory, points of view—not surprising in a work of such originality by a young contributor. Elsewhere I have examined the effect—or lack of effect—on that paper of three contemporary physicists: H. A. Lorentz,[17] Henri Poincaré,[18] and August Föppl. Here we may ask in what sense and to what extent Einstein's initial relativity paper of 1905 was imbued with the style of thinking associated with Ernst Mach and his followers—apart from the characteristics of clarity and independence, the two traits in Mach which Einstein always praised most.

In brief, the answer is that the Machist component—a strong component, even if not the whole story—shows up prominently in two related respects: first, by Einstein's insistence from the beginning of his relativity paper that the fundamental problems of physics cannot be understood until an epistemological analysis is carried out, particularly so with respect to the meaning of the conceptions of space and time;[19] and second, by Einstein's identification of reality with what is given by sensations, the "events," rather than putting reality on a plane beyond or behind sense experience.

From the outset, the instrumentalist, and hence sensationist, views of measurement and of the concepts of space and time are strikingly evident. The key concept in the early part of the 1905 paper is introduced at the top of the third page in a straightforward way. Indeed, Leopold Infeld in his biography of Einstein called them "the simplest sentence[s] I have ever encountered in a scientific paper." Einstein wrote: "We have to take into account that all our judgments in which time plays a part are always judgments of *simultaneous events*. If for instance I say, 'that train arrived here at seven o'clock,' I mean something like this: 'The pointing of the small hand of my watch to seven and the arrival of the train are simultaneous events.' "[20]

The basic concept introduced here, one that overlaps almost entirely Mach's basic "elements," is Einstein's concept of *events* [*Ereignisse*]—a word that recurs in Einstein's paper about a dozen times immediately following this citation. Transposed into Minkowski's later formulation of relativity, Einstein's "events" are the intersections of particular "world lines," say that of the train and that of the clock. The time (*t* coordinate) of an event by itself has no operational meaning. As Einstein says: "The 'time' of an event is that which is given simultaneously with the event by a stationary clock located at the place of the event."[21] We can say that just as the *time* of an event assumes meaning only when it connects with our consciousness through sense experience (that is, when it is subjected to measurement-in-principle by means of a clock present at the same place), so also is the *place,* or space coordinate, of an event meaningful only if it enters our sensory experience while being subjected to measurement-in-principle (that is, by means of meter sticks present on that occasion at the same time).[22]

This was the kind of operationalist message which, for most of his readers, overshadowed all other philosophical aspects in Einstein's paper. His work was enthusiastically embraced by the groups who saw themselves as philosophical heirs of Mach, the Vienna Circle of neopositivists and its predecessors and related followers,[23] providing a tremendous boost for the philosophy that had initially helped to nurture it. A typical response welcoming the relativity theory as "the victory over the metaphysics of absolutes in the conceptions of space and time . . . a mighty impulse for the development of the philosophical point of view of our time," was extended by Joseph Petzoldt in the inaugural session of the *Gesellschaft für positivistische Philosophie* in Berlin, 11 November 1912.[24] Michele Besso, who had heard the message from Einstein before anyone else, had exclaimed: "In the setting of Minkowski's space-time framework, it was now first possible to carry through the thought which the great mathematician, Bernhard Riemann, had grasped: 'The space-time framework itself is formed by the events in it.' "[25]

To be sure, re-reading Einstein's paper with the wisdom of hindsight, as we shall do presently, we can find in it also very different trends, warning of the possibility that "reality" in the end is not going to be left identical with "events." There are premonitions that sensory experiences, in Einstein's later work, will not be regarded as the chief building blocks of the "world," that the laws of physics themselves will be seen to be built into the event-world as the undergirding structure "governing" the pattern of events.

Such precursors appear even earlier, in one of Einstein's first letters in the Archives. Addressed to his friend Marcel Grossmann, it is dated

14 April 1901, when Einstein believed he had found a connection between Newtonian forces and the forces of attraction between molecules: "It is a wonderful feeling to recognize the unity of a complex of appearances which, to direct sense experience, seem to be separate things." Already there is a hint here of the high value that will be placed on intuited unity and the limited role seen for evident sense experience.

But all this was not yet ready to come into full view, even to the author. Taking the early papers as a whole, and in the context of the physics of the day, we find that Einstein's philosophical pilgrimage did start on the historic ground of positivism. Moreover, Einstein thought so himself, and confessed as much in letters to Ernst Mach.

The Einstein-Mach Letters

In the history of recent science, the relation between Einstein and Mach is an important topic that has begun to interest a number of scholars. Indeed, it is a drama of which we can sketch here four stages: Einstein's early acceptance of the main features of Mach's doctrine; the Einstein-Mach correspondence and meeting; the revelation in 1921 of Mach's unexpected and vigorous attack on Einstein's relativity theory; and Einstein's own further development of a philosophy of knowledge in which he rejected many, if not all, of his earlier Machist beliefs.

Happily, the correspondence is preserved at least in part. A few letters have been found, all from Einstein to Mach. Those of concern here are part of an exchange between 1909 and 1913, and they testify to Einstein's deeply felt attraction to Mach's viewpoint, just at a time when the mighty Mach himself—forty years senior to the young Einstein whose work was just becoming widely known—had for his part embraced the relativity theory publicly by writing in the second (1909) edition of CONSERVATION OF ENERGY: "I subscribe, then, to the principle of relativity, which is also firmly upheld in my MECHANICS and WÄRMELEHRE.[26] In the first letter, Einstein writes from Bern on 9 August 1909. Having thanked Mach for sending him the book on the Law of Conservation of Energy, he adds: "I know, of course, your main publications very well, of which I most admire your book on Mechanics. You have had such a strong influence upon the epistemological conceptions of the younger generation of physicists that even your opponents today, such as Planck, undoubtedly would have been called Mach followers by physicists of the kind that was typical a few decades ago."

It will be important for our analysis to remember that Planck was Einstein's earliest patron in scientific circles. It was Planck who, in 1905, as editor of the ANNALEN DER PHYSIK, received Einstein's first relativity paper and thereupon held a review seminar on the paper in Berlin. Planck defended Einstein's work on relativity in public meetings from the beginning, and by 1913 had succeeded in persuading his German colleagues to invite Einstein to the Kaiser-Wilhelm-Gesellschaft in Berlin. With a polemical essay *Against the New Energetics* in 1896, he had made clear his position, and by 1909 Planck was one of the few opponents of Mach, and scientifically the most prominent one. He had just written a famous attack, *Die Einheit des physikalischen Weltbildes*. Far from accepting Mach's view that, as he put it, "Nothing is real except the perceptions, and all natural science is ultimately an economic adaptation of our ideas to our perceptions," Planck held to the entirely antithetical position that a basic aim of science is "the finding of a *fixed* world picture independent of the variation of time and people," or, more generally, "the complete liberation of the physical picture from the individuality of the separate intellects."[27]

At least by implication, Einstein's remarks to Mach show that he dissociated himself from Planck's view. It may also not be irrelevant that just at that time Einstein, who since 1906 had been objecting to inconsistencies in Planck's quantum theory, was preparing his first major invited paper before a scientific congress, the eighty-first meeting of the Naturforscherversammlung, announced for September, 1909, in Salzburg. Einstein's paper called for a revision of Maxwell's theory to accommodate the probabilistic character of the emission of photons—none of which Planck could accept—and concluded: "To accept Planck's theory means, in my view, to throw out the bases of our [1905] theory of radiation."

Mach's reply to Einstein's first letter is now lost, but it must have come quickly, because eight days later Einstein sends an acknowledgment:

Bern, 17 August 1909. Your friendly letter gave me enormous pleasure. . . . I am very glad that you are pleased with the relativity theory. . . . Thanking you again for your friendly letter, I remain, your student [indeed: *Ihr Sie verehrender Schüler*], A. Einstein.

Einstein's next letter was written as physics professor in Prague, where Mach before him had been for twenty-eight years. The post had been offered to Einstein on the basis of recommendations of a faction (Lampa, Pick) who regarded themselves as faithful disciples of Mach. The letter was sent out about New Year's 1911–12, perhaps just before or after Einstein's sole (and, according to Philipp Frank's account in EINSTEIN, HIS LIFE AND TIMES,[28] not very successful) visit to Mach, and after the first progress toward the general relativity theory:

. . . I can't quite understand how Planck has so little understanding for your efforts. His stand to my [general relativity] theory is also one of refusal. But I can't take it amiss; so far, that one single epistemological argument is the only thing which I can bring forward in favor of my theory.

Here, Einstein is referring delicately to the Mach Principle, which he had been putting at the center of the developing theory.[29] Mach responded by sending Einstein a copy of one of his books, probably the ANALYSIS OF SENSATIONS.

In the last of these letters to Mach (who was now seventy-five years old, and for some years had been paralyzed), Einstein writes from Zürich on 25 June 1913:

Recently you have probably received my new publication on relativity and gravitation which I have at last finished after unending labor and painful doubt. [This must have been the *Entwurf einer verallgemeinerten Relativitätstheorie und einer Theorie der Gravitation*, written with Marcel Grossmann.[30]] Next year at the solar eclipse it will turn out whether the light rays are bent by the sun, in other words whether the basic and fundamental assumption of the equivalence of the acceleration of the reference frame and of the gravitational field really holds. If so, then your inspired investigations into the foundations of mechanics—despite Planck's unjust criticism—will receive a splendid confirmation. For it is a necessary consequence that inertia has its origin in a kind of mutual interaction of bodies, fully in the sense of your critique of Newton's bucket experiment.[31]

The Paths Diverge

The significant correspondence stops here, but Einstein's public and private avowals of his adherence to Mach's ideas continue for several years more. For example, there is his well-known, moving eulogy of Mach, published in 1916.[15] In August, 1918, Einstein writes to Besso quite sternly about an apparent—and quite temporary—lapse in Besso's positivistic epistemology; it is an interesting letter, worth citing in full:

28 August 1918.
Dear Michele:

In your last letter I find, on re-reading, something which makes me angry: That speculation has proved itself to be superior to empiricism. You are thinking here about the development of relativity theory. However, I find that this development teaches something else, that it is practically the opposite, namely that a theory which wishes to deserve trust must be built upon generalizable facts.

Old examples: Chief postulates of thermodynamics [based] on impossibility of perpetuum mobile. Mechanics [based] on a grasped [*ertasteten*] law of inertia. Kinetic gas theory [based] on equivalence of heat and mechanical energy (also historically). Special Relativity on the constancy of light velocity and Maxwell's equation for the vacuum, which in turn rest on empirical foundations. Relativity with respect to uniform [?] translation is a *fact of experience.*

General Relativity: *Equivalence of inertial and gravitational mass.* Never has a truly useful and deep-going theory really been found purely speculatively. The nearest case is Maxwell's hypothesis concerning displacement current; there the problem was to do justice to the fact of light propagation. . . . With cordial greetings, your Albert. [Emphasis in the original]

Careful reading of this letter shows us that already here there is evidence of divergence between the conception of "fact" as understood by Einstein and "fact" as understood by a true Machist. The impossibility of the perpetuum mobile, the first law of Newton, the constancy of light velocity, the validity of Maxwell's equations, the equivalence of inertial and gravitational mass—none of these would have been called "facts of experience" by Mach. Indeed, Mach might have insisted that—to use one of his favorite battle words—it is evidence of "dogmatism" not to regard all these conceptual constructs as continually in need of probing re-examinations; thus, Mach had written:

. . . for me, matter, time and space are still *problems,* to which, incidentally, the physicists (Lorentz, Einstein, Minkowski) are also slowly approaching.[32]

Similar evidence of Einstein's gradual apostasy appears in a letter of 4 December 1919 to Paul Ehrenfest. Einstein writes:

I understand your difficulties with the development of relativity theory. They arise simply because you want to base the innovations of 1905 on epistemological grounds (nonexistence of the stagnant ether) instead of empirical grounds (equivalence of all inertial systems with respect to light).

Mach would have applauded Einstein's life-long suspicion of formal epistemological systems, but how strange would he have found this use of the word *empirical* to characterize the hypothesis of the equivalence of all inertial systems with respect to light! What we see forming slowly here is Einstein's view that the fundamental role played by experience in the construction of fundamental physical theory is, after all, not through the "atom" of experience, not through the individual sensation or the protocol sentence, but through some creative digest or synthesis of "*die gesammten Erfahrungstatsachen*," the *totality* of physical experience.[33] But all this was still hidden. Until Mach's death, and for several years after, Einstein considered and declared himself a disciple of Mach.

In the meantime, however, unknown to Einstein and everyone else, a time bomb had been ticking away. Set in 1913, it went off in 1921, five years after Mach's death, when Mach's The Principles of Physical Optics was published at last. Mach's preface was dated July, 1913—perhaps a few days or, at most, a few weeks after Mach had received Einstein's last, enthusiastic letter and the article on general relativity theory. In a well-known passage in the preface (but one usually found in an inaccurate translation), Mach had written:

I am compelled, in what may be my last opportunity, to cancel my views[*Anschauungen*] of the relativity theory.

I gather from the publications which have reached me, and especially from my correspondence, that I am gradually becoming regarded as the forerunner of relativity. I am able even now to picture approximately what new expositions and interpretations many of the ideas expressed in my book on Mechanics will receive in the future from this point of view. It was to be expected that philosophers and physicists should carry on a crusade against me, for, as I have repeatedly observed, I was merely an unprejudiced rambler endowed with original ideas, in varied fields of knowledge. I must, however, as assuredly disclaim to be a forerunner of the relativists as I personally reject the atomistic doctrine of the present-day school, or church. The reason why, and the extent to which, I reject [*ablehne*] the present-day relativity theory, which I find to be growing more and more dogmatical, together with the particular reasons which have led me to such a view—considerations based on the physiology of the senses, epistemological doubts, and above all the insight resulting from my experiments—must remain to be treated in the sequel [a sequel which was never published].[34]

Certainly, Einstein was deeply disappointed by this belated disclosure of Mach's sudden dismissal of the relativity theory. Some months later, during a lecture on 6 April 1922 in Paris, in a discussion with the anti-Machist philosopher Emile Meyerson, Einstein allowed in a widely reported remark that Mach was "*un bon mécanicien,*" but "*déplorable philosophe.*"[35]

We can well understand that Mach's rejection was at heart very painful, the more so as it was somehow Einstein's tragic fate to have the contribution he most cared about rejected by the very men whose approval and understanding he would have most gladly had—a situation not unknown in the history of science. In addition to Mach, the list includes these four: *Henri Poincaré*, who, to his death in 1912, only once deigned to mention Einstein's name in print, and then only to register an objection; *H. A. Lorentz,* who gave Einstein personally every possible encouragement—short of fully accepting the theory of relativity for himself; *Max Planck,* whose support of the special theory of relativity was unstinting, but who resisted Einstein's ideas on general relativity and the early quantum theory of radiation; and *A. A. Michelson,* who to the end of his days did not believe in relativity theory, and once said to Einstein that he was sorry that his own work may have helped to start this "monster."[36]

Soon Einstein's generosity again took the upper hand and resulted, from then to the end of his life, in many further personal testimonies to Mach's earlier influence.[37] A detailed analysis was provided in Einstein's letter of 6 January 1948 to Besso:

As far as Mach is concerned, I wish to differentiate between Mach's influence in general and his influence on me. . . . Particularly in the MECHANICS and the WÄRMELEHRE he tried to show how concepts arose out of experience. He took convincingly the position that these conceptions, even the most fundamental ones, obtained their warrant only out of empirical knowledge that they are in no way logically necessary. . . .

I see his weakness in this, that he more or less believed science to consist in a mere ordering of empirical material; that is to say, he did not recognized the freely constructive element in formation of concepts. In a way he thought that theories arise through *discoveries* and not through inventions. He even went so far that he regarded "sensations" not only as material which has to be investigated, but, as it were, as the building blocks of the real world; thereby, he believed, he could overcome the difference between psychology and physics. If he had drawn the full consequences, he would have had to reject not only atomism but also the idea of a physical reality.

Now, as far as Mach's influence on my own development is concerned, it certainly was great. I remember very well that you drew my attention to his MECHANICS and WÄRMELEHRE during my first years of study, and that both books made a great impression on me. The extent to which they influenced my own work is, to say the truth, not clear to me. As far as I am conscious of it, the immediate influence of Hume on me was greater. . . . But, as I said, I am not able to analyze that which lies anchored in unconscious thought. It is interesting, by the way, that Mach rejected the special relativity theory passionately (he did not live to see the general relativity theory [in the developed form]). The theory was, for him, inadmissibly speculative. He did not know that this speculative character belongs also to Newton's mechanics, and to every theory which thought is capable of. There exists only a gradual difference between theories, insofar as the chains of thought from fundamental concepts to empirically verifiable conclusions are of different lengths and complications.[38]

Antipositivistic Component of Einstein's Work

Ernst Mach's harsh words in his 1913 preface leave a tantalizing mystery. Ludwig Mach's destruction of his father's papers has so far made it impossible to find out more about the "experiments" (possibly on the constancy of the velocity of light) at which Ernst Mach hinted. Since 1921, many speculations have been offered to explain Mach's remarks.[39] They all leave something to be desired. Yet, I believe, it is not so difficult to reconstruct the main reasons why Mach ended up rejecting the relativity theory. To put it very simply, Mach had recognized more and more clearly, years before Einstein did so himself, that Einstein had indeed fallen away from the faith, had left behind him the confines of Machist empiriocriticism.

The list of evidences is long. Here only a few examples can be given, the first from the 1905 relativity paper itself: what had made it really work was that it contained and combined elements based on two entirely different philosophies of science—not merely the empiricist-operationist component, but the courageous initial postulation, in the second paragraph, of two thematic hypotheses (one on the constancy of light velocity and the other on the extension of the principle of relativity to all branches of physics), two postulates for which there was and can be no direct empirical confirmation.

For a long time, Einstein did not draw attention to this feature. In a lecture at King's College, London, in 1921, just before the posthumous publication of Mach's attack, Einstein still was protesting that the origin of relativity theory lay in the facts of direct experience:

> . . . I am anxious to draw attention to the fact that this theory is not speculative in origin; it owes its invention entirely to the desire to make physical theory fit observed fact as well as possible. We have here no revolutionary act, but the natural continuation of a line that can be traced through centuries. The abandonment of certain notions connected with space, time, and motion, hitherto treated as fundamentals, must not be regarded as arbitrary, but only as conditioned by observed facts.[40]

By June, 1933, however, when Einstein returned to England to give the Herbert Spencer Lecture at Oxford entitled *On the Method of Theoretical Physics*, the more complex epistemology that was in fact inherent in his work from the beginning had begun to be expressed. He opened this lecture with the significant sentence: "If you want to find out anything from the theoretical physicists about the methods they use, I advise you to stick closely to one principle: Don't listen to their words, fix your attention on their deeds." He went on to divide the tasks of experience and reason in a very different way from that advocated in his earlier visit to England:

> We are concerned with the eternal antithesis between the two inseparable components of our knowledge, the empirical and the rational. . . . The structure of the system is the work of reason; the empirical contents and their mutual relations must find their representation in the conclusions of the theory. In the possibility of such a representation lie the sole value and justification of the whole system, and especially the concepts and fundamental principles which underlie it. Apart from that, these latter are free inventions of the human intellect, which cannot be justified either by the nature of that intellect or in any other fashion *a priori*.

In the summary of this section, he draws attention to the "purely fictitious character of the fundamentals of scientific theory." It is this penetrating insight which Mach must have smelled out much earlier and dismissed as "dogmatism."

Indeed, Einstein, in his 1933 Spencer Lecture—widely read, as were and still are so many of his essays—castigates the old view that "the fundamental concepts and postulates of physics were not in the logical sense inventions of the human mind but could be deduced from experience by 'abstraction'—that is to say, by logical means. A clear recognition of the erroneousness of this notion really only came with the general theory of relativity."

Einstein ends this discussion with the enunciation of his current credo, so far from that he had expressed earlier:

> Nature is the realization of the simplest conceivable mathematical ideas. I am convinced that we can discover, by means of purely mathematical constructions, those concepts and those lawful connections between them which furnish the key to the understanding of natural phenomena. Experience may suggest the appropriate mathematical concepts, but they most certainly cannot be deduced from it. Experience remains, of course, the sole criterion of physical utility of a mathematical construction. But the creative principle resides in mathematics. In a certain sense, therefore, I hold it true that pure thought can grasp reality, as the ancients dreamed.[41]

Technically, Einstein was now at—or rather just past—the midstage of his pilgrimage. He had long ago abandoned his youthful allegiance to a primitive phenomenalism that Mach would have commended. In the first of the two passages just cited and others like it, he had gone on to a more refined form of phenomenalism which many of the logical positivists could still accept. He has, however, gone beyond it in the second passage, turning toward interests that we shall see later to have matured into clearly metaphysical conceptions.

Later, Einstein himself stressed the key role of what we have called thematic rather than phenomenic elements[42]—and thereby he fixed the early date at which, in retrospect, he found this need to have arisen in his earliest work. Thus he wrote in his *Autobiographical Notes* of 1946 that "shortly after 1900 . . . I despaired of the possibility of discovering the true laws by means of constructive efforts based on known facts. The longer and the more despairingly I tried, the more I came to the conviction that only the discovery of a *universal formal principle* could lead us to assured results."[43]

Another example of evidence of the undercurrent of disengagement from a Machist position is an early one: it comes from Einstein's article on relativity in the 1907 JAHRBUCH DER RADIOAKTIVITÄT UND ELEKTRONIK,[44] where Einstein responds, after a year's silence, to Walter Kaufmann's 1906 paper in the ANNALEN DER PHYSIK.[45] That paper had been the first publication in the ANNALEN to mention Einstein's work on the relativity theory, published there the previous year. Coming from the eminent experimental physicist Kaufmann, it had been most significant that this very first discussion was announced as a categorical, experimental disproof of Einstein's theory. Kaufmann had begun his attack with the devastating summary:

I anticipate right here the general result of the measurements to be described in the following: *the measurement results are not compatible with the Lorentz-Einsteinian fundamental assumption.*[46]

Einstein could not have known that Kaufmann's equipment was inadequate. Indeed, it took ten years for this to be fully realized, through the work of Guye and Lavanchy in 1916. So in his discussion of 1907, Einstein had to acknowledge that there seemed to be small but significant differences between Kaufmann's results and Einstein's predictions. He agreed that Kaufmann's calculations seemed to be free of error, but "whether there is an unsuspected systematic error or whether the foundations of relativity theory do not correspond with the facts one will be able to decide with certainty only if a great variety of observational material is at hand."[47]

Despite this prophetic remark, Einstein does not rest his case on it. On the contrary, he has a very different, and what for his time and situation must have been a very daring, point to make: he acknowledges that the theories of electron motion given earlier by Abraham and by Bucherer do give predictions considerably closer to the experimental results of Kaufmann. But Einstein refuses to let the "facts" decide the matter: "In my opinion both [their] theories have a rather small probability, because their fundamental assumptions concerning the mass of moving electrons are not explainable in terms of theoretical systems which embrace a greater complex of phenomena."[48]

This is the characteristic position—the crucial difference between Einstein and those who make the correspondence with experimental fact the chief deciding factor for or against a theory: even though the "experimental facts" at that time very clearly seemed to favor the theory of his opponents rather than his own, he finds the ad hoc character of their theories more significant and objectionable than an apparent disagreement between his theory and their "facts."[49]

So already in this 1907 article—which, incidentally, Einstein mentions in his postcard of 17 August to Ernst Mach, with a remark regretting that he has no more reprints for distribution—we have explicit evidence of a hardening of Einstein against the epistemological priority of experiment, not to speak of sensory experience. In the years that followed, Einstein more and more openly put the consistency of a simple and convincing theory or of a thematic conception higher in importance than the latest news from the laboratory—and again and again he turned out to be right.

Thus, only a few months after Einstein had written in his fourth letter to Mach that the solar eclipse experiment will decide "whether the basic and fundamental assumption of the equivalence of the acceleration of the reference frame and of the gravitational field really holds," Einstein writes to Besso in a very different vein (in March 1914), before the first, ill-fated eclipse expedition was scheduled to test the conclusions of the preliminary version of the general relativity theory: "Now I am fully satisfied, and I do not doubt any more the correctness of the whole system, may the observation of the eclipse succeed or not. The sense of the thing [*die Vernunft der Sache*] is too evident." And later, commenting on the fact that there remains up to 10 per cent discrepancy between the measured deviation of light owing to the sun's field and the calculated effect based on the general relativity theory: "For the expert, this thing is not particularly important, because the main significance of the theory does not lie in the verification of little effects, but rather in the great simplification of the theoretical basis of physics as a whole."[50] Or again, in Einstein's *Notes on the Origin of the General Theory of Relativity*,[51] he reports that he "was in the highest degree amazed" by the existence of the equivalence between inertial and gravitational mass, but that he "had no serious doubts about its strict validity, even without knowing the results of the admirable experiment of Eötvös."

The same point is made again in a revealing account given by Einstein's student, Ilse Rosenthal-Schneider. In a manuscript, *Reminiscences of Conversation with Einstein*, dated 23 July 1957, she reports:

> Once when I was with Einstein in order to read with him a work that contained many objections against his theory . . . he suddenly interrupted the discussion of the book, reached for a telegram that was lying on the windowsill, and handed it to me with the words, "Here, this will perhaps interest you." It was Eddington's cable with the results of measurement of the eclipse expedition [1919]. When I was giving expression to my joy that the results coincided with

his calculations, he said quite unmoved, "But I knew that the theory is correct"; and when I asked, what if there had been no confirmation of his prediction, he countered: "Then I would have been sorry for the dear Lord—the theory *is* correct."[52]

Minkowski's "World" and the World of Sensations

The third major point at which Mach, if not Einstein himself, must have seen that their paths were diverging is the development of relativity theory into the geometry of the four-dimensional space-time continuum, begun in 1907 by the mathematician Hermann Minkowski (who, incidentally, had had Einstein as a student in Zürich). Indeed, it was through Minkowski's semipopular lecture, "Space and Time," on 21 September 1908 at the eightieth meeting of the Naturforscherversammlung,[53] that a number of scientists first became intrigued with relativity theory. We have several indications that Mach, too, was both interested in and concerned about the introduction of four-dimensional geometry into physics (in Mach's correspondence around 1910, for example, with August Föppl); according to Friedrich Herneck,[54] Ernst Mach specially invited the young Viennese physicist Philipp Frank to visit him "in order to find out more about the relativity theory, above all about the use of four-dimensional geometry." As a result, Frank, who had recently finished his studies under Ludwig Boltzmann and had begun to publish noteworthy contributions to relativity, published the "presentation of Einstein's theory to which Mach gave his assent" under the title *Das Relativitätsprinzip und die Darstellung der physikalischen Erscheinungen im vierdimensionalen Raum.*[55] It is an attempt, addressed to readers "who do not master modern mathematical methods," to show that Minkowski's work brings out the "empirical facts far more clearly by the use of four-dimensional world lines." The essay ends with the reassuring conclusion: "In this four-dimensional world the facts of experience can be presented more adequately than in three-dimensional space, where always only an arbitrary and one-sided projection is pictured."

Following Minkowski's own papers on the whole, Frank's treatment can make it nevertheless still appear that in most respects the time dimension is equivalent to the space dimensions. Thereby one could think that Minkowski's treatment based itself not only on a functional and operational interconnection of space and time, but also—fully in accord with Mach's own views—on the primacy of ordinary, "experienced" space and time in the relativistic description of phenomena.

Perhaps as a result of this presentation, Mach invoked the names of Lorentz, Einstein, and Minkowski in his reply of 1910 to Planck's first attack, citing them as physicists "who are moving closer to the problems of matter, space, and time." Already a year earlier, Mach seems to have been hospitable to Minkowski's presentation, although not without reservations. Mach wrote in the 1909 edition of CONSERVATION OF ENERGY:[26] "Space and time are here conceived not as independent entities, but as forms of the dependence of the phenomena on one another"; he also added a reference to Minkowski's lecture of 1908.[53] But a few lines earlier, Mach had written: "Spaces of many dimensions seem to me not so essential for physics. I would only uphold them if things of thought [*Gedankendinge*] like atoms are maintained to be indispensable, and if, then, also the freedom of working hypotheses is upheld."

It was correctly pointed out by C. B. Weinberg[56] that Mach may eventually have had two sources of suspicion against the Minkowskian form of relativity theory. As was noted above, Mach regarded the fundamental notions of mechanics as problems to be continually discussed with maximum openness within the frame of empiricism, rather than as questions that can be solved and settled—as the relativists, seemingly dogmatic and sure of themselves, were in his opinion more and more inclined to do. In addition, Mach held that the questions of physics were to be studied in a broader setting, encompassing biology and psychophysiology. Thus Mach wrote: "Physics is not the entire world; biology is there too, and belongs essentially to the world picture.[57]

But I see also a third reason for Mach's eventual antagonism against such conceptions as Minkowski's (unless one restricted their application to "mere things of thought like atoms and molecules, which by their very nature can never be made the objects of sensuous contemplations"[58]). If one takes Minkowski's essay seriously—for example, the abandonment of space and time separately, with identity granted only to "a kind of union of the two"—one must recognize that it entails the abandonment of the conceptions of experiential space and experiential time; and that is an attack on the very roots of sensations-physics, on the meaning of actual measurements. If identity, meaning, or "reality" lies in the four-dimensional space-time interval *ds*, one is dealing with a quantity which is hardly *denkökonomisch,* nor one that preserves the primacy of measurements in "real" space and time. Mach may well have seen the warning flag; and worse was soon to come, as we shall see at once.

In his exuberant lecture of 1908,[58] Minkowski had announced that "three-dimensional geometry becomes a chapter in four-dimensional physics. . . . Space and time are to fade away into the shadows, and only *eine Welt an sich* will subsist." In this "world" the crucial innovation is the conception of the *"zeitartige Vektorelement,"* *ds,* defined as $(1/c)\sqrt{c^2dt^2-dx^2-dy^2-dz^2}$ with imaginary components. To Mach, the word *Element* had a pivotal and very different meaning. As we saw in Schlick's summary, elements were nothing less than the sensations and complexes of sensations of which the world consists and which completely define the world. Minkowski's rendition of relativity theory was now revealing the need to move the ground of basic, elemental truths from the plane of direct experience in ordinary space and time to a mathematicized, formalistic model of the world in a union of space and time that is not directly accessible to sensation—and, in this respect, is reminiscent of absolute space and time concepts that Mach had called "metaphysical monsters."[59]

Here, then, is an issue which, more and more, had separated Einstein and Mach even before they realized it. To the latter, the fundamental task of science was economic and descriptive; to the former, it was speculative-constructive and intuitive. Mach had once written: "If all the individual facts—all the individual phenomena, knowledge of which we desire—were immediately accessible to us, science would never have risen."[60] To this, with the forthrightness caused perhaps by his recent discovery of Mach's opposition, Einstein countered during his lecture in Paris of 6 April 1922: "Mach's system studies the existing relations between data of experience: for Mach, science is the totality of these relations. That point of view is wrong, and in fact what Mach has done is to make a catalog, not a system."[61]

We are witnessing here an old conflict, one that has continued throughout the development of the sciences. Mach's phenomenalism brandished an undeniable and irresistible weapon for the critical reevaluation of classical physics, and in this it seems to hark back to an ancient position that looked upon sensuous appearances as the beginning and end of scientific achievement. One can read Galileo in this light, when he urges the primary need of *description* for the fall of bodies, leaving "the causes" to be found out later. So one can understand (or rather, misunderstand) Newton, with his too-well-remembered remark: "I feign no hypotheses."[62] Kirchhoff is in this tradition. Boltzmann wrote of him in 1888:

> The aim is not to produce bold hypotheses as to the essence of matter, or to explain the movement of a body from that of molecules, but to present equations which, free from hypotheses, are as far as possible true and quantitatively correct correspondents of the phenomenal world, careless of the essence of things and forces. In his book on mechanics, Kirchhoff will ban all metaphysical concepts, such as forces, the cause of a motion; he seeks only the equations which correspond so far as possible to observed motions.[63]

And so could, and did, Einstein himself understand the Machist component of his own early work.

Phenomenalistic positivism in science has always been victorious, but only up to a very definite limit. It is the necessary sword for destroying old error, but it makes an inadequate plowshare for cultivating a new harvest. I find it exceedingly significant that Einstein saw this during the transition phase of partial disengagement from the Machist philosophy. In the spring of 1917 Einstein wrote to Besso and mentioned a manuscript which Friedrich Adler had sent him. Einstein commented: "He rides Mach's poor horse to exhaustion." To this, Besso—the loyal Machist—responds on 5 May 1917: "As to Mach's little horse, we should not insult it; did it not make possible the infernal journey through the relativities? And who knows—in the case of the nasty quanta, it may also carry Don Quixote de la Einsta through it all!"

Einstein's answer of 13 May 1917 is revealing: "I do not inveigh against Mach's little horse; but you know what I think about it. It cannot give birth to anything living, it can only exterminate harmful vermin."

Toward a Rationalistic Realism

The rest of the pilgrimage is easy to reconstruct, as Einstein more and more openly and consciously turned Mach's doctrine upside down—minimizing rather than maximizing the role of actual details of experience, both at the beginning and at the end of scientific theory, and opting for a rationalism that almost inevitably would lead him to the conception of an objective, "real" world behind the phenomena to which our senses are exposed.

In the essay, *Maxwell's Influence on the Evolution of the Idea of Physical Reality* (1931), Einstein began with a sentence that could have been taken almost verbatim from Max Planck's attack on Mach in 1909, cited above: "The belief in an external world independent of the perceiving subject is the basis of all natural science." Again and again, in the period beginning with his work on the general relativity theory, Einstein insisted that between experience and reason, as well as between the world of sensory perception and the objective world, there are logically unbridgeable chasms. He characterized the efficacy of reason to grasp reality by the word *miraculous*; the very terminology in these statements would have been an anathema to Mach.

We may well ask when and under what circumstances Einstein himself became aware of his change. Here again, we may turn for illumination to one of the hitherto unpublished letters, one written to his old friend, Cornelius Lanczos, on 24 January 1938:

> Coming from sceptical empiricism of somewhat the kind of Mach's, I was made, by the problem of gravitation, into a believing rationalist, that is, one who seeks the only trustworthy source of truth in mathematical simplicity. The logically simple does not, of course, have to be physically true; but the physically true is logically simple, that is, it has unity at the foundation.

Indeed, all the evidence points to the conclusion that Einstein's work on general relativity theory was crucial in his epistemological development. As he wrote later in *Physics and Reality* (1936): "the first aim of the general theory of relativity was the preliminary version which, while not meeting the requirements for constituting a closed system, could be connected in as simple a manner as possible with 'directly observed facts.' "[64] But the aim, still apparent during the first years of correspondence with Mach, could not be achieved. In *Notes on the Origin of the General Theory of Relativity*, Einstein reported:

> I soon saw that the inclusion of non-linear transformation, as the principle of equivalence demanded, was inevitably fatal to the simple physical interpretation of the coordinate—i.e., that it could no longer be required that coordinate differences [*ds*] should signify direct results of measurement with ideal scales or clocks. I was much bothered by this piece of knowledge . . . [just as Mach must have been].

The solution of the above mentioned dilemma [from 1912 on] was therefore as follows: A physical significance attaches not to the differentials of the coordinates, but only to the Riemannian metric corresponding to them.[65]

And this is precisely a chief result of the 1913 essay of Einstein and Grossmann,[66] the same paper which Einstein sent to Mach and discussed in his fourth letter. This result was the final consequence of the Minkowskian four-space representation—the sacrifice of the primacy of direct sense perception in constructing a physically significant system. It was the choice that Einstein had to make—against fidelity to a catalogue of individual operational experiences, and in favor of fidelity to the ancient hope for a unity at the base of physical theory.[67]

Enough has been written in other places to show the connections that existed between Einstein's scientific rationalism and his religious beliefs. Max Born summarized it in one sentence: "He believed in the power of reason to guess the laws according to which God has built the world."[68] Perhaps the best expression of this position by Einstein himself is to be found in his essay, *Über den gegenwärtigen Stand der Feld-Theorie*, in the FESTSCHRIFT of 1929 for Aurel Stodola:

Physical Theory has two ardent desires, to gather up as far as possible all pertinent phenomena and their connections, and to help us not only to know *how* Nature is and *how* her transactions are carried through, but also to reach as far as possible the perhaps utopian and seemingly arrogant aim of knowing why Nature is *thus and not otherwise*. Here lies the highest satisfaction of a scientific person. . . . [On making deductions from a "fundamental hypothesis" such as that of the kinetic-molecular theory,] one experiences, so to speak, that God Himself could not have arranged those connections [between, for example, pressure, volume, and temperature] in any other way than that which factually exists, any more than it would be in His power to make the number 4 into a prime number. This is the promethean element of the scientific experience. . . . Here has always been for me the particular magic of scientific considerations; that is, as it were, the religious basis of scientific effort.[69]

This fervor is indeed far from the kind of analysis which Einstein had made only a few years earlier. It is doubly far from the asceticism of his first philosophic mentor, Mach, who had written in his day book: "Colors, space, tones, etc. These are the only realities. Others do not exist."[70] It is, on the contrary, far closer to the rational realism of his first scientific mentor, Planck, who had written: "The disjointed data of experience can never furnish a veritable science without the intelligent interference of a spirit actuated by faith. . . . We have a right to feel secure in surrendering to our belief in a philosophy of the world based upon a faith in the rational ordering of this world."[71] Indeed, we note the philosophical kinship of Einstein's position with seventeenth-century natural philosophers—for example, with Johannes Kepler who, in the preface of the MYSTERIUM COSMOGRAPHICUM, announced that he wanted to find out concerning the number, positions, and motions of the planets, "why they are as they are, and not otherwise," and who wrote to Herwart in April, 1599, that, with regard to numbers and quantity, "our knowledge is of the same kind as God's, at least insofar as we can understand something of it in this mortal life."

Not unexpectedly, we find that during this period (around 1930) Einstein's non-scientific writings began to refer to religious questions much more frequently than before. There is a close relation between his epistemology, in which reality does not need to be validated by the individual's sensorium, and what he called "Cosmic Religion," defined as follows: "The individual feels the vanity of human desires and aims, and the nobility and marvelous order which are revealed in nature and in the world of thought. He feels the individual destiny as an imprisonment and seeks to experience the totality of existence as a unity full of significance."[72]

Needless to say, Einstein's friends from earlier days sometimes had to be informed of his change of outlook in a blunt way. For example, Einstein wrote to Moritz Schlick on 28 November 1930:

> In general your presentation fails to correspond to my conceptual style insofar as I find your whole orientation so to speak too positivistic. . . .I tell you straight out: Physics is the attempt at the conceptual construction of a model of the *real world* and of its lawful structure. To be sure, it [physics] must present exactly the empirical relations between those sense experiences to which we are open; but only *in this way* is it chained to them. . . . In short, I suffer under the (unsharp) separation of Reality of Experience and Reality of Being. . . .
>
> You will be astonished about the "metaphysicist" Einstein. But every four- and two-legged animal is de facto in this sense metaphysicist. [Emphasis in the original.]

Similarly, Philipp Frank, Einstein's early associate and later his biographer, reports that the realization of Einstein's true state of thought reached Frank in a most embarrassing way, at the Congress of German physicists in Prague in 1929, just as Frank was delivering "an address in which I attacked the metaphysical position of the German physicists and defended the positivistic ideas of Mach." The very next speaker disagreed and showed Frank that he had been mistaken still to associate Einstein's views with that of Mach and himself. "He added that Einstein was entirely in accord with Planck's view that physical laws describe a reality in space and time that is independent of ourselves. At that time," Frank comments, "this presentation of Einstein's views took me very much by surprise."[73]

In retrospect it is, of course, much easier to see the evidences that this change was being prepared. Einstein himself realized more and more clearly how closely he had moved to Planck, from whom he earlier dissociated himself in three of the four letters to Mach. At the celebration of Planck's sixtieth birthday, two years after Mach's death, Einstein made a moving speech in which, perhaps for the first time, he referred publicly to the Planck-Mach dispute and affirmed his belief that "there is no logical way to the discovery of these elementary laws. There is only the way of intuition" based on *Einfühlung* in experience.[74] The scientific dispute concerning the theory of radiation between Einstein and Planck, too, had been settled (in Einstein's favor) by a sequence of developments after 1911—for example, by Bohr's theory of radiation from gas atoms. As colleagues, Planck and Einstein saw each other regularly from 1913 on. Among evidences of the coincidence of these outlooks there is in the Einstein Archives a handwritten draft, written on or just before 17 April 1931 and intended as Einstein's introduction to Planck's hard-hitting article *Positivism and the Real External World*.[75] In lauding Planck's article, Einstein concludes: "I presume I may add that both Planck's conception of the logical state of affairs as well as his subjective expectation concerning the later development of our science corresponds entirely with my own understanding."[76]

This essay gave a clear exposition of Planck's (and one may assume, Einstein's) views, both in physics and in philosophy more generally. Thus Planck wrote there:

The essential point of the positivist theory is that there is no other source of knowledge except the straight and short way of perception through the senses. Positivism always holds strictly to that. Now, the two sentences: (1) *there is a real outer world which exists independently of our act of knowing* and (2) *the real outer world is not directly knowable* form together the cardinal hinge on which the whole structure of physical science turns. And yet there is a certain degree of contradiction between those two sentences. This fact discloses the presence of the irrational, or mystic, element which adheres to physical science as to every other branch of human knowledge. The effect of this is that a science is never in a position completely and exhaustively to solve the problem it has to face. We must accept that as a hard and fast, irrefutable fact, and this fact cannot be removed by a theory which restricts the scope of science at its very start. Therefore, we see the task of science arising before us as an incessant struggle toward a goal which will never be reached, because by its very nature it is unreachable. It is of a metaphysical character, and, as such, is always again and again beyond our achievement.[77]

From then on, Einstein's and Planck's writings on these matters are often almost indistinguishable from each other. Thus, in an essay in honor of Bertrand Russell, Einstein warns that the "fateful 'fear of metaphysics' . . . has come to be a malady of contemporary empiricistic philosophizing."[78] On the other hand, in the numerous letters between the two old friends, Einstein and Besso, each to the very end touchingly and patiently tries to explain his position, and perhaps to change the other's. Thus, on 28 February 1952, Besso once more presents a way of making Mach's views again acceptable to Einstein. The latter, in answering on 20 March 1952, once more responds that the facts cannot lead to a deductive theory and, at most, can set the stage "for intuiting a general principle" as the basis of a deductive theory. A little later, Besso is gently scolded (in Einstein's letter of 13 July 1952): "It appears that you do not take the four-dimensionality of reality seriously, but that instead you take the present to be the only reality. What you call 'world' is in physical terminology 'spacelike sections' for which the relativity theory—already the special theory—denies objective reality."

In the end, Einstein came to embrace the view which many, and perhaps he himself, thought earlier he had eliminated from physics in his basic 1905 paper on relativity theory: that there exists an external, objective, physical reality which we may hope to grasp—not directly, empirically, or logically, or with fullest certainty, but at least by an intuitive leap, one that is only guided by experience of the totality of sensible "facts." Events take place in a "real world," of which the space-time world of sensory experience, and even the world of multidimensional continua, are useful conceptions, but no more than that.

For a scientist to change his philosophical beliefs so fundamentally is rare, but not unprecedented. Mach himself underwent a dramatic transformation quite early (from Kantian idealism, at about age seventeen or eighteen, according to Mach's autobiographical notes). We have noted that Ostwald changed twice, once to anti-atomism and then back to atomism. And strangely Planck himself confessed in his 1910 attack on Mach[27] that some twenty years earlier, near the beginning of his own career when Planck was in his late twenties (and Mach was in his late forties), he, too, had been counted "one of the decided followers of the Machist philosophy," as indeed is evident in Planck's early essay on the conservation of energy (1887).

In an unpublished fragment apparently intended as an additional critical reply to one of the essays in the collection ALBERT EINSTEIN: PHILOSOPHER-SCIENTIST (1949), Einstein returned once more to deal —quite scathingly—with the opposition. The very words he used showed how complete was the change in his epistemology. Perhaps even without consciously remembering Planck's words in the attack on Mach of 1909 cited earlier—that a basic aim of science is "the complete liberation of the physical world picture from the individuality of the separate intellects"[27]—Einstein refers to a "basic axiom" in his own thinking:

> It is the postulation of a "real world" which so-to-speak liberates the "world" from the thinking and experiencing subject. The extreme positivists think that they can do without it; this seems to me to be an illusion, if they are not willing to renounce thought itself.

Einstein's final epistemological message was that the world of mere experience must be subjugated by and based in fundamental thought so general that it may be called cosmological in character. To be sure, modern philosophy did not gain thereby a major, novel, and finished corpus. Physicists the world over generally feel that today one must steer more or less a middle course in the area between, on the one hand, the Machist attachment to empirical data or heuristic proposals as the sole source of theory and, on the other, the aesthetic-mathematical attachment to persuasive internal harmony as the warrant of truth. Moreover, the old dichotomy between rationalism and empiricism is slowly being dissolved in new approaches.[79] Yet by encompassing in his own philosophical development both ends of this range, and by always stating forthrightly and with eloquence his redefined position, Einstein not only helped us to define our own, but also gave us a virtually unique case study of the interaction of science and epistemology.

NOTES

1. The main results of the study to this point are contained in Chapters 6 through 9.

2. These documents are mostly on deposit at the Archives of the Estate of Albert Einstein at Princeton; where not otherwise indicated, citations made here are from those documents. In studying and helping to order for scholarly purposes the materials in the Archives, I am grateful for the help received from the Trustees of the Albert Einstein Estate, and particularly from Miss Helen Dukas.

I thank the Executor of the Estate for permission to quote from the writings of Albert Einstein. I also wish to acknowledge the financial support provided by the Rockefeller Foundation for cataloguing the collection in the Archives at Princeton. The Institute for Advanced Study at Princeton and its director have been most hospitable throughout this continuing work. I am also grateful to M. Vero Besso for permission to quote from the letters of his father, Michelange Besso. All translations here are the author's unless otherwise indicated.

Early drafts of portions of this essay have been presented as invited papers at the *Tagung* of *Eranos* in Ascona (August, 1965), at the International Congress for the History of Science in Warsaw (August, 1965), and at the meeting, *Science et Synthèse*, at UNESCO in Paris (December, 1965), published in SCIENCE ET SYNTHÈSE, Paris: Gallimard, 1967.

3. This letter as well as the next two letters mentioned in the text (those of 3 April 1901 and 13 April 1901) have been published by Hans-Günther Körber, FORSCHUNGEN UND FORTSCHRITTE, 38: 75–76, 1964.

4. Albert Einstein, *Folgerungen aus den Capillaritätserscheinungen*, ANNALEN DER PHYSIK, 4:513–523, 1901.

5. The only other known attempt on Einstein's part to obtain an assistantship at that time was a request to Kamerlingh-Onnes (12 April 1901), to which, incidentally, he also seems to have received no response.

6. J. T. Merz, A HISTORY OF EUROPEAN THOUGHT IN THE NINETEENTH CENTURY, 4 vols. (Edinburgh: William Blackwood & Sons, 1904–1912; reprint ed., New York: Dover Publishing Co., 1965) Volume II, pp. 184, 199.

7. ERINNERUNGEN AN ALBERT EINSTEIN, issued by the Patent Office in Bern, about 1965 (n. d., no pagination).

8. Wilhelm Ostwald, *Chemische Energie*, LEHRBUCH DER ALLGEMEINEN CHEMIE, 2nd ed. (Leipzig: Verlag von Wilhelm Engelmann, 1893), Volume II, Part 1, pp. 1014, 1016.

9. Albert Einstein, *Autobiographical Notes*, in P. A. Schilpp, ed., ALBERT

EINSTEIN: PHILOSOPHER-SCIENTIST (Evanston, Ill.: Library of Living Philosophers, 1949), p. 15.

10. Anton Reiser, ALBERT EINSTEIN (New York: A. & C. Boni, 1930), pp. 51–52.

11. Ernst Mach, DIE MECHANIK IN IHRER ENTWICKLUNG, HISTORISCH-KRITISCH DARGESTELLT, Leipzig: 1883.

12. Mach, *op. cit.*, preface, 7th ed., 1912.

13. Moritz Schlick, *Ernst Mach, der Philosoph,* in a special supplement on Ernst Mach in the NEUE FREIE PRESSE (Vienna), 12 June 1926. Einstein himself, in a brief and telling analysis, *Zur Enthüllung von Ernst Machs Denkmal,* published in the same issue (the day of the unveiling of a monument to Mach), wrote:

"Ernst Mach's strongest driving force was a philosophical one: the dignity of all scientific concepts and statements rests solely in isolated experiences [*Einzelerlebnisse*] to which the concepts refer. This fundamental proposition exerted mastery over him in all his research and gave him the strength to examine the traditional fundamental concepts of physics (time, space, inertia) with an independence which at that time was unheard of."

14. Among many evidences of Mach's effectiveness, not the least are his five hundred or more publications (counting all editions—for example, seven editions of his THE SCIENCE OF MECHANICS in German alone during his lifetime), as well as his large exchange of letters, books, and reprints (of which many important ones "carry the dedication of their authors," to cite the impressive catalogue of Mach's library by Theodor Ackermann, Munich, No. 634 [1959] and No. 636 [1960]). A glimpse of Mach's effect on those near him was furnished by William James, who in 1882 heard Mach give a "beautiful" lecture in Prague. Mach received James "with open arms. . . . Mach came to my hotel and I spent four hours walking and supping with him at his club, an unforgettable conversation. I don't think anyone ever gave me so strong an impression of pure intellectual genius. He apparently has read everything and thought about everything, and has an absolute simplicity of manner and winningness of smile when his face lights up, that are charming." From James's letter, in Gay Wilson Allen, WILLIAM JAMES, A BIOGRAPHY (New York: Viking Press, 1967), p. 249.

The topicality of Mach's early speculations on what is now part of General Relativity Theory is attested to by the large number of continuing contributions on the Mach Principle. Beyond that, Mach's influence today is still strong in scientific thinking, though few are as explicit and forthright as the distinguished physicist R. H. Dicke of Princeton University, in his recent, technical book THE THEORETICAL SIGNIFICANCE OF EXPERIMENTAL RELATIVITY (London:

Gordon and Breach, 1964), pp. vii–viii: "I was curious to know how many other reasonable theories [in addition to General Relativity] would be supported by the same facts. . . . The reason for limiting the class of theories in this way is to be found in matters of philosophy, not in the observations. Foremost among these considerations was the philosophy of Bishop Berkeley and E. Mach. . . . The philosophy of Berkeley and Mach always lurked in the background and influenced all of my thoughts."

15. Albert Einstein, *Ernst Mach*, PHYSIKALISCHE ZEITSCHRIFT, 17: 101–104, 1916.

16. Einstein, *Autobiographical Notes*, p. 21.

17. Gerald Holton, *On the Origins of the Special Theory of Relativity*, AMERICAN JOURNAL OF PHYSICS, 28:627–636, 1960. Much of this article is incorporated in Chapter 6.

18. Gerald Holton, *On the Thematic Analysis of Science: The Case of Poincaré and Relativity*, MÉLANGES ALEXANDRE KOYRÉ (Paris: Hermann, 1964), pp. 257–268, in good part used in Chapter 6.

19. For evidences that this insistence on prior epistemological analysis of conceptions of space and time are Machist rather than primarily derived from Hume and Kant (who had, however, also been influential), see Einstein's detailed rendition of Mach's critique of Newtonian space and time, in note 15; his discussion of Mach in note 9, pp. 27–29; and in note 1.

20. Albert Einstein, *Zur Elektrodynamik bewegter Körper,* ANNALEN DER PHYSIK, 17:893, 1905.

21. *Ibid.*, p. 894.

22. Philipp Frank, *Einstein, Mach, and Logical Positivism*, in Schilpp, *op. cit.*, pp. 272–273. "The definition of simultaneity in the special theory of relativity is based on Mach's requirement that every statement in physics has to state relations between observable quantities. . . . There is no doubt that . . . Mach's requirement, the 'positivistic' requirement, was of great heuristic value to Einstein."

23. For example, see Philipp Frank, MODERN SCIENCE AND ITS PHILOSOPHY (New York: George Braziller, 1955), pp. 61–89; Viktor Kraft, THE VIENNA CIRCLE, trans. Arthur Pap (New York: Philosophical Library, 1953); Richard von Mises, ERNST MACH UND DIE EMPIRISTISCHE WISSENSCHAFTSAUFFASSUNG (1938; printed as a fascicule of the series EINHEITSWISSENSCHAFT).

24. Joseph Petzoldt, *Gesellschaft für positivistiche Philosophie*, reprinted in ZEITSCHRIFT FÜR POSITIVISTISCHE PHILOSOPHIE, 1:4, 1913. In that same speech, Petzoldt sounded a theme that became widely favored in the positivistic inter-

pretation of the genesis of relativity theory—namely, that the relativity theory was developed in direct response to the puzzle posed by the results of the Michelson experiment.

In his interesting essay, *Das Verhältnis der Machschen Gedankenwelt zur Relativitätstheorie*, published as an appendix to the eighth German edition of Ernst Mach's DIE MECHANIK IN IHRER ENTWICKLUNG (Leipzig: F. A. Brockhaus, 1921), pp. 490–517, Petzoldt faithfully attempts to identify and discuss several Machist aspects of Einstein's relativity theory:

(1) The theory "in the end is based on the recognition of the coincidence of sensations; and therefore it is fully in accord with Mach's world-view, which may be best characterized as a relativistic positivism" (p. 516).

(2) Mach's works "produced the atmosphere without which Einstein's Relativity Theory would not have been possible" (p. 494), and in particular Mach's analysis of the equivalence of rotating reference objects in Newton's bucket experiment prepared for the next step, Einstein's "equivalence of relatively moving coordinate systems" (p. 495).

(3) Mach's principle of economy is said to be marvelously exhibited in Einstein's succinct and simple statements of the two fundamental hypotheses. The postulate of the equivalence of inertial coordinate systems deals with "the simplest case thinkable, which now also serves as a fundamental pillar for the General Theory. And Einstein chose also with relatively greatest simplicity the other basic postulate [constancy of light velocity]. . . . These are the foundations. Everything else is logical consequence" (pp. 497–498).

25. Letter of Besso to Einstein, 16 February 1939. Among many testimonies to the effect of Einstein on positivistic philosophies of science, see P. W. Bridgman, *Einstein's Theory and the Operational Point of View*, in Schilpp, *op. cit.*, pp. 335–354.

26. Ernst Mach, HISTORY AND ROOT OF THE PRINCIPLE OF THE CONSERVATION OF ENERGY (Chicago: The Open Court Publishing Co., 1911), p. 95, translation by Philip E. B. Jourdain of the second edition (Ernst Mach, DIE GESCHICHTE UND DIE WURZEL DES SATZES VON DER ERHALTUNG DER ARBEIT, Leipzig: J. A. Barth, 1909). For a brief analysis of Mach's various expressions of adherence as well as reservations with respect to the principle of relativity, see Hugo Dingler, DIE GRUNDGEDANKEN DER MACHSCHEN PHILOSOPHIE (Leipzig: J. A. Barth, 1924), pp. 73–86. Friedrich Herneck (*Nochmals über Einstein und Mach*, PHYSIKALISCHE BLÄTTER, 17:275, 1961) reports that Frank wrote him he had the impression during a discussion with Ernst Mach around 1910 that Mach "was fully in accord with Einstein's special relativity theory, and particularly with its philosophical basis."

27. Republished in Max Planck, A SURVEY OF PHYSICAL THEORY (New

York: Dover Publications, 1960), p. 24. We shall read later a reaffirmation of this position, in almost exactly the same words, but from another pen.

After Mach's rejoinder (*Die Leitgedanken meiner naturwissenschaftlichen Erkenntnislehre und ihre Aufnahme durch die Zeitgenossen*, SCIENTIA, 7:225, 1910), Planck wrote a second, much more angry essay, *Zur Machschen Theorie der physikalischen Erkenntnis*, VIERTELJAHRSCHRIFT FÜR WISSENSCHAFTLICHE PHILOSOPHIE, 34:507, 1910. He ends as follows: "If the physicist wishes to further his science, he must be a Realist, not an Economist [in the sense of Mach's principle of economy]; that is, in the flux of appearances he must above all search for and unveil that which persists, is not transient, and is independent of human senses."

28. Philipp Frank, EINSTEIN, HIS LIFE AND TIMES, trans. George Rose, ed. and rev. Suichi Kusaka, New York: Alfred A. Knopf, 1947.

29. Later Einstein found that this procedure did not work; see Albert Einstein, *Notes on the Origin of the General Theory of Relativity*, IDEAS AND OPINIONS, trans. Sonja Bargmann (London: Alvin Redman, 1954), pp. 285–290; and other publications. In a letter of 2 February 1954 to Felix Pirani, Einstein writes: "One shouldn't talk at all any longer of Mach's principle, in my opinion. It arose at a time when one thought that 'ponderable bodies' were the only physical reality and that in a theory all elements that are not fully determined by them should be conscientiously avoided. I am quite aware of the fact that for a long time, I, too, was influenced by this fixed idea."

30. Albert Einstein and Marcel Grossman, *Entwurf einer verallgemeinerten Relativitätstheorie und einer Theorie der Gravitation*, ZEITSCHRIFT FÜR MATHEMATIK UND PHYSIK, 62:225–261, 1913.

31. For a further analysis and the full text of the four letters see Friedrich Herneck, *Zum Briefwechsel Albert Einsteins mit Ernst Mach*, FORSCHUNGEN UND FORTSCHRITTE, 37:239–243, 1963, and *Die Beziehungen zwischen Einstein und Mach, documentarisch Dargestellt*, WISSENSCHAFTLICHE ZEITSCHRIFT DER FRIEDRICH-SCHILLER-UNIVERSITÄT JENA, MATHEMATISCH-NATURWISSENSCHAFTLICHE REIHE, 15:1–14, 1966; and Helmut Hönl, *Ein Brief Albert Einsteins an Ernst Mach*, PHYSIKALISCHE BLÄTTER, 16:571–580, 1960. Many other evidences, direct and indirect, have been published to show Mach's influence on Einstein prior to Mach's death in 1916. For example, recently a document has been found which shows that in 1911 Mach had participated in formulating and signing a manifesto calling for the founding of a society for positivistic philosophy. Among the signers, together with Mach, we find Joseph Petzoldt, David Hilbert, Felix Klein, Georg Helm, Sigmund Freud, and Einstein. (See Herneck, *Nochmals über Einstein und Mach*, p. 276.)

32. Ernst Mach, *Die Leitgedanken meiner naturwissenschaftlichen Erk-*

enntnislehre und ihre Aufnahme durch die Zeitgenossen, PHYSIKALISCHE ZEITSCHRIFT, 11:605, 1910. (Emphasis added.) To be sure, Mach was not always a "Machist" himself.

33. See Einstein, *Time, Space, and Gravitation* (1948), OUT OF MY LATER YEARS (New York: Philosophical Library, 1950), pp. 54–58. Einstein makes the distinction between constructive theories and "theories of principle." Einstein cites, as an example of the latter, the relativity theory, and the laws of thermodynamics. Such theories of principle, Einstein says, start with "empirically observed general properties of phenomena." See also *Autobiographical Notes*, p. 53.

34. Ernst Mach, DIE PRINZIPIEN DER PHYSIKALISCHEN OPTIK, Leipzig: J. A. Barth, 1921. The English edition is THE PRINCIPLES OF PHYSICAL OPTICS, trans. John S. Anderson and A. F. A. Young (London: Methuen, 1926); reprint ed., New York: Dover Publications, 1953), pp. vii–viii.

35. Einstein, in *Séance du 6 avril, 1922: La Théorie de la relativité*, BULLETIN DE LA SOCIÉTÉ FRANÇAISE DE PHILOSOPHIE, 22:91–113, 1922. In his 1913 preface rejecting relativity, Mach expressed himself perhaps more impetuously and irascibly than he may have meant. Some evidence for this possibility is in Mach's letters to Petzoldt. In early 1914, Mach wrote: "I have received the copy of the positivistic ZEITSCHRIFT which contains your article on relativity; I liked it not only because you copiously acknowledge my humble contributions with respect to that theme, but also in general." And within a month, Mach writes—rather more incoherently—to Petzoldt: "The enclosed letter of Einstein [a copy of the last of Einstein's four letters, cited above] proves the penetration of positivistic philosophy into physics; you can be glad about it. A year ago, philosophy was altogether sheer nonsense. The details prove it. The paradox of the clock would not have been noticed by Einstein a year ago."

I thank Dr. John Blackmore for drawing my attention to the Mach-Petzoldt letters, and Dr. H. Müller for providing copies from the Petzoldt Archive in Berlin.

36. R. S. Shankland, *Conversations with Einstein*, AMERICAN JOURNAL OF PHYSICS, 31:56, 1963.

37. A typical example is his letter of 18 September 1930 to Armin Weiner: ". . . I did not have a particularly important exchange of letters with Mach. However, Mach did have a considerable influence upon my development through his writings. Whether or to what extent my life's work was influenced thereby is impossible for me to find out. Mach occupied himself in his last years with the relativity theory, and in a preface to a late edition of one of his works even spoke out in rather sharp refusal against the relativity theory. However, there can be no doubt that this was a consequence of a lessening

ability to take up [new ideas] owing to his age, for the whole direction of thought of this theory conforms with Mach's, so that Mach quite rightly is considered as a forerunner of general relativity theory. . . ."

I thank Colonel Bern Dibner for making a copy of the letter available to me from the Archives of the Burndy Library in Norwalk, Connecticut. Among other hitherto unpublished letters in which Einstein indicated his indebtedness to Mach, we may cite one to Anton Lampa, 9 December 1935: ". . . You speak about Mach as about a man who has gone into oblivion. I cannot believe that this corresponds to the fact since the philosophical orientation of the physicists today is rather close to that of Mach, a circumstance which rests not a little on the influence of Mach's writings."

Moreover, practically everyone else shared Einstein's explicitly expressed opinion of the debt of relativity theory to Mach; thus Hans Reichenbach wrote in 1921, "Einstein's theory signified the accomplishment of Mach's program." (*Der gegenwärtige Stand der Relativitätsdiskussion,* Logos, 10:333, 1922.) Even Hugo Dingler agreed: [Mach's] criticism of the Newtonian conceptions of time and space served as a starting point for the relativity theory. . . . Not only Einstein's work, but even more recent developments, such as Heisenberg's quantum mechanics, have been inspired by the Machian philosophy." Hugo Dingler, *Ernst Mach*, Encyclopedia of the Social Sciences, ed. Edwin R. A. Seligman and Alvin Johnson (New York: Macmillan Co., 1933), Volume 9, p. 653. And H. E. Hering wrote an essay whose title is typical of many others: *Mach als Vorläufer des physikalischen Relativitätsprinzips*, Kölner Universitätszeitung, 1:3–4, January 17, 1920. I thank Dr. John Blackmore for a copy of the article.

38. In his article, *Zur Enthüllung von Ernst Machs Denkmal*, in the special supplement of Neue freie Presse cited in note 13, Einstein—then already disenchanted for some time with the Machist program—wrote immediately after the portion quoted in note 13:

"Philosophers and scientists have often criticized Mach, and correctly so, because he erased the logical independence of the concepts vis-à-vis the 'sensations,' [and] because he wanted to dissolve the Reality of Being, without whose postulation no physics is possible, in the Reality of Experience . . . ."

There were additional sources, both published and unpublished, on the detailed aspects of the relation between Einstein and Mach, which, for lack of space, cannot be summarized here.

39. For example, by Einstein himself, by Joseph Petzoldt, and by Hugo Dingler. I assign relatively little weight to the possibility that the rift grew out of the difference between Einstein and Mach on atomism. Herneck provides the significant report that according to a letter from Philipp Frank, Mach was personally influenced by Dingler, whom Mach had praised in the 1912

edition of the MECHANIK and who was from the beginning an opponent of relativity theory, becoming one of the most "embittered enemies" of Einstein (Herneck, *Die Beziehungen zwischen Einstein und Mach*, p. 14; see note 31). The copies of letters from Dingler to Mach in the Ernst-Mach-Institute in Freiburg indicate Dingler's intentions; nevertheless, there remains a puzzle about Dingler's role which is worth investigating. It is significant that in his 1921 essay (see note 24) Petzoldt devotes much space to a defense of Einstein's work against Dingler's attacks. See also Joachim Thiele, *Analysis of Mach's Preface*, in NTM, SCHRIFTENREIHE FÜR GESCHICHTE DER NATURWISSENSCHAFTEN, TECHNIK UND MEDIZIN (Leipzig), 2, no. 6: 10–19, 1965.

40. Albert Einstein, *Über Relativitätstheorie*, MEIN WELTBILD (Amsterdam: Querido Verlag, 1934), pp. 214–220; republished as *On the Theory of Relativity*, IDEAS AND OPINIONS, pp. 246–249. Herneck has given the texts of similar discussions on phonographic records by Einstein in 1921 and even in 1924; cf. Herneck's transcriptions of Einstein, *Zwei Tondokumente Einsteins zur Relativitätstheorie*, FORSCHUNGEN UND FORTSCHRITTE, 40: 133–134, 1966.

41. Quotations from Einstein, *Zur Methodik der theoretischen Physik*, MEIN WELTBILD, pp. 176–187, as reprinted in translation in *On the Method of Theoretical Physics*, IDEAS AND OPINIONS, pp. 270–276, except for correction of mistranslation of one line. There are a number of later lectures and essays in which the same point is made. See, for example, the lecture, *Physics and Reality* (1936, reprinted in IDEAS AND OPINIONS, pp. 290–323), which states that Mach's theory of knowledge is insufficient on account of the relative closeness between experience and the concepts which it uses; Einstein advocates going beyond this "phenomenological physics" to achieve a theory whose basis may be further removed from direct experience, but which in return has more "unity in the foundations." Or see *Autobiographical Notes*, p. 27: "In the choice of theories in the future," he indicates that the basic concepts and axioms will continue to "distance themselves from what is directly observable."

Even as Einstein's views developed to encompass the "*erlebbare, beobachtbare*" facts as well as the "*wild-spekulative*" nature of theory, so did those of many of the philosophers of science who also had earlier started from a strict Machist position. This growing modification of the original position, partly owing to "the growing understanding of the general theory of relativity," has been chronicled by Frank, for example, in *Einstein, Mach, and Logical Positivism*, in Schilpp, *op. cit.*, pp. 271–286.

42. For a discussion of thematic and phenomenic elements in theory construction, see my article, *The Thematic Imagination in Science*, in SCIENCE AND CULTURE, ed. Gerald Holton (Boston: Houghton Mifflin Co., 1965), and incorporated here in the Introduction and Chapter 1.

43. Einstein, *Autobiographical Notes*, p. 53. Emphasis added. On pp. 9–11, Einstein describes what may be a possible precursor of this attitude in his study of geometry in his childhood.

44. Einstein, *Über das Relativitätsprinzip und die aus demselben gezogenen Folgerungen*, JAHRBUCH DER RADIOAKTIVITÄT UND ELEKTRONIK, 4:411–462, 1907.

45. Walter Kaufmann, *Über die Konstitution des Elektrons*, ANNALEN DER PHYSIK, 19:487–553, 1906.

46. *Ibid.*, p. 495. Emphasis in original.

47. See note 44, p. 439.

48. *Ibid.* Shortly after Kaufmann's article appeared, Max Planck (*Die Kaufmannschen Messungen der Ablenkbarkeit der β-Strahlen in ihrer Bedeutung für die Dynamik der Elektronen*, PHYSIKALISCHE ZEITSCHRIFT, 7:753–761, 1906) took it on himself publicly to defend Einstein's work in an analysis of Kaufmann's claim. He concluded that Kaufmann's data did not have sufficient precision for his claim. Incidentally, Planck tried to coin the term for the new theory that had not yet been named: "*Relativtheorie.*"

49. It should be remembered that Poincaré, with a much longer investment in attempts to fashion a theory of relativity, was quite ready to give in to the experimental "evidence." See note 18.

50. Carl Seelig, ALBERT EINSTEIN (Zurich: Europa Verlag, 1954), p. 195.

51. Einstein, *Einiges über die Entstehung der allgemeinen Relativitätstheorie*, MEIN WELTBILD, pp. 248–256, reprinted in translation as *Notes on the Origin of the General Theory of Relativity*, IDEAS AND OPINIONS, pp. 285–90.

52. "*Da könnt' mir halt der liebe Gott leid tun, die Theorie stimmt doch.*" This semi-serious remark of a person who was anything but sacrilegious indeed illuminates the whole style of a significant group of new physicists. P. A. M. Dirac, *The Evolution of the Physicist's Picture of Nature*, SCIENTIFIC AMERICAN, 208:46–47, 1963 speaks about this, with special attention to the work of Schrödinger, a spirit close to that of his friend, Einstein, despite the ambivalence of the latter to the advances in quantum physics. We can do no better than quote *in extenso* from Dirac's account:

"Schrödinger worked from a more mathematical point of view, trying to find a beautiful theory for describing atomic events, and was helped by deBroglie's ideas of waves associated with particles. He was able to extend deBroglie's ideas and to get a very beautiful equation, known as Schrödinger's wave equation, for describing atomic processes. Schrödinger got this equation by pure thought, looking for some beautiful generalization of deBroglie's ideas and not by keeping close to the experimental development of the subject in the way Heisenberg did.

"I might tell you the story I heard from Schrödinger of how, when he first got the ideas for this equation, he immediately applied it to the behavior of the electron in the hydrogen atom, and then he got results that did not agree with experiment. The disagreement arose because at that time it was not known that the electron has a spin. That, of course, was a great disappointment to Schrödinger, and it caused him to abandon the work for some months. Then he noticed that if he applied the theory in a more approximate way, not taking into account the refinements required by relativity, to this rough approximation, his work was in agreement with observation. He published his first paper with only this rough approximation, and in this way Schrödinger's wave equation was presented to the world. Afterward, of course, when people found out how to take into account correctly the spin of the electron, the discrepancy between the results of applying Schrödinger's relativistic equation and the experiment was completely cleared up.

"I think there is a moral to this story, namely, that it is more important to have beauty in one's equations than to have them fit experiment. If Schrödinger had been more confident of his work, he could have published it some months earlier, and he could have published a more accurate equation. The equation is now known as the Klein-Gordon equation, although it was really discovered by Schrödinger before he discovered his nonrelativistic treatment of the hydrogen atom. It seems that if one is working from the point of view of getting beauty in one's equations, and if one has really a sound insight, one is on a sure line of progress. If there is not complete agreement between the results of one's work and experiment, one should not allow oneself to be too discouraged, because the discrepancy may well be due to minor features that are not properly taken into account and that will get cleared up with further developments of the theory. That is how quantum mechanics was discovered . . ." (pp. 46–47).

53. Published several times—for example, by B. G. Teubner, Leipzig, 1909.

54. Friedrich Herneck, *Zu einem Brief Albert Einsteins an Ernst Mach*, PHYSIKALISCHE BLÄTTER, 15:564, 1959. Frank's remark is reported by Herneck in *Ernst Mach und Albert Einstein*, SYMPOSIUM AUS ANLASS DES 50. TODESTAGES VON ERNST MACH, ed. Frank Kerkhof (Freiburg: Ernst Mach Institut, 1966), pp. 45–61.

55. Philipp Frank, *Das Relativitätsprinzip und die Darstellung der physikalischen Erscheinungen im vierdimensionalen Raum*, ZEITSCHRIFT FÜR PHYSIKALISCHE CHEMIE, 74:466–495, 1910.

56. C. B. Weinberg, *Mach's Empirio-Pragmatism in Physical Science* (Thesis, Columbia University, 1937).

57. Mach, *Die Leitgedanken meiner naturwissenschaftlichen Erkenntnislehre und ihre Aufnahme durch die Zeitgenossen*, SCIENTIA, 7:225.

58. Ernst Mach, SPACE AND GEOMETRY (Chicago: Open Court Publishing Co., 1906), p. 138. Mach's attempts to speculate on the use of n-dimensional spaces for representing the configuration of such "mere things of thought"—the derogatory phrase also applied to absolute space and absolute motion in Newton—are found in his first major book, CONSERVATION OF ENERGY (first edition, 1872).

59. Cf. Joseph Petzoldt, *Verbietet die Relativitätstheorie Raum und Zeit als etwas Wirkliches zu denken?* VERHANDLUNGEN DER DEUTSCHEN PHYSIKALISCHEN GESELLSCHAFT, No. 21–24 (1918), pp. 189–201. Here, and again in his 1921 essay (note 24), Petzoldt tries to protect Einstein from the charge—for example, made by Sommerfeld—that space and time no longer "are to be thought of as real."

60. Mach, CONSERVATION OF ENERGY, p. 54.

61. See note 35; also reported in *Einstein and the Philosophies of Kant and Mach*, NATURE, 112:253, 1923.

62. That Einstein did not so misunderstand Newton can be illustrated, for example, in a comment reported by C. B. Weinberg: "Dr. Einstein further maintained that Mach, as well as Newton, tacitly employs hypotheses—not recognizing their non-empirical foundations." (Weinberg, *op. cit.*, p. 55.) Dingler analyzed some of the nonempirical foundations of relativity theory in KRITISCHE BEMERKUNGEN ZU DEN GRUNDLAGEN DER RELATIVITÄTSTHEORIE (Leipzig: S. Hirzel, 1921).

63. Cited by Robert S. Cohen in his very useful essay *Dialectical Materialism and Carnap's Logical Empiricism*, THE PHILOSOPHY OF RUDOLF CARNAP, ed. P. A. Schilpp (La Salle, Ill.: Open Court Publishing Co., 1963), p. 109. I am also grateful to Professor Cohen for a critique of parts of this paper in earlier form.

64. Albert Einstein, *Physics and Reality*, JOURNAL OF THE FRANKLIN INSTITUTE, 221:313–347, 1936.

65. Einstein, *Notes on the Origin of the General Theory of Relativity*, pp. 288, 289.

66. See note 30, pp. 230–231.

67. I am not touching in this essay on the effect of quantum mechanics on Einstein's epistemological development; the chief reason is that while from his "heuristic" announcement of the value of a quantum theory in 1905, Einstein remained consistently skeptical about the "reality" of the quantum

theory of radiation, this opinion only added to the growing realism stemming from his work on general relativity theory. In the end, he reached the same position in quantum physics as in relativity; cf. his letter of 7 September 1944 to Max Born: "In our scientific expectations we have become antipodes. You believe in the dice-playing God, and I in the perfect rule of law in an objectively existing world which I try to capture in a wildly speculative way." (Reported by Max Born, *Erinnerungen an Einstein*, UNIVERSITAS, ZEITSCHRIFT FÜR WISSENSCHAFT, KUNST UND LITERATUR, 20:795–807, 1965.)

68. Max Born, *Physics and Relativity*, PHYSICS IN MY GENERATION (London: Pergamon Press, 1956), p. 205.

69. Albert Einstein, *Über den gegenwärtigen Stand der Feldtheorie*, FESTSCHRIFT PROF. DR. A. STODOLA ZUM 70. GEBURTSTAG, ed. E. Honegger (Zurich and Leipzig: Orell Füssli Verlag, 1929), pp. 126–132. I am grateful to Professor Cornelius Lanczos and Professor John Wheeler for pointing out this reference to me.

70. Dingler, DIE GRUNDGEDANKEN DER MACHSCHEN PHILOSOPHIE, p. 98.

71. Max Planck, THE PHILOSOPHY OF PHYSICS, trans. W. H. Johnson (New York: W. W. Norton & Co. 1936), pp. 122, 125.

72. Albert Einstein, *Religion and Science*, THE NEW YORK TIMES MAGAZINE, 9 November 1930; cf. MEIN WELTBILD, p. 39, and COSMIC RELIGION (New York: Covici-Friede, 1931), p. 48.

Possible reasons for Einstein's growing interest in these matters, partly related to the worsening political situation at the time, are discussed in Frank, EINSTEIN, HIS LIFE AND TIMES. (See note 28). It is noteworthy that while Einstein was quite unconcerned with religious matters during the period of his early scientific publications, he gradually returned later to a position closer to that at a very early age, when he reported he had felt a "deep religiosity. . . . It is quite clear to me that the religious paradise of youth . . . was a first attempt to free myself from the chains of the 'merely personal.'" *Autobiographical Notes*, in Schilpp, *op. cit.*, pp. 3, 5. For further remarks, see Chapter 3.

73. Frank, EINSTEIN, HIS LIFE AND TIMES, p. 215. Einstein's change of mind was, of course, not acceptable to a considerable circle of previously sympathetic scientists and philosophers. See, for example, P. W. Bridgman, *Einstein's Theory and the Operational Point of View* in Schilpp, *op. cit.*, pp. 335–354.

An interesting further confirmation of Einstein's changed epistemological position became available after this paper was first published. Werner Heisenberg, in PHYSICS AND BEYOND (New York, Harper & Row, 1971), pp. 62–66, writes about his conversation with Einstein concerning physics and philosophy.

See, for example, p. 63—a portion of a conversation set in 1925–1926:

"But you don't seriously believe," Einstein protested, "that none but observable magnitudes must go into physical theory?"

"Isn't that precisely what you have done with relativity?" I asked in some surprise. "After all, you did stress the fact that it is impermissible to speak of absolute time, simply because absolute time cannot be observed; that only clock readings, be it in the moving reference system or the system at rest, are relevant to the determination of time."

"Possibly I did use this kind of reasoning," Einstein admitted, "but it is nonsense all the same. Perhaps I could put it more diplomatically by saying that it may be heuristically useful to keep in mind what one has actually observed. But in principle, it is quite wrong to try founding a theory on observable magnitudes alone. In reality the very opposite happens. It is the theory which decides what we can observe."

See also Heisenberg's account of Einstein's critique of Mach, *ibid.*, pp. 63–66.

74. Originally entitled *Motiv des Forschens* (in ZU MAX PLANCKS 60. GEBURTSTAG [Karlsruhe: Müller, 1918]), reprinted in translation by James Murphy, as a preface to Max Planck, WHERE IS SCIENCE GOING? (London: Allen & Unwin, 1933), pp. 7–12. In an earlier appreciation of Planck in 1913, Einstein had written only very briefly about his epistemology, merely lauding Planck's essay of 1896 against energetics, and not mentioning Mach. [Einstein's important 1918 essay is analyzed in Chapter 9.]

75. Planck, *Positivism and External Reality*, INTERNATIONAL FORUM, 1, No. 1:12–16; 1, No. 2:14–19, 1931.

76. Einstein sent his introduction to the editor of the journal on 17 April 1931, but it appears to have come too late for inclusion.

77. Planck, *Positivism and External Reality*, pp. 15–17. Emphasis in original.

78. Albert Einstein, *Bemerkungen zu Bertrand Russells Erkenntnis-Theorie*, in P. A. Schilpp, ed., THE PHILOSOPHY OF BERTRAND RUSSELL (Evanston, Ill.: Library of Living Philosophers, 1944), p. 289.

79. Toward the end, Einstein himself acknowledged a similar point in his *Remarks Concerning the Essays Brought Together in This Co-operative Volume*, in Schilpp, ALBERT EINSTEIN, pp. 679–680: " 'Einstein's position . . . contains features of rationalism and extreme empiricism. . . .' This remark is entirely correct. . . . A wavering between these extremes appears to me unavoidable."

Gerald Holton

Einstein and the Cultural Roots of Modern Science

THE ROOTS OF SCIENCE IN THE CULTURAL SOIL

THE FRUITS OF SCIENTIFIC RESEARCH are nourished by many roots, including the earlier work of other scientists. Significantly, Albert Einstein himself characterized his work as the "Maxwellian Program."[1]

But the imagination of scientists often draws also on another, quite different, "extrascientific" type of source. In Einstein's own intellectual autobiography, he asserted that reading David Hume and Ernst Mach had crucially aided in his early discoveries.[2] Such hints point to one path that historical scholarship on Einstein, to this day, has hardly explored—tracing the main *cultural roots* that may have helped shape Einstein's scientific ideas in the first place, for example, the literary or philosophic aspect of the cultural milieu in which he and many of his fellow scientists grew up.[3] To put the question more generally, as Erwin Schrödinger did in 1932: To what extent is the pursuit of science *milieubedingt*, where the word *bedingt* can have the strict connective sense of "dependent on," the more gentle and useful meaning of "being conditioned by," or, as I prefer, "to be in resonance with"? In short, the main thrust of this essay is to explore how the cultural milieu in which Einstein found himself resonated with and conditioned his science.

Gerald Holton is Mallinckrodt Professor of Physics and Professor of History of Science Emeritus at Harvard University. *This essay is based on the Robert and Maurine Rothschild Distinguished Lecture in the History of Science, Harvard University, April 8, 1997.*

There are major studies of such milieu resonances for earlier scientists: for example, the effect of the neo-Platonic philosophy on the imagination of seventeenth-century figures such as Kepler and Galileo; the theological interests that affected Newton's work; the adherence to *Naturphilosophie* that supported the discoveries of Oersted, J. R. Mayer, and Ampère; or the connection between the religious beliefs of the Puritan period and the science of the day, described in the apt metaphor that concludes Robert K. Merton's famous 1938 monograph, "The *cultural soil* of seventeenth century England was peculiarly fertile for the growth and spread of science."[4]

But there have thus far been few attempts to take up the influence of the cultural milieu on the scientific advances of twentieth-century physical scientists. The best known is that of Paul Forman, who more than two decades ago tried to interpret aspects of some scientists' presentations of quantum mechanics chiefly as their response to the sociopolitical malaise in the Weimar Republic[5]—although that work has been vigorously disputed by John Hendry, Stephen Brush, and more recently by Kraft and Kroes.[6] An example of a different sort is in an area in which Max Jammer and I have published, namely, the study of the extent to which Niels Bohr's introduction of the complementarity principle into physics was influenced by his delight in Søren Kierkegaard's philosophical writings, by his courses taken under the philosopher Harald Høffding, and also, as he claimed, by his reading of William James.[7]

But so far, there have been few such investigations in the wider, intellectual-cultural direction. I have long thought (and taught) that the full understanding of any particular scientific advance requires attention to both content and context, employing the whole orchestra of instruments, so to speak, playing out the many interacting components, without which there cannot be a full description or understanding of a case. But this is rarely done, though a middle ground exists between the extremes of internalistic study of the text alone, on one end, and constructivist externalism on the other. Moreover, in tracing the contributions of twentieth-century physical scientists themselves, the bridge from the *humanistic* parts of culture to the scientific ones—which carried much traffic in the past—has narrowed and become fragile. That is a deplorable loss, and one that deserves our attention.

The specific case of Einstein demands such attention for at least two reasons. First, it may serve as an example for studying other major twentieth-century scientists whose work has been nourished by subterranean connections to elements of the humanistic tradition. Second, it will help us resolve an intriguing paradox that has plagued scholars concerned with the source and originality of Einstein's creativity.

A PERSONAL INTERLUDE

While it is fashionable for scholars to hide assiduously the private motivations and circumstances that initiated a specific research program, on this occasion it will be useful to sketch the personal trajectory that caused me to become aware of the puzzling, paradoxical aspects of Einstein's early work.

I can fix the moment at which I was first drawn into this field of research. When the news of Einstein's death on April 18, 1955, reached our physics department, my colleagues proposed a local commemoration of Einstein's life and work. Although my own research was chiefly in experimental high-pressure physics, I had also begun to write on topics in the history of science, and so my assignment was to present how Einstein's work had been analyzed by modern historians of science. Little did I know that this suggestion would start me on a search that eventually would change profoundly my life as a scholar.

First, I discovered to my dismay that practically nothing had been done by modern historians to study seriously Einstein's *scientific* contributions—their roots, their structure, their development, their wider influence. This was in striking contrast to the volume and distinction of scholarship on the work of scientists of earlier periods, which had examined and assessed the legacy of such giants as George Sarton, Otto Neugebauer, Joseph Needham, Marjorie Nicolson, Robert Merton, Alexander Koyré, Hélène Metzger, Ludwig Fleck, and others—not to speak of *their* ancestors, such as Pierre Duhem and Ernst Mach. I seemed to be in virgin territory. Even among the many Einstein biographies, there were few serious sources.[8]

Figure 1: Albert Einstein at age nineteen. (By permission of the Einstein Archive, Hebrew University, Jerusalem.)

In truth, at the time of Einstein's death he was still deeply respected, but chiefly by way of ancestral piety and for his courageous political opposition at the time to McCarthyism, the arms race, and the Cold War. Scientists generally regarded him as having become an obstinate seeker who had wasted his last decades pursuing in vain his program of finding a unified field theory; as he told a friend, "At Princeton they regard me as the village idiot." Even his general relativity theory began to be widely taught again only after his death. In his last years, he had become a ghostly figure—a long way from the image of the vigorous young man, ready for a brilliant career (figure 1).

Today, four decades later, this perception has vastly changed. To be sure, many bubbles are bursting from the deluge of trendy journalism, whose motto in writing on major figures is well summarized in a recent essay on Herman Melville that carried the headline "Forget the Whale—the Big Question is: Did He Beat His Wife?"[9] But among the people at large Einstein's image is perhaps more ubiquitous that ever; from professional science historians, there is now an increasing flood of good scholarship on Einstein, especially since a team of researchers at Boston University has begun to publish the volumes of Einstein's *Collected Papers*, with their extraordinarily valuable editorial comments providing further stimuli for research.

None of this could have been foreseen in 1955. In retrospect, I regret not having the wit, as I was drawn into this field, to quote Marie Curie. When asked why she took up the study of radioactivity, she is said to have replied, "Because there was no bibliography." But as the historian Tetsu Hiroshige later commented, somebody had to take a "first step" in research on Einstein; eventually, it helped launch an industry analogous to the long-established ones on Newton or Darwin.[10]

That first step came in the form of a trip to the Institute for Advanced Study in Princeton to look for documents on which to base some original remarks at the memorial meeting. The key to access would be Helen Dukas, not only a trustee of Einstein's estate, but active in Einstein's household from 1928 as his secretary and later as general marshallin in the household—knowledgeable about much of his life and work, she was the untiring translator of his drafts into English and, as it turned out, endowed with an encyclopedic memory of the details of Einstein's vast correspondence that had passed through her hands.

Elsewhere I have described something of my first encounter.[11] In the bowels of Fuld Hall at the Institute was a large vault, similar to those in banks. The heavy door was partly open, and inside, illuminated dimly by a lamp on her desk, was Helen Dukas, still handling correspondence, among twenty or so file drawers that turned out to contain Einstein's scientific correspondence and manuscripts.

Once I had calmed her inborn suspicion about strangers and was allowed to have access to the files, I found myself in a state of indescribable exhilaration, in a fantastic treasure house—the kind of which most historians dream. Those documents, almost all unpublished, were arranged in a chaotic state through which only Miss Dukas knew her way with ease; they seemed to breathe

the life of the great scientist and his correspondents from all points of the compass, a rich mixture of science and philosophical speculation, of humor and dead-serious calculations.

Eventually, during two stays at the Institute, I induced Miss Dukas to help reorganize the papers into an archive suitable for scholarly research, to have a *catalogue raisonné* made, and by and by to add to the files at the Institute what she called "the more personal correspondence," which she had kept at Einstein's Mercer Street home. The whole lot, now numbering about 45,000 documents, has since been transferred by Einstein's will to the library at Hebrew University in Jerusalem. Represented in that collection are most major physicists in Europe and abroad who were living at that time, as well as authors, artists, statesmen, and the wretched of the earth, seeking help. The collection is indeed a microscope on half a century of history.

It is an amazingly diverse correspondence. Take, for example, the letters exchanged during just one of Einstein's immensely busy and creative periods (1914–18); they indicate a wide spectrum of interests among the correspondents—mostly scientists—even if gauged just by the references made to the works of major scientific, literary, and philosophical figures, including Ampère, Boltzmann, Hegel, Helmholtz, Hertz, Hume, Kant, Kirchhoff, Mach, Poincaré, and Spinoza. And one word repeatedly appeared in the correspondence—*Weltbild*, only faintly translatable as "worldpicture" or "worldview." Initially I hardly knew how important this concept, and these authors, would become in understanding Einstein's whole research program.

But to return to my mission at the time. How to proceed? In that mountain of papers at Princeton, the question of which problem I would use to start on a historical study was almost irrelevant; wherever one looked, there were exciting possibilities. For example, what role did experiments play in the genesis of the special relativity theory? Like practically everyone else, I had thought that the Michelson-Morley experiment of 1886 was the *crucial* influence that led Einstein to the relativity theory. (Indeed, I had just recently published a textbook on physics that had said so.) I had read that opinion everywhere: Robert Millikan, for example, after describing the Michelson-Morley experiment, simply concluded with the sentence, "Thus was born the special theory of relativity."[12]

But looking at samples of Einstein's correspondence, it turned out to be not so simple. One such warning occurs in his letter of February 9, 1954, to F. G. Davenport: "One can therefore understand why in my personal struggle Michelson's experiment played no role or at least no decisive role." Indeed, I later found that Einstein had repeated his stance over and over again.[13] He had typically gone his own way, relying on well-established, much older findings—experiments by Faraday, Bradley, and Fizeau—saying, "They were enough."[14] The haunting question suggests itself: what helped young Einstein make the leap when other, more established physicists could have done it so much earlier?

Another example of a key document in the files was a copy of a letter dated April 14, 1901, from Einstein to his friend and fellow student Marcel Grossmann, the existence of which was known from Seelig's biography.[15] Its eye-opening content will become clearer when we later reread that letter in the context of others in the archive. Here we need only the key sentence, in which the twenty-three-year-old beginner announces the overarching theme that would guide him through the rest of his career: "*It is a wonderful feeling to recognize the unity* [Einheitlichkeit] *of a complex of appearances, which, to direct sense experience, seem to be separate things.*"

OUTLINING THE PARADOX

These hasty first glimpses of the products of a creative mind seemed puzzling, incoherent, contradictory to me at first. They also seemed to reinforce the paradox I have mentioned before, which, in its simplest form, runs like this: It is not difficult to document that, from the start, Einstein proudly rebelled against main conventions in science as well as the social and political norms of his time. But it can be shown that at the same time he also was deeply devoted to large parts of the existing cultural canons. Was this dichotomy a hindrance, or could it possibly be a clue to understanding Einstein's uniqueness in a new way?

As to Einstein's rebelliousness, that is easily summarized in its various forms—where "rebelliousness" is a shorthand term for such traits as disobedience or insubordination to authority, a tendency to be revolutionary, obstinately nonconformist, dissident, defiant, and, in a phrase he applied to himself, "stubborn as a mule." That image of Einstein is embedded both in the public perception and throughout the literature. For example, an Einstein biography written jointly by the mathematician Banesh Hoffmann (who once worked with Einstein) and Helen Dukas herself is entitled *Albert Einstein, Creator and Rebel.*[16] Lewis Feuer, in his 1974 book *Einstein and the Generations of Science*, presented an Einstein whose whole attitude in life and science was shaped by the countercultural milieu of the throng of young revolutionaries of every sort who lived in Zurich and Bern around the turn of the century.[17] Even the *New York Times* seemed to view the confirmation of the predictions of Einstein's general relativity theory as a grave social threat. On November 16, 1919, under the title "Jazz in Scientific World," the newspaper reported at length that Charles Poor, a professor of celestial mechanics at Columbia University, thought Einstein's success showed that the spirit of unrest of that period had "invaded science," and the *Times* added its own warning: "When is space curved? When do parallel lines meet? When is a circle not a circle? When are the three angles of a triangle not equal to two right angles? Why, when Bolshevism enters the world of science, of course."[18]

But concentrating only on that aspect of Einstein overlooks an entirely different aspect of his persona, namely, Einstein as a *cultural traditionalist*, even within the limits set by his innate skepticism. If it can be proven that these opposites are combined in Einstein (as I shall show), his type of rebellion would be far from the modern image of our twentieth-century rebels in art, poetry, politics, parts of academe, or folklore—rebels who typically reject the social-political conventions of the bourgeoisie along with its cultural canon. Moreover, we shall see how Einstein's assertion of obstinate nonconformity enabled him to clear the ground ruthlessly of obstacles impeding his great scientific advance, even though the program of that advance itself ran along one of the oldest traditionalist lines. Skepticism, while necessary, was not enough to build the Temple of Isis, to use the metaphor that had long been current among German scientists.[19] Einstein made that crystal clear in a famous letter to his friend Michele Besso, who had urged him now to apply Ernst Mach's skepticism, as he had earlier, in attacking the infernal difficulties of quantum physics. Einstein replied: "You know what I think about it. [Mach's way] cannot give birth to anything living; it can only exterminate harmful vermin."[20]

We shall document Einstein's rebellious image in more detail; then examine the contradictory element; and demonstrate how the paradoxical tension between Einstein's rebellious image and his contradictory side was put to constructive use in his work. In particular, we shall examine the influence of this tension as he adopted with daring courage a set of personal presuppositions that had a history reaching back to antiquity—but for which, as he put it to Max Born, moral support came only from his own "little finger."[21] For these courageous presuppositions, on which his early success depended, he could and did draw on supporting allies—little noticed so far but far more powerful even than Einstein's little finger—i.e., ideas he had absorbed through his cultural roots, from what Merton had called, in another context, the "cultural soil" of the time.[22] In the end, we will be able to understand Einstein's program, method, and results in a new way.

AN EXCURSION INTO TERMINOLOGY

Here, a side excursion into the terminology and social stratification of Einstein's milieu is necessary. When talking about "the cultural roots of Einstein's science"—and especially today, when various definitions of "culture" are violently battling for primacy among anthropologists—a brief summary is needed of key concepts operative in the German context at the time of young Einstein's formation, in order to understand the framework within which he and his work found their place, as well as the class to which he belonged, including the aspirations of that class. The main concepts that are relevant here are *Kultur* and its companions, *Zivilisation* and *Bildung*, as well as the two composite notions of *Kulturträger* and *Bildungsbürgertum*.

The German language distinguished more sharply between *Kultur* and *Zivilisation* than did the English or French languages between their equivalents.[23] Although both *Kultur* and *Zivilisation* were generally understood in German-speaking Europe as supra-individual, collective phenomena, typically *Zivilisation* focused on the material and technological side, while *Kultur*—as first adapted in the German context by Johann Gottfried Herder—referred to the spiritual and value-related products. In extreme cases, *Zivilisation* was identified with superficial "French reason," *Kultur* with deep "German soul."[24]

At the level of the individual, the term *Bildung* (loosely translated as "intellectual formation," "self-refinement," or "education") referred to the process through which a person could acquire the attitudes and products of *Kultur*. In turn, the nation's *Kultur* as a whole was sustained—and advanced at its upper, creative level—by such *gebildete* individuals. *Bildung* thus meant much more than job-related training; it defined an ideal of human development. And a chief tool for the young to acquire *Bildung* at its best, albeit for only a small fraction of the population, was by beginning one's study in the *Gymnasium*, the neo-humanistic secondary school for ages ten to eighteen or so. The students were expected to be quite thoroughly acquainted with the great German poets and thinkers (the *Dichter und Denker*) as well as classics from other cultures, especially of antiquity.

Happily, the team now preparing Einstein's *Collected Papers* has found the curricula at Einstein's Munich schools as well as at the high school in Aarau. A quick scan of a few mandatory parts of the canon gives a good impression of how the young minds of Einstein and his cohorts were meant to be shaped. Initially there are readings from the Bible; then Latin enters at age ten, and Greek at age thirteen; Caesar's *Gallic Wars* and Ovid's *Metamorphoses* are read; then, under the supervision of his only beloved teacher, Ferdinand Ruess, poems by Uhland, Schiller, Goethe, and others; Goethe's prose poem "Hermann and Dorothea" is studied along with Xenophone's "Anabasis"; and next year, more Schiller, Herder, Cicero, Virgil. At Aarau, Einstein encounters more of the classics in German, French, and Italian; a typical entry for his course in German in 1896 reads: "History of literature from Lessing to the death of Goethe. Read

Götz von Berlichingen . . . ," and the list ends with Iphigenia and Torquato Tasso.

Such knowledge also was intended to contribute to forging a common bond between the *gebildete* individuals raised on similar *Gymnasium* curricula throughout German-speaking countries, regardless of the particular professional discipline they were later to study at the universities, whether law, medicine, the humanities, or science—a preparation for the common understanding of that class in their conversations, letters, and popular lectures, across specialties and even in their intimate personal relations.

But while the *Gymnasium* placed heavy emphasis on Latin and Greek and other aspects of "pure" *Bildung*, it had little concern for the kind of practical knowledge offered in other types of German secondary schools without such attention to classical languages, for instance, the so-called *Realschulen* (where Einstein's father Hermann and uncle Jakob, headed for electrical engineering, had received their secondary education). Needless to say, those other schools were considered, with a dose of snobbism, to be culturally less valuable; their graduates were generally not considered for university training and hence unlikely to achieve the status of *Kulturträger*.

Here it is crucial to understand a subtlety in the German concept *Kulturträger*. The term had a double meaning: both carrier and pillar of *Kultur*. On the one hand, *gebildete* individuals—chiefly the graduates of *Gymnasium* who had gone on to the universities—were seen as personally carrying or even embodying *Kultur*, living among its products, and, in the case of the most outstanding ones, advancing the *Kultur*. On the other hand, as a group they functioned also as the chief supporters (*Träger*, or "pillars") of the nation's collective project of *Kultur*. Although the term *Kulturträger* itself became generally popular only after World War I, it was a key concept earlier, as the following episode illustrates. In 1910, a bill in Prussia proposed a change in the three-tiered electoral law so that *Kulturträger* be expressly favored; they would be put into a smaller pool of voters "above the class for which their wealth would qualify them," so that their votes would count more.[25]

At the level of social stratification, most of the *Kulturträger* could be identified as belonging to what has been called the *Bildungsbürgertum* (the educated members of the bourgeoisie). The sociologist Karl Mannheim usefully distinguished two components in the modern bourgeoisie. From the beginning, he wrote, it had two kinds of social roots:"on the one hand the owners of capital, on the other those whose only capital consisted in their education."[26] In nineteenth-century Germany, the latter formed the *Bildungsbürgertum*; their social ranks were symbolized by the certificates they had attained during the process of *Bildung* and often also by a position within the hierarchies of the civil service. *Bildungsbürger* worked predominantly in professions that required university training, as physicians, lawyers, and clergy, as well as teachers and professors and other higher officials in government service.

Variants of this social stratum of the *Bildungsbürgertum* existed in many countries, but its social clout was particularly strong in nineteenth-century Germany. First, in the context of its relatively backward economy at the time, the importance of serving in the governments of the multitude of German territories large or small favored the prominence of the *Bildungsbürgertum* over the economic bourgeoisie. Second, in the absence of a nation-state and a centralized economy, German nationalism focused on *Kultur* as the basis of the nation. What held the conception of Germany together was perhaps chiefly the cultural and scholarly output of its poets and dramatists, thinkers, composers, and, eventually, its scientists. One thinks here of Goethe and Schiller, Friedrich Gottlieb Klopstock, Gotthold Ephraim Lessing, Johann Gottfried Herder, Friedrich Hölderlin and Johann Joachim Winckelmann (the prophets of Hellenism), Friedrich Schleiermacher, Friedrich Schelling, Friedrich Schlegel, Immanuel Kant, Schopenhauer, and Nietzsche, as well as Bach, Haydn, Mozart, Schubert, and Beethoven.

Thus, the academic elite among the *Kulturträger* had fundamentally a twofold mission. One was to help secure, through their scholarship, the foundation of German nationhood—though, for most of them, this also involved keeping their distance from participation in political life—and so they tended to be looked up to by those who did not, or not yet, qualify for that rank. The other was to help prepare a cadre of *gebildete* individuals, high-level functionaries who were, to adapt Fritz Ringer's terminology, "Mandarins."[27]

It is ironic that whenever Einstein, after becoming world famous, traveled abroad to lecture, an official from the local German embassy or consulate would secretly report to the foreign office in Berlin on how Einstein had behaved and how he had been received. A typical account, now available, would state that Einstein had behaved well enough, and Germany would be wise to use him to conduct what one report calls "*Kulturpropaganda*."[28] In short, he might yet be put to use as a Mandarin.

As Mannheim noted, there existed among the *Kulturträger* themselves a small group of "free-floating" (*freischwebende*) intellectuals who led marginal existences, lacked a well-defined anchor in society, and had rather critical and even rebellious inclinations. They could not or would not share the staid material comforts of the *Bildungsbürger* and disliked the whole business of "climbing up to the next rung of social existence."[29] At this point we can connect these concepts with the status and hopes of the Einstein family, asking what young Einstein's place was within the cultural-social order of the time.

The Einsteins could trace their origins in southern Germany to the seventeenth century.[30] On the male side of the family, they had largely come from the small town of Buchau, in Swabia, which in midcentury had some two thousand inhabitants, of whom a few hundred were Jews. On the maternal side, the origins were chiefly in the similarly small Swabian town of Jebenhausen. Einstein's maternal grandfather Julius Koch left Jebenhausen for Cannstadt near Stuttgart and became quite wealthy through the grain trade. Einstein's mother, Pauline, thus belonged to the bourgeoisie chiefly by virtue of capital. His father Hermann's preparation in technical school and technical trade—like that of his brother and business partner, the engineer Jakob—also did not quite qualify them as part of *Bildungsbürgertum*, and certainly not as *Kulturträger*, though one may doubt that Hermann ever gave any thought to that. But at last the family tree had sprouted, in the form of Albert Einstein, a promise to grow into that higher social region—if only the bright lad would behave as he should!

We can now reformulate the paradoxical tension of Einstein's tendency toward social-political rebelliousness and his adherence to the products of *Kultur*. Was he just one of these rootless, rebellious intellectuals, reneging on his mission as a *Kulturträger*, or did his sympathies lie with the true carriers and pillars of national culture? To make the question more graphic, imagine a scene in which Einstein first stands accused of being a free-floating intellectual intent on undermining authority, and then is defended from that charge. The testimonials offered by either side will aid in understanding better the motivations behind Einstein's behavior—and his science.

CHRONOLOGY OF A CURIOUS REBELLION

A prosecuting attorney would find it easy to establish, both by chronology as well as psychosociological profile, a portrait of Einstein as a rebellious individual throughout his life. I have no competence or deep interest in searching for the possible causes; but as to the documentable facts, many details are well known, and the pattern they form is persuasive. Einstein made his obstinacy known almost from birth, refusing to speak until about age two and a half, or, as Erik Erikson remarked, until he could begin to speak sensibly in whole sentences.[31] When Albert reached school age, his penchant for defiance took a different form. In her memoir, his sister Maja reported that in opposition to his thoroughly secular home environment, young Albert decided to become a religious Jew and accordingly "obeyed in all particulars the religious commands," including the dietary ones.[32] But after he had advanced to the Munich Luitpold Gymnasium and encountered the state-prescribed, compulsory courses on Jewish religion there, Albert's interest in Judaism came to an abrupt end. His reading in scientific books led him, as he put it in his autobiography, to the conviction that organized religious education left him "with the impression that youth is intentionally being deceived by the state through lies." He now turned to a "positively fanatic [orgy of] free thinking," having formed a "suspicion against every kind of authority"; he found solace in what he later called his "holy little book of Geometry," which was given to him as a present for self-study—a first hint of where his destiny would lead him.[33]

But as one would expect, he found school life too regimented for his taste, and he dropped out of the *Gymnasium* at fifteen and a half, surely much to the relief of some of his teachers. About a year later, he renounced his citizenship as well. When he moved for his final year of high school to Aarau, Switzerland, he arrived as a thoroughly alienated youth, having left his school, his country, and his family; he even failed in his first attempt to enroll at the Swiss Polytechnic Institute. Once Albert got into the Polytechnic he continued his "in your face" rebelliousness, to the point that when speaking to his main professor, Heinrich Friedrich Weber—on whom his career might well depend—he refused to use the obligatory title and obstinately called him just "Herr Weber." In turn, Weber did nothing to help him in his job search later.

Einstein's lifestyle at the time was distinctly bohemian.[34] He lived on the margins of bourgeois society economically, socially, and (by the standards of the day and the place) morally; he lived together with his fellow student Mileva Marić, who bore their first child before they were married in 1903. To be sure, they passionately loved each other, and as their letters show, they were of one mind in railing against the "philistine" life and conventions they saw all around them.

Even in Einstein's great paper of 1905 on relativity, one can find many touches of that self-confident defiance and seeming arrogance, not only with respect to accepted ideas in science, but also to accepted style and practice. Thus the paper contained none of the expected footnote references or credits, only a mention of his friend Michele Besso, a person who of course would be unknown among research physicists.

We shall come back to that magical first period of Einstein's brilliance. Einstein, who often characterized himself as a gypsy, at first found only temporary teaching jobs, and those tended to end abruptly and noisily; finally, after the intercession of the father of his friend Marcel Grossmann, he found refuge at the Patent Office. By 1909, he began to be sought after by universities and in 1914 accepted a call to the University in Berlin and the Prussian Academy, chiefly to gain freedom from teaching and other obligations. In fact, he managed to avoid turning out more than a single Ph.D. of his own during his lifetime.[35] As director of the Institute for Physics, his record shows that his model of leadership was to pay minimal attention to his directorial duties, even to the recruitment of new members or to drawing up regulations.[36] "Red tape," he explained, "encases the spirit like the bands of a mummy." When he first met John D. Rockefeller, Jr., the two men compared notes on how to get things done. "I put my faith in organization," Rockefeller said; "I put my faith in intuition," came Einstein's reply.[37]

When war broke out in August of 1914, ninety-three of the chief intellectuals of Germany published a manifesto with the significant title "Appeal to the World of Culture," supporting the military. Einstein, for his part, supported a pacifist counterdeclaration entitled "Appeal to the Europeans"; however, it was never published, having attracted a grand total of only four signatures. But throughout the war Einstein never made a secret of his pacifist and cosmopolitan attitude, and in an increasingly hostile Germany he took care to express publicly his support for the founding of a Jewish state in Palestine. He also made it plain that he regarded himself again as a Jew and indeed as a religious person; of course, as shown in several essays in his book *Ideas and Opinions*, his idea of religion was contrary to any religious establishment. It was a Spinozistic pantheism that he called "cosmic religion," and he put his position simply and seriously in one of his letters: "I am a deeply religious unbeliever."[38]

After his move to America when World War II broke out, the authorities kept Einstein uninformed about nuclear research. On the contrary, he was carefully monitored by the military and the FBI, which considered him a security risk. The FBI files on Einstein are voluminous; J. Edgar Hoover apparently was personally convinced that Einstein had to be watched—the physicist's whole history showed that here was a really dangerous rebel.

THE SELECTIVE REVERENCE FOR TRADITION

One could add even more weight to the side of the balance that measures Einstein's iconoclastic nature. But if now the defense attorney for the accused is given some moments for rebuttal, a counterargument might be introduced by noting that Einstein's rebelliousness was only half the story; the other half was his selective reverence for tradition. Indeed, the counsel for the defense might well urge us to consider it a hallmark of genius to tolerate and perhaps even relish what seems to us such apparent contradiction.

For there were significant limits to the offenses cited. For example, one might be more lenient about Einstein's leaving his *Gymnasium* early, since he preferred reading classics of science and literature on his own. After all, the school system was by no means beloved by all its pupils—not least because it devoted itself not only to educational goals but also to political indoctrination. Although there were variations among school systems in different parts of Germany, an official Prussian publication was typical in setting forth the plan and aims for the upper schools of 1892, when Einstein was a *Gymnasiast*. It announced, "Instruction in German is, next to that in religion and history, ethically the most significant in the organism of our higher schools. The task to be accomplished here is extraordinarily difficult and can be properly met only by those teachers who warm up the impressionable hearts of our youths for the German language, for the destiny of the German people, and the German spiritual greatness. And such teachers must be able to rely on their deeper understanding of our language and its history, while also being borne up by enthusiasm for the treasures of our literature, and being filled with patriotic spirit."[39] Clearly, *Bildung* and *Kultur* were here instrumentalized in the service of the state. To young Einstein, it smelled of militarism.

Moreover, when the public and his fellow scientists later hailed him as the great scientific revolutionary, Einstein always took pains to deny this label. He emphasized over and over again that his work was firmly embedded in the tradition of physics and had to be considered an evolution of it, rather than a revolution. He would have been appalled to know that a few years after his death a philosopher would assert that a wall of incommensurability existed between the world of Newton and the world of Einstein.

But points such as these pale in comparison to a central one: Einstein's lifelong interest in and devotion to the European literary and philosophical cultural tradition, and especially to German literary and philosophical *Kultur*. That allegiance, in which his science was clearly embedded, had been fostered early in his childhood. While the classics of music were offered in their home by his mother, Einstein's father would assemble the family in the evening around the lamplight to read aloud from works by such writers as Friedrich Schiller or Heinrich Heine.[40] The family perceived itself as participating in the movement of general *Bildung* in this way, the uplifting of mind, character, and spirit that characterized the rising portion of the *Bürgertum*. This was especially true for its Jewish segments. *Kultur* advocated and legitimized emancipation, and also provided a vehicle of social assimilation.

After all, during his scientifically most creative and intense period in Bern, Einstein formed with two young friends an "academy" for the self-study of scientific, philosophical, and literary classics. We have the list of the books they read and discussed at their meetings, which sometimes convened several times a week: Spinoza, Hume, Mach, Avenarius, Karl Pearson, Ampère, Helmholtz, Riemann, Dedekind, Clifford, Poincaré, John Stuart Mill, and Kirchhoff, as well as Sophocles and Racine, Cervantes and Dickens.[41] They would not have wanted to be *ignorant* of the cultural milieu, even if they did not necessarily agree with all they read.

To illuminate the point with but a single example: We know that Albert at the tender age of thirteen was introduced to Immanuel Kant's philosophy, starting with the *Critique of Pure Reason*, through his contacts with a regular guest at the Einstein home, Max Talmey.[42] He reread Kant's book at the age of sixteen and enrolled in a lecture course on Kant while at the Technical Institute in Zurich.[43] He wrote a lengthy book review of a philosopher's analysis of Kant, and at the Institute in Princeton his favorite topic of discussion with his friend Kurt Gödel was, again, Kant.[44] Einstein surely knew of the overwhelming influence of Kant on, for example, the late-nineteenth-century philosophers arguing against materialism.

All this, typically, did not make Einstein a Kantian at all. While sympathizing with Kantian categories—and very likely to remember that Kant had listed "Unity" as the first of his categories[45]—Einstein objected to the central point of Kant's transcendental idealism by denying the existence of the synthetic a priori, arguing: "[W]e do not conceive of the 'categories' as unalterable (conditioned by the nature of the understanding) [as Kant did], but as (in the logical sense) free conventions. They appear to be a priori only insofar as thinking without the positing of categories and of concepts in general would be as impossible as is breathing in a vacuum."[46] The essential point for him was, again, freedom, the "free play" of the individual imagination, within the empirical boundaries the world has set for us.

Thus Einstein's reverence was carefully selective, even while his outreach into the traditional cultural environment was enormous. He loved books, and they were his constant companions. A list of only those books found in the Einstein household that had been published up to 1910 includes the works of Aristophanes, Boltzmann, Ludwig Büchner, Cervantes, William Clifford, Dante, Richard Dedekind, Dickens, Dostoyevski, Frederick Hebbel, the collected works of Heine (two editions), Helmholtz, Homer, Alexander von Humboldt (both the collected works and his *Cosmos*), many books of Kant, Lessing, Mach, Nietzsche, Schopenhauer, Sophocles, Spinoza, and, for good measure, Mark Twain.[47] But what looms largest are the collected works of Johann Wolfgang von Goethe: a thirty-six-volume edition and another of twelve volumes, plus two volumes on his optics, one on the exchange of letters between Goethe and Schiller, and also a separate volume of the tragedy *Faust*, which will become a significant part of our story.

Some of those books have such early dates of publication that they may have been heirlooms; others must have been lost in the turmoil of the various migrations and separations. But this list, though only a part of the total library, indicates roughly what an aspiring member of the culture-carrying class would want to know about. And their schooling had prepared them, willing or not, to take such exemplars of higher culture seriously, not least as preparation for school examinations.

Einstein's required courses in high school were mentioned earlier; at the Polytechnic Institute, where Einstein was training to be a high-school physics teacher, he took all the obvious required science courses, including differential equations, analytical geometry, and mechanics—although what he most wanted to learn about, Maxwell's electromagnetism, he had to study on his own. In his first year, he enrolled in two additional optional courses, one on the philosophy of Kant, as noted earlier, and one entitled "Goethe, Werke and Weltanschauung." No doubt—he had been captured.

I think we now have at least an outline of the gestalt of young Einstein's complex intellectual-cultural inner life and an idea of his perception of his quite individualistic place among the *Kulturträger*—to which his whole education, both through compulsory or private reading, had carried him and where in fact he found a satisfying spiritual home.

TOWARD A VERDICT

The opposing evidences—Einstein's rebelliousness and his attention to tradition—having now been presented, is not the obvious conclusion that in Einstein we are dealing with a sort of split personality? The answer is no; we have seen two different perspectives of one coherent mental structure that uses the apparently conflicting parts to support each other.

The bonds between the apparent opposites are of three kinds. The first lies in the presence of an alternative subcurrent in the *Kultur* itself. As I have hinted, *Kultur* carried within itself a strain that we may call a "tradition of rebellion," which made it in fact potentially unstable and volatile. The anti-Enlightenment Sturm und Drang and Romantic products of the earlier period had become canonized and remained part of the tradition-bound, late nineteenth-century *Kultur*; the ideal of the active, creative, unbounded individual continued to be championed. Employing the evocative phrase Max Weber had used in a different context, such a person had to accept the plain and simple duty "to find and obey the demon that holds the fibers of one's very life"—to strive for authenticity and intensity of feeling, even heroism and sacrifice.[48] The purest expression of individuality was embodied in the genius, who led an often marginal, tormented, and, by conventional standards, failing or even demonic existence but who nonetheless saw and created things far beyond the reach of comfortable philistines.[49] Those philistines were the enemies for the Sturm und Drang authors, as they were for Einstein.

These two strains in *Kultur*, the rebellious and the traditional, often occurred in a complementary manner. Those formed by this *Kultur* were prepared to flout convention, while at the same time revering the outstanding cultural figures of all times. Though willing to dissent, they also understood themselves as loyal members of a supratemporal community of exceptional minds that existed in a universe parallel to that of the philistine masses. This mixture was not considered contradictory, although note must be taken here of what history was to record later in blood-stained letters: When these elements of rebelliousness later broke away from their stabilizing counterparts in culture, they flamed up for a time in twentieth-century Germany into the transformation and destruction of *Kultur* itself—as Einstein and so many others were to experience. But during his formative years, this complementary nature of *Kultur* still functioned, and it was precisely what Einstein needed for his work and life.

The second of the three bonds connecting those seemingly contradictory aspects of Einstein lies, unsurprisingly, in his approach to physics, both in his manner of radically clearing obstacles and in how he achieved his insights with the aid of tools from the traditional culture. Looking at his papers and letters, one can almost watch the seemingly centrifugal tendencies of Einstein's spirit being used and tamed to his service. I found the first hint in the letter he wrote in the spring of 1905 to his friend Carl Habicht.[50] In a single paragraph, Einstein poured out an accounting of major works he was then completing. First on his list is what is now known as the discovery of the quantum nature of light, as evidenced in the photoelectric effect. Another was his prediction and detailed explanation of a random, zigzag movement of small bodies in suspension that are large enough to be seen through a microscope, in which he traced the cause in exact detail to the bombardment of these visible bodies by the invisible submicroscopic chaos of molecules. (The existence of such motion, referred to as Brownian movement, was known.) And the last of the papers-in-progress he referred to was what became the original presentation of Einstein's relativity, identifying that work to Habicht only as an evolutionary act, a "modification of the teachings of space and time." To achieve that, in the published paper he casually discarded the ether, which had been preoccupying the lives of a large number of prominent physicists for more than a century, with the nonchalant remark that it was "superfluous"; dismissed the ideas of the absolutes of space, time, and simultaneity; showed that the basic differences between the two great warring camps, the electromagnetic and mechanistic worldviews, were easily dissolved into a new, relativistic one; and finally, as an afterthought, derived $E=mc^2$.

Each of these papers, completed in 1905, is a dazzling achievement, and, what is more, they always have seemed to be in three completely different fields. But I could not rid myself of the thought that behind their obvious differences something common was motivating these articles, published rapidly, one right after the other. Something was missing in that exuberant letter to Habicht.

An important lead was found at last in an unpublished letter Einstein had written to Max von Laue in January 1952, which indicates the hidden connection.[51] To put it very briefly, Einstein's study of Maxwell's theory, which had led him to the theory of relativity, had also convinced him that radiation has an atomistic (that is to say, quantum) structure, exhibiting fluctuation phenomena in the radiation pressure, and that these fluctuation should show up in the Brownian movement of a tiny suspended mirror. Thus the three separate fireworks—relativity, the quantum, and Brownian movement—had originated in a common cartridge.

Moreover, once this is understood, Einstein's approach to the problem in each of these diverse papers could be recognized as having essentially the same style and components. Unlike most other physicists of the time, Einstein did not start with a review of puzzling new experimental facts, the latest news from the laboratory, but rather by stating his dissatisfaction with what seemed to him asymmetries or other incongruities that others would dismiss as being merely aesthetic in nature. He then proposed a principle of great generality, analogous to the axioms Euclid had placed at the head of that "holy" geometry book. Then Einstein showed in each case how to remove, as one of the deduced consequences, his original dissatisfaction; at the end, briefly and in a seemingly offhand way, he proposed a few experiments that would bear out the predictions following from his theory. Once more there was only one Einstein, not three.

Most significant, the fundamental motivation behind each paper was really the very same one he had announced five years earlier in the letter to Marcel Grossmann in which he revealed what would become his chief preoccupation in science for the rest of his life: "To recognize the unity of a complex of appearances which . . . seem to be separate things." Thus, the paper on the quantum nature of light begins with a typical sentence: "There is a deep formal difference between the theoretical understanding which physicists have about gases and other ponderable bodies, and Maxwell's theory of electromagnetic processes in the so-called vacuum."[52] That is to say, energy of palpable bodies is concentrated, and not infinitely divisible; but as a light wave spreads out, its energy at a given point constantly decreases.

Why should atomicity not apply to both matter and light energy? The Brownian movement article declared that if there is chaotic motion, spontaneous fluctuation in the microcosm of classical thermodynamics, it must also show up in the macrocosm of visible bodies. And the relativity paper in effect removed the old barriers between space and time, energy and mass, electromagnetic and mechanistic worldviews. In the end, all these papers endeavored to bring together and unify apparent opposites, removing the illusory barriers between them.

THEMATIC PRESUPPOSITIONS

The longer I studied the papers and correspondence of this scientist, the more impressed I became by his courage to place his confidence, often against all available evidence, in a few fundamental guiding ideas or presuppositions, which he called "categories" in a non-Kantian sense, i.e., freely chosen. In studying other major scientists, I have repeatedly found the same courageous tendency to place one's bets early on a few nontestable but highly motivating presuppositions, which I refer to as themata. In Einstein's case, an example of themata would be simplicity, harking back to Newton's first rule of philosophy: "Nature is pleased with simplicity, and affects not the pomp of superfluous causes."[53] Einstein wrote veritable hymns to the concept of simplicity as a guide in science, and he exemplified it in his own lifestyle.[54]

Another of his thematic presuppositions was symmetry, a concept he introduced into physics in 1905, considering it basic—when most of his readers surely wrote it off as an aesthetic, optional choice. It has since become one of the fundamental ideas in modern physics. Yet another thema was his belief in strict Newtonian causality and completeness in the description of natural phenomena, which explains why Einstein could not accept as final Niels Bohr's essentially probabilistic, dice-playing universe. Einstein's utter belief in the continuum was another such thema, as in the field concepts that enchanted him from the moment he saw his first magnet compass in boyhood.

There are a few more themata to which he also clung obstinately. But beyond that, we must ask a key question: Because the themata are not a priori or innate but choosable, are those that are selected chosen at random from some infinite set of possible themata? That I do not believe. *Or are the themata so confidently held because they are reinforced by, and in resonance with, the scientist's cultural milieu?* That was the initial question here, but now it can be tested in a real case.

For that purpose, one thema that was the most important to Einstein—that of unity, unification, wholeness—will serve as the prototypical example to answer the question whether themata in science may be reinforced by the cultural milieu.[55] Einstein's dedication to the presupposition of finding unities in Nature at work is evidenced in the motivation for his three great papers of 1905. As he put it in a letter of 1916 to the astronomer Willem de Sitter, he felt always driven by "my need to generalize" (*mein Verallgemeinerungsbedürfnis*).[56] That need continued uninterrupted from his first paper on capillarity to his last ones on finding a general unified field theory that would join gravity and electromagnetism, and even provide a new interpretation of quantum phenomena—as may yet happen, although along a path different from his.[57] In between, that preoccupation had led him from the special theory to what he at first called typically the *verallgemeinerte*, the generalized theory of relativity.

That self-imposed, unquenchable desire to find unifying theories had possessed many other scientists (for example, Alexander von Humboldt, who celebrated in 1828 the "deep feeling for a unity of Nature"); however, this presupposition sometimes led Einstein astray, as had Galileo's analogous obsession with the primacy of circular motion. To be sure, some splendid science is done by researchers who seem to have no need of thematic presuppositions, as I have found in other case studies. Nor do I want to paint all German scientists as having been caught up in the dream of unity; for example, as Pauline Mazumdar's study of German immunologists showed, there were "Pluralists" among them to oppose the "Unitarians."[58]

But my subject is Einstein, and it is clear that his thematic acceptance of unity or wholeness was one of the demons that had got hold of the central fiber of his soul. He even lent his name—along with thirty-two other scholars from a great variety of fields, ranging from David Hilbert and Ernst Mach to Jacques Loeb, Sigmund Freud, Felix Klein, and Ferdinand Tönnies—to the publication, as

Aufruf!

Eine umfassende Weltanschauung auf Grund des Tatsachenstoffes vorzubereiten, den die Einzelwissenschaften aufgehäuft haben, und die Ansätze dazu zunächst unter den Forschern selbst zu verbreiten, ist ein immer dringenderes Bedürfnis vor allem für die Wissenschaft geworden, dann aber auch für unsere Zeit überhaupt, die dadurch erst erwerben wird, was wir besitzen.

Doch nur durch gemeinsame Arbeit vieler kann das erreicht werden. Darum rufen wir alle philosophisch interessierten Forscher, auf welchen wissenschaftlichen Gebieten sie auch betätigt sein mögen, und alle Philosophen im engeren Sinne, die zu haltbaren Lehren nur durch eindringendes Studium der Tatsachen der Erfahrung selbst zu gelangen hoffen, zum Beitritt zu einer Gesellschaft für positivistische Philosophie auf. Sie soll den Zweck haben, alle Wissenschaften untereinander in lebendige Verbindung zu setzen, überall die vereinheitlichenden Begriffe zu entwickeln und so zu einer widerspruchsfreien Gesamtauffassung vorzudringen.

Um nähere Auskunft wende man sich an den mitunterzeichneten Herrn Dozent M. H. Baege, Friedrichshagen b. Berlin, Waldowstraße 23.

E. Dietzgen, Fabrikbesitzer u. philos. Schriftsteller Bensheim.
Prof. Dr. Einstein, Prag
Prof. Dr. Forel Yvorne.
Prof. Dr. Föppl, München.
Prof. Dr. S. Freud, Wien
Prof. Dr. Helm, Geh. Hofrat, Dresden.
Prof. Dr. Hilbert, Geh. Reg.-Rat. Göttingen.
Prof. Dr. Jensen, Göttingen.
Prof. Dr. Jerusalem, Wien.
Prof. Dr. Kammerer, Geh. Reg.-Rat. Charlottenburg.
Prof. Dr. B. Kern, Obergeneralarzt u. Inspekteur der II. Sanitäts-Inspektion. Berlin.
Prof. Dr. F. Klein, Geh. Reg.-Rat. Göttingen.
Prof. Dr. Lamprecht, Geh. Hofrat, Leipzig.
Prof. Dr. v. Liszt, Geh. Justizrat, Berlin.
Prof. Dr. Loeb, Rockefeller-Institute. New-York.
Prof. Dr. E. Mach, Hofrat. Wien.
Prof. Dr. G. E. Müller, Geh. Reg.-Rat. Göttingen.
Dr. Müller-Lyer, München.
Josef Popper, Ingenieur, Wien.
Prof. Dr. Potonié, Königl. Landesgeologe. Berlin.
Prof. Dr. Rhumbler, Hann.-Münden.
Prof. Dr. Ribbert, Geh. Medizinalrat. Bonn.
Prof. Dr. Roux, Geh. Medizinalrat. Halle a. S.
Prof. Dr. F. C. S. Schiller, Corpus Christi College. Oxford.
Prof. Dr. Schuppe, Geh. Reg.-Rat. Breslau.
Prof. Dr. Ritter v. Seeliger, München.
Prof. Dr. Tönnies, Kiel.
Prof. Dr. Verworn, Bonn.
Prof. Dr. Wernicke, Oberrealschuldirektor u. Privat-Dozent. Braunschweig.
Prof. Dr. Wiener, Geh. Hofrat. Leipzig
Prof. Dr. Th. Ziehen, Geh. Medizinalrat. Wiesbaden.

M. H. Baege, Dozent d. Freien Hochschule Berlin Friedrichshagen.
Prof. Dr. Petzoldt, Oberlehrer u. Priv.-Dozent. Spandau.

Figure 2: Appeal for the formation of the Gesellschaft für positivistische Philosophie. (Courtesy of Wilhelm-Ostwald-Archiv, Deutsche Akademie der Wissen-schaften zu Berlin.)

early as 1912, of a public manifesto (*Aufruf*, figure 2) calling for the establishment of a new society aiming to develop, across *all* branches of scholarship, one set of unifying ideas and unitary conceptions. As the *Aufruf* put it in its second paragraph, the new Society's aim would be "to join all fields of learning [*alle Wissenschaften*] together in an organic association, to develop everywhere the unifying ideas, and thus to advance to a non-contradictory comprehensive conception."[59]

Yet if it was allegiance to a few themata that supported Einstein in launching into uncharted territory, often with the barest encouragement from the phenomena, *what provided the courage to adopt these themata*, and to stick with them through thick and thin? This is where the various strands we have pursued will converge, where we make closest contact with the "cultural soil" that helped to feed his scientific imagination, for one can show the resonance between Einstein's thematic belief in unity in *science* and the belief in the primacy of unity contained in certain *literary works* to which he had allegiance. While here I can demonstrate the case for only one of his themata, and for one set of major literary works, the case made is more general and applies not only to this particular scientist.

THE CULTURAL ROOTS OF UNITY—A POET POINTS THE WAY

So far, we have noted that Einstein drew on the work of other scientists, on the tools of his trade that he assembled during his education—so joyfully by himself, less so in his schooling. We have discussed his personal attitude as a *gebildete* individual, who refused to be a mere functionary of the state and kept his freedom of imagination and destiny. Other useful suggestions for pieces of the puzzle have also been proposed, for example, the interesting point made by Robert Schulmann and Jürgen Renn that Einstein's reading in popular scientific books as a boy consisted largely of ones that did not dwell on details but instead provided an overview of science as a coherent corpus of understanding, and that this experience predisposed him early to fasten upon the big questions rather than the small pieces.[60]

All this was necessary; but it was not enough. His wide reading in humanistic works beyond science—where the *Bildung* during his formative years was to lead to continued self-refinement through study of the "best works," analogous to Matthew Arnold's concept of culture—hinted at what else was needed to understand his particular genius.[61] From the list of icons of high culture at the time who greatly impressed Einstein, I must focus on just one author, indeed one who, with Friedrich Schiller, was among the most universally revered: Johann Wolfgang von Goethe.[62] Since Goethe is today certainly not on everyone's mind, I will attempt to convey in a few words his unimaginable influence at the time, not merely on educated Germans in general, but on German scientists in particular.[63]

There are two major parts to that influence. One was the fact that Goethe was arguably Germany's most accomplished and productive poet. He began his long and fruitful career when, as we noted, Germany was not a modern state. Indeed, in many ways it was backward compared with Britain and France; it was politically impotent, a motley assembly of about three hundred fragments, large and small, within the dying Holy Roman Empire. In 1775, when the twenty-six-year-old Goethe arrived in Weimar, it was still an impoverished duchy, and his own youthful presence there was possibly one of its biggest assets. His skill, intelligence, and humanity had begun to show itself even in his first, fiery works that were still linked to the Sturm und Drang tradition, for example, the irreverent revolutionary drama *Götz von Berlichingen*, written at age twenty-four, and the romantic novel *The Sorrows of Young Werter*, written one year later. The *Götz* drama was based on a legendary early sixteenth-century German knight, a bold and impudent adventurer who made it known to all, in strong language, that he was beholden to no one but God, Kaiser Maximilian, and his own independent self. (I find it delightful that during Einstein's final *Matura* examination, his essay in the subject of German was on *Götz*, the very embodiment of the independent individual spirit.[64])

Goethe, too, was a complex of apparent opposites. In his early works he had established himself as the foremost German spokesman for the Sturm und Drang movement, the forerunner of the Romantic revolt, while still adhering to Enlightenment ideas (one of those contradictions when viewed from our level below). And he was still in his twenties when he began work on the first part of his *Faust*, the tragedy into which he poured his superb poetic skills and all the varied and mutually antagonistic aspects of his maturing soul. It was, like much of his writings, part of a "great confession," but it had an especially strong grip on the German imagination, on the upward-striving bourgeoisie as well as the elite; the nearest analog that comes to mind is the indelible impression of Dante's epic on intellectuals in Italy. As G. H. Lewes remarked, the Faust tragedy "has every element: wit, pathos, wisdom, farce, mystery, melody, reverence, doubt. magic, and irony."[65]

In his early period, Goethe himself, like his Faust, accepted the dictum "To live, not to learn." But this rebellion took a special form as he matured, similar to Einstein's own. Goethe's "central tenet" was the belief in individuality or individualism: one was a free person, defying some of the social conventions but at the same time revering the geniuses of history and legend, which for him (according to Goethe's biographers) included the original Dr. Johann Faustus of the sixteenth century, Prometheus, Spinoza, Mohammed, Caesar, and the original knight Götz von Berlichingen.[66] Like Spinoza, Goethe saw God and Nature as two aspects of the same basic reality, and in that belief, too, he shared the spirit of Einstein and other scientists. Among German *Kulturträger*, Goethe became a fascinating and inexhaustible part of their imaginative lives.[67]

I will return to that point in a moment. But it must be noted that a second aspect of Goethe's power was his position as a serious and productive scientist on certain topics, such as the investigation of the subjective impression of color; the discovery, in his first scientific paper, of the presence of an intermaxilliary bone in man; his early version of what Ernst Haeckel later called an evolutionary mechanism; his concept of the metamorphosis of plants, and other such matters. Thus Goethe has an honored place even in the modern *Dictionary of Scientific Biography*, and despite the huge controversy about others of his contributions, especially on the theory of colors (the *Zur Farbenlehre* of 1810), his scientific activities—totaling fourteen volumes of the Weimar edition of his collected works—added to his standing as a figure representing the best of culture in all its dimensions.

To be sure, Goethe's science was chiefly that of the poet-philosopher. For example, one early "scientific" essay, entitled "Study after Spinoza," begins with the sentence "The concept of being and of completeness is one and the same"; from this, Goethe goes on to ponder the meaning of the infinite.[68] But significantly, the main point of that work was to argue for the primacy of unity in scientific thinking, and for the wholeness "in every living being." The sorry and misguided war he waged for over four decades against Newton's ideas, especially on color theory, must be understood in terms of Goethe's philosophical and poetic beliefs. For example, the quantification and subdivision of natural phenomena, he thought, missed the whole point of the organic unity of man and nature in the explanation of phenomena, particularly for what he regarded as qualities, such as colors. This is a prominent aspect of much of Goethe's whole corpus: the theme of unity, wholeness, the interconnection of all parts of nature. Those are main conceptions that informed both his science and his epics. As one of his commentators has mentioned, "The nature of the entire cycle [is this]: unity in duality."[69] It pervaded even his belief in the existence of an original, archetypal plant (*Urpflanze*), an archetypal man, and so on—all part of what has been called the Ionian Fallacy, looking for one overarching explanation of the diversity of phenomena.[70] Even at age eighty-one, two years before his death, he was immensely excited by news that in France, the biologist St. Hilaire had associated himself with the concept of unity at the base of biology, and he exclaimed:

> What is all intercourse with Nature, if we merely occupy ourselves with individual material parts, and do not feel the breath of the spirit which prescribes to every part its direction, and orders or sanctions every deviation by means of an inherent law! I have exerted myself in this great question for fifty years. At first I was alone, then I found support, and now at last, to my great joy, I am surpassed by congenial minds.[71]

Much has been written about the interest among scientists in various aspects of Goethe's work, and not only in Germany. A list of such scientists would contain names such as Johann Bernhard Stallo, Wilhelm Ostwald, the physiologist Arnold Adolphe Berthold, the neurophysiologists Rudolf Magnus and Emile du Bois-Reymond, the botanist Gottlieb Haberlandt, the physical chemist Gustav Tammann, the bacteriologist Robert Koch, the psychologist Georg Elias Müller, and the English scientist William Henry Fox Talbot. A curious case is that of Nicola Tesla, who, although not German by descent, was so caught up in the German style of *Bildung* that he claimed, and sometimes demonstrated, that he knew the whole of Goethe's *Faust* by heart—all 12,110 lines.[72]

Of course not everyone shared Tesla's enthusiasm. Many a scientist had to give lip service to Goethe's dominance while actually fighting for a down-to-earth, pragmatic, properly experimental style of thought. But wherever these readers turned, from their school days on, they, like Einstein, were likely to encounter Goethe and so were liable to absorb and sympathize with that central point in Goethe's work, the longing for unity, for wholeness, for the interconnectivity of all parts of nature. As Walter Moore put it in his biography of Erwin Schrödinger, "All German-speaking youth [were] imbued with the spirit of Goethe. . . . They have absorbed in their youth Goethe's feeling for the unity of Nature."[73] Fragments of Goethe's poetry could be encountered routinely, not only in the popular lectures of other *Kulturträger* or in the exhortations of politicians, but even in the lectures and textbooks on science itself, in the writings of physicists such as Hermann von Helmholtz, Erwin Schrödinger, Wilhelm Wien, and Max Born. Thus Arnold Sommerfeld, in the third volume of his *Lectures on Theoretical Physics*, sends his readers on the general relativity theory off with a quotation from *Faust*, part II.[74]

My favorite example of that ubiquity occurs on two pages of a textbook by one of Einstein's own scientific predecessors, one whom in 1900 he had called "quite magnificent."[75] Ludwig Boltzmann's *Vorlesungen über Maxwells Theorie der Elektricität und des Lichtes* was published in two parts (1891 and 1893), each preceded by a short epigraph. Boltzmann could count on every German reader to recognize the origin of the lines he quoted there, for they referred to the early pages of Goethe's *Faust* tragedy. My free translation of the first passage is: "That I may no longer, with sour labor, have to teach others that which I do not know myself"; Boltzmann does not even have to add the next, most celebrated and programmatic lines of *Faust*: "*and that I may perceive what holds the world together in its innermost.*"

Boltzmann's second epigraph refers to the passage given in italics below where Faust has just opened the book of Nostradamus, seeking even there a guide to the force that holds the world together; he gazes at the wondrous "Sign of the Macrocosm" and exclaims:

> Ha! as I gaze what rapture suddenly
> begins to flow through all my senses! . . .
> *Did some god inscribe these signs*
> *that quell my inner turmoil,*
> *fill my poor heart with joy,*
> *and with mysterious force unveil*
> *the natural powers all about me?*
> Am I a god? I see so clearly now!
> In these lines' perfection I behold
> creative nature spread out before my soul. . . .
> How all things interweave as one
> and work and live each in the other.[76]

By referring to the God-like signs Boltzmann meant of course to indicate Maxwell's equations, the summary of Maxwell's synthesis of electricity, magnetism, and optics. The equations relating the electric and magnetic field terms are indeed stunningly beautiful in their simplicity, scope, and symmetry, particularly when written in modern form:[77]

$$\text{curl } E = -\frac{1}{c}\frac{\delta B}{\delta t} \qquad \text{div } E = 0$$

$$\text{curl } B = \frac{1}{c}\frac{\delta E}{\delta t} \qquad \text{div } B = 0$$

But back to the enchanted Boltzmann. It is quite significant that in both epigraphs Boltzmann's version of Goethe's lines are in fact just a bit wrong.[78] He too was no doubt quoting from memory, going back to school days. Used constantly, such verses tend to be taken for granted and get fuzzy at the edges. Boltzmann's errors are really one sign that Goethe's lines have become part of common culture.

But never mind. We must dig a bit deeper to see why such literary allusions were so meaningful to the scientific reader. Consider the context of those lines, near the beginning of the first part of the *Faust* tragedy. Having painfully worked his way through every major specialty, Faust's thirst for knowledge at its

Figure 3: Rembrandt's etching, called "Dr. Faustus" (detail). From L. Münz, ed., *Rembrandt's Etchings* (London: Phaidon Press, 1972).

deepest level had not been satisfied by these separate (let us say, reductionist) studies—any more than were the signers of the 1912 *Appeal* for unity throughout all sciences and scholarship. Even if he has to turn to the realm of the magical, Faust must discover the secret of the world's coherence. Nostradamus's book offers him the blinding revelation in terms of the Sign of the Macrocosmos, that ancient symbol of the connection between the part and the whole, man and nature (figure 4). This is why Boltzmann connects the passage to Maxwell's equations, which express the synthesis of large parts of physics.[79]

The main point here is the strong resonance between the Goethean or Faustian drive toward a unified fundamental understanding of nature, symbolized by the Sign of the Macrocosmos, and that of the analogous ambition of Boltzmannian scientists and their pupils: the search for one single, totally coherent worldpicture, a *Weltbild* encompassing all phenomena. Physical science, too, yearned to progress by the discovery of ever fewer, ever more encompassing fundamental concepts and laws, so that one might achieve at last what Max Planck called, in the title of his 1908 essay, "Die Einheit des physikalischen Weltbildes."[80] Indeed, some physical scientists still work toward the day when one single equation, one world equation, will be found that will subsume all the diversity of physical phenomena. Then the Sign of the Macrocosmos will indeed stand before our gaze.

Einstein, starting with his very first publication in 1901 on capillarity, was committed to an early stage of such a Faustian plan. In that paper he tried to remove a duality between Newtonian gravitation, which directs the motion of macroscopic objects downward, and capillary action, which drives the molecules of the submicroscopic world of the liquid upward. In its way this was also a search for the commonality between the macrocosm of observable gravitation and the microcosm of molecular motions. Here was a case where, he thought, apparently opposite phenomena could be brought into a common vision. Even though Einstein later dismissed the physics he had used in that first paper as juvenilia, he never turned his back on the inherent goal.

Perhaps its most eloquent expression appears in his address of 1918, "Principles of Research," given in honor of Max Planck.[81] There he solemnly states that into the shaping of a coherent worldview every serious artist, philosopher, or scientist, each in his own way, "places the center of gravity of his emotional life." Einstein called that search for a worldpicture "*the supreme task*" of the physicist—the task "to arrive at those universal elementary laws from which the cosmos can be built up. . . ."

The intensity of the impulse toward a unified *Weltbild*, so typical for many German scientists of the time—even while specialization was rising all around them—was not confined to them. David Cassidy has noted that

> the "unifying spirit," as it was called, pervaded much of central European thought at the turn of the century. German idealism,

> neo-Romanticism, and historicism, stretching from Immanuel Kant and Georg Wilhelm Hegel to Benedetto Croce and Wilhelm Dilthey, each pointed to some sort of transcendent higher unity, the existence of permanent ideas or forces that supersede or underlie the transient, ephemeral world of natural phenomena, practical applications, and the daily struggle of human existence. The scholar, the artist, the poet, the theoretical physicist all strove to grasp that higher reality, a reality that because of its permanence and transcendence must reveal ultimate "truth" and, hence, serve as a unifying basis for comprehending, for reacting to, the broader world of existence in its many manifestations.[82]

I can only add in closing that movements as different as turn-of-the-century Monism, and later the "Unity of Science" movement, were closely related to this set of aims and ideas. And as Anne Harrington has shown in her recent book, the "'holistic' biological impulse" in early nineteenth-century Germany later flourished with the assistance of our poet. As she put it, "Goethe's resulting aesthetic-teleological vision of living nature would subsequently function as one of the later generations' recurrent answers to the question of what it 'meant' to be a holistic scientist in the grand German style."[83]

At this end of Einstein's century, many excellent scientists and some philosophers are ready to settle for a hierarchical or "disunited" science rather than participate in the pursuit of overarching unities.[84] To them, the self-imposed task of those earlier culture-carriers in search of grand unifications appears perhaps overreaching, and even discussing it as a historic fact may be written off as nostalgia. Moreover, Ernst Mayr and E. O. Wilson have long insisted that for modern evolutionary biologists and naturalists the chief guiding concept should be diversity rather than unity. Perhaps Henry Adams was right when he wrote that after the nineteenth century the course of all history will be away from unity, and toward multiplicity and fragmentation.[85]

Yet the fundamental motivation of Einstein's program has helped to keep alive the modern idea of a search for a physical theory that will encompass all phenomena, from gravitation through nuclear science (a path that Einstein had not explored).

The ascent to that Mt. Everest is now taking various forms among different camps, along different routes. The physics journals and even the daily papers are witnesses; and the International School of Physics has announced a physics seminar at Lake Como with the title, in part, "A Probe of Nature's Grand Design." I have little doubt that hovering there above the audience will be a throng of kindred ghosts, including Kant, Maxwell, Boltzmann, and Einstein, and of course, among the poets, Goethe, with Faust himself next to him; and, way in back, the Greek philosopher Thales of Miletus in Ionia, who twenty-six hundred years ago had launched that Ionian dream, the thema that all things are made of *one* essence. All of those forebears had tilled and seeded the cultural soil of their time and, in turn, in their different ways, had been nourished and reinforced by it.

ASHES INTO THE WINDS

When death approached to claim Einstein in April of 1955, his last acts were still fully in character. He remained strong-willed to the end, obstinately adhering to his ways. He had recently signed a manifesto with Bertrand Russell and others, intending to bring together the international community of scientists as a unifying counterweight against the divisive, national ambitions then rampant during the arms race. For seven years, Einstein had known that a growing intestinal aneurysm of his aorta might rupture at any time, but he had refused any major operation when it still might have averted the threat. He explained his uncomplaining state of mind to his stepdaughter Margot by saying simply, "I have done my thing here." At about one o'clock in the morning, as the aneurysm burst, he suddenly spoke once more, but the night nurse did not understand German.

Einstein's requests concerning his last rest also bore all the marks of his lifelong struggle for simplicity and against ordinary convention. There was to be no funeral—only a few family members and friends gathering at the crematorium. No speeches, no flowers, not even music. No gravestone. But as Einstein's ashes were dispersed into the winds, an old friend and fellow émigré felt moved to recite a few verses of poetry, ending with these lines:

He gleams like some departing meteor bright,
Combining, with his own, eternal light.

As it happened, the poem had been written a century and a half earlier, by the grief-stricken Goethe on the occasion of the death of his friend Friedrich Schiller. A great circle had closed. Symbolically, Einstein's lifelong comrades had helped him, once more, to move across those illusory divisions between space, time, and cultures.

ACKNOWLEDGMENTS

In addition to thanking Robert and Maurine Rothschild for their support to the History of Science department, I also wish to acknowledge several colleagues whom I have consulted on aspects of this work, including Gordon Craig, Frederick Gregory, Roald Hoffmann, Robert Schulmann, S. S. Schweber—none of whom are responsible for possible errors—and above all, Gerhard Sonnert, who provided essential and dedicated help throughout. I am grateful to the Andrew W. Mellon Foundation for support of a research project of which this essay is part.

ENDNOTES

[1]It is symbolic that among the framed portraits he kept in his Princeton home there were only three scientists, each of whom pursued a great synthesis in physics—Newton, Faraday, and Maxwell.

[2]Albert Einstein, "Autobiographical Notes," in *Albert Einstein: Philosopher-Scientist*, ed. Paul Arthur Schilpp (Evanston, Ill.: Library of Living Philosophers, 1949), 53.

[3]Looking in the opposite direction, e.g., how cultural elements later were affected *by* relativity theory rather than initially helping to shape it, we know how certain of Einstein's publications were interpreted to affect the culture of his and our time, misguided though most of these attempts have been—as Steven Weinberg recently warned—such as the transfer of relativity concepts into anthropology, ethics, religion, literature, and to the so-called relativism haunting other fields. Einstein himself was perturbed by popular misunderstanding of his theory. He would have preferred if his theory—which Max Planck and Max Abraham, not Einstein himself, had named in 1906 the "theory of relativity"—had become known as the "theory of invariance" instead. Einstein, letter to E. Zschimmer, 30 September 1921; cf. Gerald Holton, *Einstein, History, and Other Passions: The Rebellion Against Science at the End of the Twentieth Century* (Reading, Mass.: Addison-Wesley,

1996), 131–132. See also Steven Weinberg, "Sokal's Hoax," *New York Review of Books*, 8 August 1996, 11–15.

[4]Robert K. Merton, *Science, Technology and Society in Seventeenth Century England* (New York: H. Fertig, 1970; first published 1938), 238.

[5]Paul Forman, "Weimar Culture, Causality, and Quantum Theory, 1918–1927: Adaption by German Physicists and Mathematicians to a Hostile Intellectual Environment," *Historical Studies in the Physical Sciences* 3 (1971): 1–115.

[6]John Hendry, "Weimar Culture and Quantum Causality," *History of Science* 18 (1980): 155–180; Stephen G. Brush, "The Chimerical Cat: Philosophy of Quantum Mechanics in Historical Perspective," *Social Studies of Science* 10 (1980), 393–447; P. Kraft and P. Kroes, "Adaption of Scientific Knowledge to an Intellectual Environment: Paul Forman's 'Weimar Culture, Causality, and Quantum Theory, 1918–1927,'" *Centaurus* 27 (1984), 76–99.

[7]Max Jammer, *The Conceptual Development of Quantum Mechanics* (New York: McGraw-Hill, 1966); Gerald Holton, *Thematic Origins of Scientific Thought: Kepler to Einstein* (Cambridge, Mass.: Harvard University Press, 1973, 1988).

[8]Specifically: Philipp Frank, *Einstein: Sein Leben und seine Zeit* (Munich: Paul List, 1949), published in English as *Einstein: His Life and Times*, tr. George Rosen, ed. Shuichi Kusaka (New York: Alfred A. Knopf, 1947); Anton Reiser (pseud. of Rudolf Kayser), *Albert Einstein: A Biographical Portrait* (New York: Albert and Charles Boni, 1930); and Carl Seelig, *Albert Einstein: Eine Dokumentarische Biographie* (Zurich: Europa Verlag, 1954). One must include as well as Einstein's own fascinating intellectual autobiography, in Schilpp, *Albert Einstein: Philosopher-Scientist*.

[9]*New York Times Sunday Magazine*, 15 December 1996.

[10]Tetsu Hiroshige, "The Ether Problem, the Mechanistic World View, and the Origin of the Theory of Relativity," *Historical Studies in the Physical Sciences* 7 (1976): 3–82.

[11]Gerald Holton, *Einstein, History, and Other Passions*, 174–175.

[12]Gerald Holton, *Introduction to Concepts and Theories in Physical Science* (Cambridge, Mass.: Addison-Wesley, 1952), 506; Robert A. Millikan, "Albert Einstein on His Seventieth Birthday," *Reviews of Modern Physics* 21 (1949), 343–344.

[13]Holton, *Thematic Origins of Scientific Thought*, ch. 8 and 477–480.

[14]Cited in Robert S. Shankland, "Conversations with Albert Einstein," *American Journal of Physics* 31 (1963):47–57.

[15]Seelig, *Albert Einstein: Eine Dokumentarische Biographie*, 61–62.

[16]Banesh Hoffmann, with Helen Dukas, *Albert Einstein: Creator and Rebel* (New York: Viking, 1972).

[17]Lewis S. Feuer, *Einstein and the Generations of Science* (New York: Basic Books, 1974).

[18]*New York Times*, 16 November 1919, 8.

[19]At the 1832 meeting of the Association of German Scientists and Physicians in Vienna, for instance, a hymn proclaimed, "Laßt uns vereint der Isis Tempel bauen / Der Göttin, welcher keine andre gleich / Die rätselhaft so nahe uns und ferne / Im Sandkorn thront wie dort im Flammensterne." See H. Schipperges, *Weltbild und Wissenschaft: Eröffnungsreden zu den Naturforscher-versammlungen 1822 bis 1972* (Hildesheim: H. A. Gerstenberg, 1976).

[20]Einstein to Besso, 13 May 1917; from Albert Einstein and Michele Besso, *Correspondance, 1903–1955*, tr. and intro. Pierre Speziali (Paris: Hermann, 1972), 114.

[21]Albert Einstein, Hedwig Born, and Max Born, *Briefwechsel, 1916–1955* (Munich: Nymphenburger Verlagshandlung, 1969), 215.

[22]Robert K. Merton, *Science, Technology, and Society in Seventeenth Century England*, 238.

[23]For a classic exposition of the contrast between *Kultur* and *Zivilisation*, see N. Elias, *Über den Prozess der Zivilisation*, 2 vols. (Basel: Verlag Haus zum Falken, 1939), vol. 1, 1–42.

[24]See, for example, *Brockhaus Enzyklopädie*, 1990.

[25]Bruno Gebhard, *Handbuch der Deutschen Geschichte*, 8th ed. (Stuttgart: Union Verlag, 1962), vol. 3, 305.

[26]Karl Mannheim, *Ideology and Utopia: An Introduction to the Sociology of Knowledge* (translation of *Ideologie und Utopie*, 1929; tr. Lewis Wirth and Edward Shils) (New York: Harcourt, Brace & World, 1970), 156. By 1843 Karl Marx had noted (in his *Kritik des Hegelschen Staatsrechts*) that "Geld und Bildung" were the main criteria for social differentiation in the *bürgerliche* society; Marx, *Karl Marx, Friedrich Engels, Werke*, ed. Institut für Marxismus-Leninismus beim ZK der SED (Berlin: Dietz Verlag, 1957), vol. 1, 203-333. For a useful summary of the *Bildungsbürgertum*, see F. Gregory, "Kant, Schelling, and the Administration of Science in the Romantic Era," *Osiris* (second series) 5 (1989):17–35.

[27]In the original sense of the educated Mandarinate that served the Chinese Empire and were chiefly concerned with administering or furthering the political and social needs of the state authorities. Fritz K. Ringer, *The Decline of the German Mandarins: The German Academic Community, 1890–1933* (Cambridge, Mass.: Harvard University Press, 1969).

[28]Christa Kirsten and Hans-Jürgen Treder, eds., *Albert Einstein in Berlin, 1913–1933*, 2 vols. (Berlin: Akademie-Verlag, 1979), vol. 1, 207. This document collection contains reports to the German foreign office from German diplomats in The Hague, Oslo, Copenhagen, Paris, Buenos Aires, Tokyo, Madrid, Montevideo, Rio de Janeiro, Chicago, New York, and Vienna. See vol. 1, 225–240.

[29]K. Mannheim, *Ideologie und Utopie* (Frankfurt: Verlag G. Schulte-Bulmke, 1969: first published in 1929), 221–222.

[30]For Albert Einstein's family tree, see Aron Tänzer, "Der Stammbaum Prof. Albert Einsteins," *Jüdische Familien-Forschung: Mitteilungen der Gesellschaft für jüdische Familienforschung* 7 (1931): 419–421.

[31]Erik Erikson, "Psychoanalytic Reflections on Einstein's Centenary," in Gerald Holton and Yehuda Elkana, eds., *Albert Einstein: Historical and Cultural Perspectives* (Princeton, N.J.: Princeton University Press, 1982), 151–173.

[32]Maja Winteler-Einstein, "Albert Einstein—Beitrag für sein Lebensbild," *Collected Papers: The Collected Papers of Albert Einstein* (Princeton University Press, 1987), vol. 1, xlviii–lxvi.

[33]Einstein, "Autobiographical Notes," 3–5.

[34]Nevertheless, Helen Dukas insisted that Einstein's lifestyle in Zurich and Bern was "anything but 'bohemian,'" as noted in L. Pyenson, *The Young Einstein: The Advent of Relativity* (Bristol: Adam Hilger, 1985), 77, note 9. Pyenson would have characterized Einstein not as a rebel but a stranger or marginal man (60–61).

[35]Seelig, *Albert Einstein: Eine Dokumentarische Biographie*, 125. The graduate student's name was Hans Tanner; Einstein supervised Tanner's dissertation while a professor at Zurich University.

[36]Giuseppe Castagnetti and Hubert Goenner, "Directing a Kaiser-Wilhelm-Institut: Albert Einstein, Organizer of Science?" paper given at the Boston University Colloquium for Philosophy of Science, 3 March 1997.

[37]Cited in Otto Nathan and Heinz Norden, eds., *Einstein on Peace* (New York: Schocken, 1968; reprint of 1960 edition), 157.

[38]Letter to Hans Muehsam, 30 March 1954, Einstein Archive 38–434; cited in *The Quotable Einstein*, ed. Alice Calaprice (Princeton, N.J.: Princeton University Press, 1996), 158. See also Einstein's declaration of his religiosity in Harry Graf Kessler, *Tagebücher 1918–1937*, ed. Wolfgang Pfeiffer-Belli (Frankfurt: Insel-Verlag, 1961), 521–522, and in Hubert Goenner and Giuseppe Castagnetti, "Albert Einstein as Pacifist and Democrat During World War I," *Science in Context* 9 (1996): 348–349. See also Albert Einstein, *Ideas and Opinions* (New York: Dell, 1954).

[39]Ministerium der geistlichen, Unterrichts- und Medizinalangelegenheiten, "Lehrpläne und Lehraufgaben für die höheren Schulen, nebst Erläuterungen und Ausführungsbestimmungen" (Berlin: Wilhelm Hertz, 1892), 20.

[40]Reiser, *Albert Einstein: A Biographical Portrait*, 26. Toward the end of his life, when Einstein's sister Maja visited him in Princeton (as Einstein wrote to Besso), both would spend their time together reading "Herodotus, Aristotle, Russell's *History of Philosophy*, and many other interesting books." Einstein and Besso, *Correspondance*; see also Albrecht Fölsing, *Albert Einstein: Eine Biographie* (Frankfurt: Suhrkamp, 1993), 819. One might add here that Heine was often excluded from the "official" cultural canon, especially outside Jewish circles, because of his religious background and his affiliation with French and revolutionary ideas.

[41]See the introduction by Maurice Solovine to Albert Einstein, *Letters to Solovine* (New York: Philosophical Library, 1987), 8–9. Auguste Comte is

notably absent from Einstein's reading list or exchanges. Comte remained relatively unknown in the German-speaking parts of Europe at the turn of the century. German translations of his works were slow to appear; see the chronology in Auguste Comte, *Rede über den Geist des Positivismus*, tr. and intro. I. Fetscher (Hamburg: Felix Meiner Verlag, 1994; originally published in 1844), xliii–xliv. In 1914, none other than Wilhelm Ostwald translated Comte's *Prospectus des travaux scientifiques nécessaires pour réorganiser la société*, almost a century after it was first published in 1822.

[42]Max Talmey, *The Relativity Theory Simplified, and the Formative Period of its Inventor* (New York: Falcon Press, 1932), 164.

[43]Seelig, *Albert Einstein: Eine Dokumentarische Biographie*, 17. The course was Professor Stadler's lecture course on "Die Philosophie I. Kants"; see *Collected Papers: The Collected Papers of Albert Einstein* (multiple vols.; Princeton, N.J.: Princeton University Press, 1987–), vol. 1, 364.

[44]Albert Einstein, "Elsbachs Buch: Kant und Einstein," *Deutsche Literaturzeitung* 1 (n.f.), 1685–1692.

[45]Immanuel Kant, *Critique of Pure Reason*, tr. Norman Kemp Smith (London: Macmillan, 1929), 113.

[46]Albert Einstein, "Remarks Concerning the Essays Brought Together in this Co-Operative Volume," in *Albert Einstein: Philosopher–Scientist*, 674.

[47]This data base for all books remaining after his death was compiled by NHK (Japan Broadcasting Corporation) and is scheduled to be published. A reading list of additional books may be found in Abraham Pais, *'Subtle is the Lord . . .': The Science and the Life of Albert Einstein* (Oxford: Oxford University Press, 1982).

[48]Max Weber, *Wissenschaft als Beruf* (Berlin: Duncker & Humboldt, 1967), 37; cf. Isaiah Berlin, *The Crooked Timber of Humanity*, ed. Henry Hardy (New York: Vintage Books, 1992), 213–216.

[49]As Fritz Stern shrewdly observed in a passage mentioning both Goethe and Einstein: "A genius could also be seen as a public nuisance. . . ." Stern, *Dreams and Delusions: The Drama of German History* (New York: Alfred A. Knopf, 1987). See also the first chapter of this work, on "Einstein's Germany." On the uses and abuses of Goethe by German ideologues, as well as on how Einstein's view of himself as a Jew differed from others (e.g., Fritz Haber), see Stern, *The Politics of Cultural Despair: A Study in the Rise of the Germanic Ideology* (Berkeley: University of California Press, 1961).

[50]Seelig, *Albert Einstein: Eine Dokumentarische Biographie*, 88–89.

[51]The letter was printed in Holton, *The Advancement of Science, and its Burdens: The Jefferson Lecture and Other Essays* (Cambridge: Cambridge University Press, 1986), 65; it is also hinted at in a letter to Max Born, cited in Born, "Physics and Relativity," *Helvetica Physica Acta, Supplementum IV* (1956), 249.

[52]*Collected Papers*, vol. 2, 150.

[53]Issac Newton, *Mathematical Principles of Natural Philosophy* (translation of *Philosophiae naturalis principia mathematica*), 2 vols., original translation by

Andrew Motte (1729), revised translation by Florian Cajori (Berkeley, Calif.: University of California Press, 1962), vol. 2, 398.

[54]Holton, *The Advancement of Science, and its Burdens*, 15.

[55]The theme of unity and unification also played an important role in biology, as Vassiliki Smocovitis has documented in her *Unifying Biology: The Evolutionary Synthesis and Evolutionary Biology* (Princeton, N.J.: Princeton University Press, 1996). William Morton Wheeler commented (as cited in Smocovitis, 109) that it might take "a few super-Einsteins" to unify biology, using Einstein as the icon of the theme of unification.

[56]Cited in Holton, *The Advancement of Science, and its Burdens*, 86.

[57]Abraham Pais, "*Subtle is the Lord . . .*", 9.

[58]Pauline Mazumdar, *Species and Specificity: An Interpretation of the History of Immunology* (Cambridge: Cambridge University Press, 1995).

[59]Gerald Holton, *Science and Anti-Science* (Cambridge, Mass.: Harvard University Press, 1993), 12–15.

[60]Jürgen Renn and Robert Schulmann, "Introduction," in *Albert Einstein—Mileva Marić: The Love Letters*, ed. Renn and Schulmann (Princeton, N.J.: Princeton University Press, 1992), xi–xxviii.

[61]See Matthew Arnold, *Culture and Anarchy*, ed. Samuel Lipman (New Haven, Conn.: Yale University Press, 1994; first published in 1869).

[62]Einstein kept sculptured busts of both Goethe and Schiller in his Berlin home. F. Herneck, *Einstein privat: Herta W. erinnert sich an die Jahre 1927 bis 1933* (Berlin: Buchverlag Der Morgen, 1978), 47–48.

[63]In the latter part of the nineteenth and in the early twentieth century, it was quite common to assemble "best book" lists of the outstanding works of literature. In 1911, Heinrich Falkenberg compiled such a bibliography, "Listen der besten Bücher," in the *Zeitschrift für Bücherfreunde*; it comprised forty-six entries. The earliest such bibliography was Johann Neukirch's *Dichterkanon* of 1853; in Neukirch's compilation, as well as in the subsequent ones, Goethe played a dominant role. Around 1906, the Viennese bookseller Hugo Heller polled a number of intellectuals about their choice of the "ten best books." A selection of the responses was printed in the *Jahrbuch deutscher Bibliophilen und Literaturfreunde*, ed. H. Feigl (Zurich: Amalthea-Verlag, 1931), 108–127. As one might expect, Goethe figured prominently in these replies, both explicitly and implicitly. At that time, the consensus about the classic literary canon was so strong that it almost went without saying. Much has changed since then. Some ninety years later, the German weekly *Die Zeit* again asked a group of German intellectuals about the literary canon. This time they were to nominate only three to five works that they thought German *Gymnasiasten* had to read. Goethe, and particularly his *Faust*, still received numerous nominations, but now many respondents lamented the almost complete erosion of the classic literary canon. Indeed, this project of *Die Zeit* was intended to help resurrect a canon that had clearly faded. The report started with the statement that nowadays "up to 90 percent of those who begin to study German at a university do not know *Faust*"—

which to the earlier generations of *Bildungsbürger* would have sounded utterly unbelievable. *Die Zeit*, 16 May 1997.

[64]*Collected Papers*, vol. 1, 26–27.

[65]George Henry Lewis, *The Life of Goethe*, 3rd ed. (London: Smith, Elder and Co., 1875).

[66]Henry C. Hatfield, *Goethe: A Critical Introduction* (New York: New Directions, 1963), 28.

[67]And of course not only in German-speaking countries; to cite a single example, Ralph Waldo Emerson taught himself German specifically in order to read Goethe's works. See Robert D. Richardson, *Emerson: The Mind on Fire* (Berkeley, Calif.: University of California Press, 1995).

[68]Johann Wolfgang von Goethe, *Goethes Werke* (Hamburger Ausgabe), 4th ed. (Hamburg: Christian Wegner Verlag, 1962), vol. 13, 7–10.

[69]Hatfield, *Goethe: A Critical Introduction*, 114.

[70]Isaiah Berlin, *Concepts and Categories* (New York: Viking Press, 1979).

[71] Lewes, *The Life of Goethe*, 558.

[72]On the other hand, there can be no doubt, of course, that many of the *Bildungsbürger* and of those aspiring to their ranks rampantly quoted from this and all other classics merely to demonstrate their membership in the educated elite. Such people were greatly helped by Georg Büchmann's *Geflügelte Worte: Der Zitatenschatz des deutschen Volkes*, 27th ed. (Berlin: Haude & Spenersche Buchhandlung, 1926), a best-selling compilation of classic quotations and lengthier excerpts that was first published in 1864, and went through 27 editions by 1926. See Wolfgang Frühwald, "Büchmann und die Folgen: Zur sozialen Funktion des Bildungszitates in der deutschen Literatur," in *Bildungsbürgertum im 19. Jahrhundert*, part II: *Bildungsgüter und Bildungswissen*, ed. Reinhart Koselleck (Stuttgart: Klett-Cotta, 1990), 197–219.

[73]Walter Moore, *Schrödinger: Life and Thought* (Cambridge: Cambridge University Press, 1989), 47.

[74]Arnold Sommerfeld, *Electrodynamics* (volume three of his *Lectures on Theoretical Physics*, tr. Edward Ramberg) (New York: Academic Press, 1952), 311.

[75]In a September 1900 letter to Marić. *Collected Papers*, vol. 1, 260.

[76]Taken from Stewart Atkins's prose translation, *Johann Wolfgang von Goethe: Faust I & II* (Cambridge, Mass.: Suhrkamp/Insel Publishers Boston, 1984), lines 430–431, 434–441, 447–448.

[77]Maxwell's equations in empty space are taken from E. M. Purcell, *Electricity and Magnetism*, 2d. ed. (New York: McGraw–Hill, 1985), 331. I am fairly sure no physics text would connect them today with Dr. Faust, who is under a dark cloud these days for his various transgressions. See Roger Shattuck, *Forbidden Knowledge: From Prometheus to Pornography* (New York: St. Martin's Press, 1996).

[78]In Boltzmann's epigraphs, his rendering of Goethe's lines was, for Part I of his treatise: "So soll ich denn mit saurem Schweiss/Euch lehren, was ich selbst nicht weiss." For Part II, Boltzmann wrote: "War es ein Gott, der diese Zeichen schrieb, / Die mit geheimnissvoll verborg'nem Trieb / Die Kräfte der Natur um mich enthüllen / Und mir das Herz mit stiller Freude füllen."

[79]There is no authoritative picture of how Goethe imagined that heavenly Sign of the Macrocosmos, since no stage directions for it appear in the text of *Faust*. There is, of course, a good amount of literature on that question; see, e.g., Ernst Beutler, ed., *Johann Wolfgang von Goethe: Die Faustdichtungen* (Munich: Winkler Verlag, 1977), 754–757; Heinrich O. Proskauer, ed., *Goethes Faust: Erster Teil* (Basel: Zbinden Verlag), 1982; Rudolf Steiner, *Geisteswissenschaftliche Erläuterungen zu Goethes Faust* (Freiburg: Novalis-Verlag, 1955), vol. 1, 25–27 and *Die Rätsel in Goethes "Faust": exoterisch und esoterisch* (Dornach, Switzerland: Rudolf Steiner Verlag, 1981); and Erich Trunz's two editions, *Goethes Faust* (Hamburg: Christian Wegner Verlag, 1949), 496–497, and *Goethe—Faust* (Munich: C. H. Beck, 1986), 517–518). But we know at least the image that seems to have satisfied Goethe himself: in the 1790 edition of volume 7 of his writings (which included *Faust*), he commissioned as a frontispiece a version of an etching by Rembrandt, which historically was known as representing "Dr. Faust," named after the original sixteenth-century legendary figure (as shown in figure 4).

[80]Max Planck, "Die Einheit des physikalischen Weltbildes," in his *Vorträge und Erinnerungen* (Darmstadt: Wissenschaftliche Buchgesellschaft, 1970), 28–51.

[81]Originally entitled "Motiv des Forschens," the address was published under the rather unfortunate title "Prinzipien der Forschung" in Einstein's *Mein Weltbild* (Frankfurt: Ullstein Bücher 1955, first published in 1934), 107–110. This led to "Principles of Research" in the English translation of the address in *Ideas and Opinions* (New York: Dell, 1954), 219–222. While the second quote from this address is here taken directly from the published English translation, the first is my own translation of the original German text (in *Mein Weltbild*).

[82]David Cassidy, *Einstein and Our World* (Atlantic Highlands, N.J.: Humanities Press, 1995), 14.

[83]Anne Harrington, *Reenchanted Science: Holism in German Culture from Wilhelm II to Hitler* (Princeton, N.J.: Princeton University Press, 1996), 5, 10.

[84]See, for example, Peter Galison and David J. Stump, eds., *The Disunity of Science: Boundaries, Contexts, and Power* (Stanford, Calif.: Stanford University Press, 1996) and Ian Hacking, "Disunified Sciences," in Richard Q. Elvee, ed., *The End of Science? Attack and Defense* (Nobel Conference XXV; St. Peter, Minn.: Gustavus Adolphus College, 1992), 33–52.

[85]Henry Adams, *The Education of Henry Adams: An Autobiography* (Boston: Houghton Mifflin, 1918).

Henri Poincaré's Criticism of *Fin De Siècle* Electrodynamics*

Olivier Darrigol†

Il faudra donc un jour ou l'autre modifier nos idées en quelque point important et briser le cadre où nous cherchons à faire rentrer à la fois les phénomènes optiques et les phénomènes électriques. Poincaré, *L'Éclairage Électrique*, October 1895.

At the turn of the century Poincaré diagnosed a severe crisis in mathematical physics. The diseased theory was Lorentz's promising theory of electrons. The symptoms were violations of two general principles of mechanics, the principle of relativity and the principle of reaction, when applied to matter alone. And the pathogenic agent was Fizeau's experiment on the partial dragging of light waves, which apparently no theory could explain without developing the symptoms. Poincaré was isolated in his worries. Other physicists believed that ether had enough materiality to provide the missing reference and reaction; or they denied that the principles of mechanics should be relevant at all.

The principal aim of this paper is to explain the origins and coherence of Poincaré's position. Very briefly, Poincaré favoured a 'physics of principles' in which the compatibility of theories with general principles was more important than the completeness of their physical picture. Poincaré early adopted a skeptical attitude toward ether, and identified progress with a gradual elimination of this concept. Accordingly, he believed that ether should not be considered when applying the principles of relativity and reaction. Moreover, he realized that a violation of the reaction principle, when applied to matter

*The following abbreviations are used: *AJP*: *American Journal of Physics*; *AP*: *Annalen der Physik*; *AHES*: *Archive for History of Exact Sciences*; *BJHS*: *British Journal for the History of Science*; *BSM*: *Bulletin des Sciences Mathématiques*; *CR*: Académie des Sciences, Paris, *Comptes Rendus Hebdomadaires des Séances*; *EE*: *L'Éclairage Électrique*; *EO*: Poincaré, *Electricité et Optique. La Lumière et les Théories Électrodynamiques* (Sorbonne lectures of 1888, 1890, and 1899, edited by J. Blondin and E. Néculcéa; Paris, 1901); *FS*: Poincaré, *The Foundations of Science* (New York, 1929), transl. of *La Science and l'Hypothèse* (Paris, 1902), *La Valeur de la Science* (Paris, 1905), and *Science et Méthode* (Paris, 1908); *HSPS*: *Historical Studies in the Physical Sciences*; *PO*: *Oeuvres de Henri Poincaré*, 11 vols (Paris, 1954); *PZ*: *Physikalische Zeitschrift*; *RSPA*: *Revue Générale des Sciences Pures et Appliquées*; *SH*: Poincaré, *La Science et l'Hypothèse* (Paris, 1902); *TM*: Poincaré, *Electricité et Optique. I. Les Théories de Maxwell et la Théorie Électromagnétique de la Lumière* (Sorbonne lectures of spring 1888, edited by J. Blondin; Paris, 1890).

†CNRS: REHSEIS, 27 rue Damesne, 75013 Paris, France.

Received 4 *July* 1994; *in revised form* 1 *November* 1994.

alone, implied grave paradoxes; and he perceived an intimate connection between the principle of relativity and the principle of reaction, so that the two had to stand or fall together.

In retrospect we may judge that Poincaré's solid trust in the relativity principle was a major source of progress. Poincaré himself drew great profits from this attitude. As is well known, in 1905 he obtained the exact form of the Lorentz transformation for coordinates, fields, and currents; he made Lorentz's theory fully compatible with the relativity principle; and he used the Lorentz invariance as a constraint to be imposed on other, non-electromagnetic interactions, including the so-called Poincaré stress of the electron, and gravitation.

Less well known is Poincaré's discovery in 1900 that Lorentz's 'corresponding states', for which the field equations in a moving system become identical to those in a system at rest, were the *apparent states* recorded by moving observers. This result, which Poincaré proved to first order in velocity, involved an essential insight: Lorentz's local time—a pure mathematical fiction from Lorentz's point of view—was the time measured by observers moving with the system, if these observers synchronized their clocks with light signals and ignored their motion.

As is even less known, in his Sorbonne lectures of 1906–1907 Poincaré generalized this result, and introduced the Lorentz transformation as the transformation giving the apparent space and time relations for moving observers. He also showed that the relativistic dynamics of the electron resulted from the requirement of Lorentz invariance, without any assumptions regarding the structure of the electron. These achievements depended on Poincaré's faith in the validity and regulative power of the relativity principle.

The reaction principle had a different fate. By 1904 Poincaré was willing to admit that it did not hold for matter alone. In 1905 he understood why, at the electronic level, the violation did not contradict the relativity principle. This consideration suggested that the connection between the two principles which he had earlier perceived, and the related macroscopic paradoxes, were only spurious. To the contrary, in 1906 Einstein showed that Poincaré's paradoxes could be solved only if a revolutionary assumption was made: the mass of a macroscopic body had to depend on its energy content.

Previous studies of Poincaré's contributions to electrodynamics have focused on his *Dynamique de l'Électron* of 1905–1906, and usually aimed at showing the contrast with Einstein's relativity theory.[1] The present study pays less attention

[1] Cf. C. Cuvaj, *A History of Relativity. The Role of Henri Poincaré and Paul Langevin*, Ph. D., Yeshiva University (1970); 'Henri Poincaré's mathematical contributions to relativity and the Poincaré stresses', *AJP* **36** (1968), 1102–1113; 'Note on Poincaré and relativity', *AJP* **38** (1970), 774–775; S. Goldberg, 'Henri Poincaré and Einstein's theory of relativity', *AJP* **37** (1969), 498–513; 'Poincaré's silence and Einstein's relativity: the role of theory and experiment in Poincaré's physics', *BJHS* **5** (1970), 73–84; G. Holton, 'Poincaré and relativity' (1964), in *Thematic Origins of Scientific*
continued opposite

to these aspects, but establishes the continuities and ruptures of Poincaré's approach from his Sorbonne lectures on Maxwell's theory of spring 1888 to his addresses on the *mécanique nouvelle* of 1909–1910.[2] Knowing that Poincaré analyzed and expounded all the major electrodynamic theories of his time, a necessary condition for the present study was a sufficient knowledge of the original theories. Here I have benefited from recent works by Jed Buchwald, Bruce Hunt, Walter Kaiser and Daniel Siegel.[3]

The first section of this paper describes how Poincaré read the works of major electrodynamicists, and traces the role of general principles in his criticism. The second section discusses Poincaré's conception of the principles of relativity and reaction, and the paradox which he demonstrated in 1895. The third section analyzes the contents and immediate reception of a very significant memoir of 1900 in which Poincaré sharply aggravates the contradictions of Lorentz's theory. In the fourth and final section I show how Poincaré, relying on Lorentz's improved electron theory of 1904 and on his own conception of the relativity principle, contributed to solve the crisis he had earlier detected. I briefly describe the analogies and differences with Einstein's approach, and

footnote 1 continued from previous page
Thought: Kepler to Einstein (Cambridge, 1973), pp. 185–195; G. H. Keswani and C. W. Kilmister, 'Intimations of relativity: Relativity before Einstein', *British Journal for the Philosophy of Science* **34** (1983), 343–354; A. I. Miller, 'A study of Henri Poincaré's "Sur la dynamique de l'électron" ', *AHES* **10** (1973), 207–328; A. I. Miller, 'On some other approaches to electrodynamics in 1905', in H. Woolf (ed.), *Some Strangeness in the Proportion. A Centennial Symposium to Celebrate the Achievements of Albert Einstein* (Reading, 1980), pp. 66–91; *Albert Einstein's Special Theory of Relativity. Emergence (1905) and Early Interpretation (1905–1911)* (Reading, 1981); M. Paty, 'The scientific reception of relativity in France', in T. Glick (ed.), *The Comparative Reception of Relativity* (Dordrecht, 1987), pp. 113–167; *Einstein Philosophe* (Paris, 1993); K. Schaffner, 'Space and time in Lorentz, Poincaré, and Einstein: Divergent approaches of the special theory of relativity', in P. K. Machamer and R. G. Turnbull (eds), *Motion and Time, Space and Matter: Interrelations in the History and Philosophy of Science* (Columbus, 1976), pp. 465–507; C. Scribner, 'Henri Poincaré and the principle of relativity', *AJP* **32** (1964), 672–678.

[2]Relevant discussions are found in M. Atten, 'Les Théories Électromagnétiques en France. 1870–1900. La Contribution des Mathématiciens, des Physiciens et des Ingénieurs à la Construction de la théorie de Maxwell', Thèse de doctorat, Paris (1992); T. Hirosige, 'Electrodynamics before the theory of relativity', *Japanese Studies in the History of Science* **5** (1966), 1–49; H. Stein, ' "Subtler forms of matter" in the period following Maxwell', in G. N. Cantor and M. J. S. Hodge (eds), *Conceptions of Ether. Studies in the history of Ether Theories. 1740–1900* (Cambridge, 1981), pp. 309–340; C. Cuvaj, *op. cit.*, note 1; S. Goldberg, *op. cit.*, note 1; A. I. Miller, *Albert Einstein*, *op cit.*, note 1; A. I. Miller, 'On some other approaches', *op. cit.*, note 1.

[3]J. Buchwald, *From Maxwell to Microphysics. Aspects of Electromagnetic Theory in the Last Quarter of the Nineteenth Century* (Chicago, 1985); B. Hunt, *The Maxwellians* (Cornell, 1991); W. Kaiser, 'Helmholtz's instrumental role in the formation of classical electrodynamics', in D. Cahan (ed.), *Hermann von Helmholtz and the Foundations of Nineteenth-century Science* (Berkeley, 1993), pp. 374–402; D. Siegel, *Innovation in Maxwell's Electromagnetic Theory: Molecular Vortices, Displacement Current, and Light* (Cambridge, 1991). See also E. Whittaker, *A History of the Theories of Aether and Electricity*, 2 Vols. (London, 1951); O. Darrigol, 'The electrodynamic revolution in Germany as documented by early German expositions of ' "Maxwell's theory" ', *AHES* **45** (1993), 189–280; 'The electrodynamics of moving bodies from Faraday to Hertz', *Centaurus* **36**, (1993), 245–360; 'The electron theories of Larmor and Lorentz: A comparative study', *HSPS* **24** (1994), 265–336; 'Helmholtz's electrodynamics and the comprehensibility of nature', in L. Krüger (ed.), Universalgenie Helmholtz. Rückblick nach 100 Jahren (Berlin, 1944), pp. 216–242.

I relate Einstein's and Langevin's discoveries of the mass–energy relation to Poincaré's paradoxes of 1900.

Through this paper I use rationalized electrostatic units, and the vector notation. It should be known, however, that Poincaré used Maxwell's electromagnetic units, and never abandoned the Cartesian notation. Regarding the principle of relativity, I have respected the variability of Poincaré's usage. A few words may help to prevent confusions.

In 1904 Poincaré defined 'the principle of relativity' as the principle according to which 'the laws of physical phenomena must be the same for a stationary observer as for an observer carried along in a uniform motion of translation; so that we have not and cannot have any means of discerning whether or not we are carried along in such a motion' (*FS*, p. 300). In earlier writings Poincaré used the expression 'principle of relative motion', sometimes meaning the same as the principle of relativity, but often meaning something different: that the laws of physics should only depend on the relative motion of the various bodies of the system. In the latter acceptance, all observers are equivalent, including accelerated ones. Some phenomena (electrodynamic induction), and some electromagnetic theories (Weber's and Hertz's) meet this condition. Of course, Newton's mechanics does not, so that in general Poincaré could only require that the laws of physics should not depend on absolute velocity. This is not yet the modern relativity principle. Ether must first be erased from the list of bodies with respect to which relative motion is defined. This crucial step leads to what Poincaré first called 'the principle of relative motion when applied to matter *alone*' (his emphasis), and later 'the relativity principle'.

1. Surveying Electrodynamic Theories

Late nineteenth century mathematicians were commonly interested in natural phenomena. Yet, the extent and diversity of Poincaré's involvement in physics was highly exceptional. In 1886 he assumed the chair of 'Physique mathématique et calcul des probabilités' at the Sorbonne, and then in 1896, that of 'Astronomie mathématique et mécanique céleste'. From 1904 to 1908, he was the titular professor of general astronomy at the Ecole Polytechnique, and from 1902 he taught theoretical electricity at the Ecole Professionnelle Supérieure des Postes et des Télégraphes. He was a member of the Bureau des Longitudes and belonged to the editorial board of the leading French electrical journal, *L'Éclairage Électrique*. He published eleven books on mathematical physics (originating mostly in his Sorbonne lectures), and more than sixty papers on various physics subjects, not including those on celestial and analytical mechanics.[4]

[4]Cf. E. Lebon, *Henri Poincaré. Biographie. Bibliographie Analytique des Ecrits* (Paris, 1909); Atten, *op. cit.*, note 2. On Poincaré's physics studies at the lycée and at the Ecole Polytechnique, cf. S. Walter, 'Henri Poincaré's student notebooks, 1870–1878', communication at the Poincaré congress in Nancy, on 18 May, 1994.

Poincaré's interests in physics ranged from the mathematically abstract to the commercially practical. He discussed the existence of solutions for the differential equations of mathematical physics. He criticized and developed the leading physical theories of his time. He played a role in analyzing important new experiments, for instance Hertz's on electromagnetic waves, Röntgen's on X-rays, and Crémieu's on electrical convection. His speculation about a relationship between fluorescence and X-rays led to Becquerel's discovery of radioactivity. A Polytechnician, an Ingénieur des Mines, and a patriot, he kept an active interest in technological questions. For instance, he gave designs for telegraphic devices and electric motors. He often played the role of the modern consultant, as his numerous contributions to *L'Éclairage Électrique* demonstrate. Finally, he excelled in popularizing science, and wrote a best-seller on wireless telegraphy.[5]

Here I will only discuss Poincaré's contributions to the theory of electricity. One should keep in mind, however, that he did not always subordinate his interest in physics to theoretical or mathematical questions. As Langevin noted, Poincaré occasionally found in physicists' works the inspiration for his mathematical work. Much more frequently, he used his mathematical superiority to extract the kernel from physical theories and ease access to their practical consequences. In the mathematical thicket of the new field theories, where others easily got lost, he moved with the greatest ease and grace and went straight to the essentials.[6]

The Sorbonne Lectures

An interest in foreign theories, especially British ones, was not very common among French physicists. There were a few exceptions, like Alfred Potier, who arranged the translation of Maxwell's *Treatise*, and Eleuthère Mascart, who devoted a whole chapter to Maxwell' s theory in the popular 'Mascart–Joubert'. Moreover, some engineers-physicists were familiar with Faraday's and Maxwell's works. Their favorite journal, *L'Éclairage Électrique*, published translations and critical expositions of numerous foreign memoirs, including the most theoretical ones (for example, Hertz's and Helmholtz's). Telegraphic engineers had an interest in Maxwell's theory of electricity even before Hertz's experiments with electric waves.[7]

[5]Cf. P. Langevin, 'Le physicien', in V. Volterra, J. Hadamard, P. Langevin, P. Boutroux, *Henri Poincaré. L'Oeuvre Scientifique et Philosophique* (Paris, 1914); L. de Broglie, 'Henri Poincaré et les théories de la physique', in *PO*, Vol. 11, pp. 62–71; M. G. Darrieus, 'Contributions de Henri Poincaré à l'électrotechnique', in *PO*, Vol. 11, pp. 132–139; W. Wien, 'Die Bedeutung Henri Poincaré's für die Physik', in *PO*, Vol. 11, pp. 242–246.

[6]Langevin, *op. cit.*, note 5.

[7]Cf. Atten, *op. cit.*, note 2; 'Les télégraphistes français et Maxwell', *Recherches Sur l'Histoire des Télécommunications* **2** (1988), 7–40; Poincaré, 'Préface' of A. Potier, *Mémoires Sur l'Electricité et l'Optique* (Paris, 1912), pp. V–X. Potier was a *répétiteur* at the Polytechnique when Poincaré was a student (I thank S. Walter for this information).

In the spring of 1888 Poincaré taught a course on 'Maxwell's theories and the electromagnetic theory of light'. Hertz's experiments could not have been the incentive, since they were concluded a few months later. Most likely, Poincaré regarded this topic as a continuation of his previous lectures on 'the mathematical theories of light'.[8]

The way Poincaré read Maxwell and other authors may be illuminated by a comment he once made on his handling of sources:

> When I read a memoir, I am used to first glance at it quickly so as to get an idea of the whole, and then to return to the points which seem obscure to me. I find it more convenient to redo the demonstrations rather than [going through] those of the author. My demonstrations may be much inferior in general, but for me they have the advantage of being mine.

The method is highly recommended to mathematicians and physicists (although not to historians). When applied to a physical theory, it has at least three virtues. First, it helps in clarifying, completing, and correcting formal developments. Second, it may reveal in full light the essential character of some features of the investigated theory. Third, it may imply creative misunderstandings, prompt reinterpretations, and initiate new approaches.[9]

As I have discussed elsewhere, Poincaré's exposition of Maxwell had all these qualities. He corrected some mathematical mistakes of the *Treatise*, and frequently clarified Maxwell's derivations. He highlighted Maxwell's use of the Lagrangian method, in which he saw the inauguration of a more abstract version of mechanical reductionism. The mathematical physicist of the future would no longer require explicit mechanisms, but solely the possibility of a mechanism. Along with this insight, Poincaré misrepresented what British Maxwellians took to be the essence of Maxwell's theory: the reduction of electrification and conduction to ether processes. As a victim of Maxwell's ambiguous wordings, he failed to understand the basic notions of electric displacement and strains in ether, and declared them contradictory. In his view Maxwell used mutually incompatible pictures within the same theory, and avoided contradiction only by not combining these pictures.[10]

In 1889, 1892 and 1899, Poincaré lectured on the electrodynamics of other foreign masters.[11] When several theories compete on the same ground, most teachers tend to dwell on their favorite, and to neglect the others. Poincaré, the

[8]Poincaré, *Electricité et Optique. I. Les Théories de Maxwell et la théorie électromagnétique de la Lumière* (Sorbonne lectures of 1888, J. Blondin (ed.); Paris, 1890); *Théorie Mathématique de la Lumière* (Sorbonne lectures of 1887–1888, J. Blondin (ed.); Paris, 1889). Cf. Darrigol, 'Revolution', *op. cit.*, note 3.

[9]Poincaré to Mittag-Leffler, 5 Feb 1889, in *PO*, Vol. 11, p. 69.

[10]Darrigol, 'Revolution', *op. cit.*, note 3.

[11]Poincaré, *Electricité et Optique. Les Théories de Helmholtz et les Expériences de Hertz* (Sorbonne lectures of 1889–1890, B. Bruhnes (ed.); Paris, 1891); *Les Oscillations Electriques* (Sorbonne lectures of 1892–1893, C. Maurain (ed.); Paris, 1894); *Electricité et Optique. La Lumière et les Théories Electrodynamiques* (Sorbonne lectures of 1888, 1890 and 1899, J. Blondin and E. Néculcéa (eds); Paris, 1901).

superior, magnanimous mathematician, did not do so. He looked at all these theories with an apparently equal interest, and with moderate judgments of value. This attitude prevented dogmatism, as Poincaré explained in the introduction to his lectures on mathematical optics:

> The theories proposed to explain optical phenomena by the vibrations of an elastic medium are very numerous and equally plausible. It would be dangerous to confine oneself to one of them; one would thus be prone to a blind, and therefore misleading confidence in this theory.'

Poincaré's openness squares well with his view that once successful theories never completely die: they contain true relations (*rapports vrais*) that may reappear in subsequent theories. The comparison of contemporary, competing theories eases the extraction of these true relations.[12]

Poincaré successively expounded the theories of Maxwell, Helmholtz, Hertz, Larmor, and Lorentz, with equal thoroughness and respect. Not fearing intellectual strabismus and ignoring incommensurability, he multiplied comparisons and connections between these theories. For example, he was first to provide a detailed, complete proof that Maxwell's theory was 'nothing but a limiting case of Helmholtz's theory'. He showed how Hertz's equations could be derived from Maxwell's. He examined the relation between Larmor's theory and Neumann's optical theory. He found the modifications of Hertz's equations that led to Lorentz's microscopic and macroscopic field equations. And he showed how Larmor 'appropriated Lorentz's assumptions and combined them with his own'.[13]

Poincaré essentially based these comparisons on the differential equations of the theories, which he expressed in standardized notation (usually Maxwell's) and units (electromagnetic ones), and arranged in a transparent manner. Such emphasis on mathematical equipment could obscure the contrast between different theories. For example, a true Maxwellian would have rejected the statement that Maxwell's theory was 'a limiting case of Helmholtz's theory' as undoing Maxwell's reform of the concept of electricity. For Poincaré, it was sufficient that the physical predictions, the *rapports vrais*, should agree in the two theories.

Yet, Poincaré sensed the individuality of each theory and tried to convey the inventor's style. His Maxwell '[did] not try to construct a unique, definitive, and well-ordered edifice, but rather seem[ed] to erect a great number of provisional and independent constructions, among which communication was difficult and sometimes impossible'. He portrayed Hertz as a rigorist who 'only admitted the equations established by Maxwell, left aside Maxwell's classical text, regarding

[12]Poincaré, 'Lumière', *op. cit.*, note 8, p. II; 'Sur les rapports de la physique expérimentale et de la physique mathématique', in *Rapports Présentés au Congrès International de Physique de 1900*, Vol. 1 (Paris, 1900), pp. 1–29, also in *SH*, Chapters 9 and 10, see pp. 173–176. Cf. D. Stump, 'Henri Poincaré's philosophy of science', *Studies in History and Philosophy of Science* **20** (1989), 335–363.

[13]Poincaré, *Théories de Helmholtz, op. cit.*, note 11, p. 83; *EO*, pp. 345–352, 587, 627 (quotation).

it as obscure, and tried, by setting the final equations in advance, to make a theory that leads to them'. He saw in Larmor a clever reactionary who 'sought a common mechanical explanation for light and electricity, and [for this purpose] returned to [older] elastic theories [of optics]'. Finally, his Lorentz was the champion of a new microscopic vision that eliminated true magnetism and conduction currents and replaced them with Amperean currents and electronic flow. In his eyes diversity was a virtue: 'Each has his characteristic aptitudes, and these aptitudes should be diverse, else would the scientific concert resemble a quartet where every one wanted to play the violin'.[14]

What made Poincaré's lectures especially attractive was his luminous style of exposition. His mathematical derivations were supremely elegant, and his French beautifully clear. He was very careful to state all necessary hypotheses, but avoided the excessive abstraction of a purely axiomatic presentation. For pedagogical reasons he preferred an inductive presentation, and provided expressive analogies for the most difficult statements. The appeal of such qualities was not limited to French readers. The Maxwellian Andrew Gray applauded: 'Here are to be found exemplified that order and harmony which renders the work of the best French mathematical writers so exquisitely clear, and that artistic charm which is so seldom seen in the writings of scientific men of other nationalities.'[15]

Reviews

Poincaré's lectures were edited by several of his students and published in four volumes. No reader questioned Poincaré's mathematical superiority, with the exception of Joseph Bertrand, who perceived a grave lack of rigor in the 1890 lectures. In the eyes of the stiff, towering Academician, Maxwell's *Treatise* was a dangerous incongruity and its French supporters betrayed the Amperean tradition. Bertrand also detected mistakes in Helmholtz's memoirs, mostly out of his own ignorance of the German idiom. Fortunately, some French physicists had a more open attitude. Marcel Brillouin wrote very detailed reviews of Poincaré's lectures, with enthusiastic comments such as: 'There is no book that expounds a greater number of interesting theories, that compares and scrutinizes them with more penetration; that, in sum, gives more to think about to the geometer-physicist who wishes to embrace in the same coherent theory a vast collection of phenomena'.[16]

The British had mixed feelings. Gray, an undogmatic Maxwellian, saw some advantage to Poincaré's departure from Maxwell's conception of the

[14]*TM*, p.IV; *EO*, pp. 344, 583, 422; *SH*, p. 4.

[15]A. Gray, review of *TM*, in *Nature* **44** (1891), 296–299.

[16]J. Bertrand, review of Poincaré, *TM*, in *Journal des Savants* (1891), 742–748; cf., e.g., Helmholtz, 'Sur le mouvement le plus général d'un fluide', *CR* **47** (1868), 222–215, 754–757; M. Brillouin, review of Poincaré, *TM*, and *Théories de Helmholtz, op. cit.*, note 11, in *BSM* **15** (1891), 129–146; review of *EO* in *BSM* **25** (1901), 118–127, see p. 127.

displacement current. He admired the depth of Poincaré's views on mechanical explanation, and he was altogether satisfied with his rendering of 'the wild and native beauty of Maxwell's work'. But the soul of the Maxwellian group, George Francis FitzGerald, protested in *Nature*'s columns. Among other infidelities, Poincaré attributed two incompatible properties, incompressibility and elasticity, to Maxwell's imaginary fluid, whereas, for Maxwell, elasticity belonged to the *ether*, of which the incompressible fluid was only one component. Poincaré had simply lost the essence of the Maxwellian conception of electricity. With a touch of irony FitzGerald concluded: 'As a more mathematical work the book is admirable and clear'.[17]

In 1901 FitzGerald passed away with his idea of a strict Maxwellian orthodoxy. The new prince of British electrodynamics, Joseph Larmor, welcomed Poincaré's lectures with some ceremony: 'It is a matter of congratulation that an analyst of M. Poincaré's vast command of all the resources of modern mathematics finds time not only to apply his genius to a thorough revision of mathematical astronomy, but also to survey the field of general physics as he has done in this interesting volume'. However, Larmor said very little about the contents of the lectures, and gave instead an update of his own ideas. Paradoxically, he regarded Poincaré's predilection for mathematical completeness as preventing the communication between different theories:

> The distinction [...] between the crystallised mathematical and the fluent physical point of view is at the root of what is a prominent characteristic of the writer's criticism. The development of electrodynamics appears as split up in so many independent and largely irreconciliable theories [...], whereas on the view which works by models and general ideas rather than by formulas there is but one theory of electrodynamics.

As noted above, Poincaré's intention was just the opposite. The misunderstanding shows just how uncongenial French mathematical physics could remain for a British *fin de siècle* physicist.[18]

The Germans had less difficulty with Poincaré's style. Two of their heroes, Helmholtz and Hertz, made excellent use of the hypothetico-deductive manner in mathematical physics. With respect to mechanical explanation, Poincaré's views were similar to those of the older Helmholtz. Germans widely shared Poincaré's difficulties in understanding Maxwell's concepts of charge and current. They promptly translated Poincaré's lectures on Maxwell and Helmholtz, giving them much attention and praise. One reviewer characterized Poincaré's presentation of Helmholtz as not quite faithful, but he had only

[17]Gray, *op. cit.*, note 14; G. F. FitzGerald, 'M. Poincaré and Maxwell', letter to the editor, *Nature* **45** (1982), 532–533, see p. 533.

[18]J. L., review of *EO*, in *Nature* **64** (1901), 273–276, see p. 275. The identification of the author as Joseph Larmor is mine.

compliments for the rest of the lectures. As I have shown elsewhere, traces of Poincaré's influence are visible in several German electrodynamic texts of the 1890s.[19]

The Physics of Principles

In his discussion of Maxwell's views on mechanical explanation, Poincaré distinguished between phenomenological laws and mechanical interpretation. The laws are concerned with 'the parameters which can be directly reached and measured by experience', and with the differential equations giving their evolution, whereas the mechanical interpretation reduces the laws to the Newtonian motion of ordinary matter and hypothetical fluids. Poincaré proved the following theorem: 'If a phenomenon admits a mechanical explanation, there exist an infinity of other mechanical explanations that account equally well for all aspects revealed by experience'. In the face of such arbitrariness, he recommended the principle of least action, which implied the possibility of a mechanical representation without giving it. As mentioned, he regarded this transformation of the mechanistic ideal as Maxwell's greatest achievement.[20]

In general, Poincaré perceived an inevitable evolution from the Newtonian or Laplacian ideal to more general principles:

> A day arrived when the conception of central forces no longer appeared sufficient [...]. What was done then? The attempt to penetrate into the detail of the structure of the universe, to isolate the pieces of this vast mechanism, to analyze one by one the forces which put them in motion, was abandoned, and we were content to take as guides certain general principles the express object of which is to spare us this minute study.

Poincaré's list of useful principles included the two principles of thermodynamics, the principle of action and reaction (Newton's third law), the principle of relativity, the principle of conservation of mass, and the principle of least action. All the principles, except Carnot's, could be extracted from mechanics. Unlike the complete set of Newton's laws, however, their application did not require a full reduction to matter and motion. They compromised effectively between the older Newtonian ideal and pure phenomenology: 'The application of these [...] general principles to the different physical phenomena is sufficient for our learning of them all that we could reasonably hope to know of them'.[21]

[19]Wangerin, abstracts of Poincaré, *TM*, and *Théories de Helmholtz, op. cit.*, note 11, in *Jahrbuch über die Fortschritte der Mathematik* **23** (1891), 1108–1110, **24** (1892), 1041–1042. The German translation of *Electricité et Optique* was made by W. Jäger and E. Gumlich, and appeared in 1891. Cf. Darrigol, 'Revolution', *op. cit.*, note 3.

[20]*TM*, pp. IX, XIV.

[21]Poincaré, 'L'état actuel de la physique mathématique', *BSM* **28** (1904), 302–324, see pp. 305–307, also in *FS*, pp. 299, 301. Poincaré's views resembled those which Helmholtz held late in his life: cf. Darrigol, 'Helmholtz's electrodynamics', *op. cit.*, note 3.

The principles were no more true a priori than the laws of mechanics. Following Kelvin, Helmholtz and Planck, Poincaré gave them an inductive origin:

> The principles are results of experiments boldly generalized; but they seem to derive from their very generality a high degree of certainty. In fact, the more general they are, the more frequent are the opportunities to check them, and the verifications multiplying, taking the most varied, the most unexpected forms, end by no longer leaving place for doubt.

In *Science and Hypothesis*, Poincaré argued that after the initial stage of experimental generalization, the fundamental principles of mechanics become completely irrefutable, that they act as definitions or conventions. The refutation of a principle can never be asserted on a given system, if only because the system is never perfectly isolated. The left-over entities can always save the principle.[22]

In practice, however, Poincaré avoided nominalism. Even though a principle could never be strictly refuted, it could become sterile, it could cease to control phenomena.[23] This could happen, for instance, if invisible entities were introduced to save the principle. As we shall see below, Poincaré denied that ether should be taken into account while applying the principles of relative motion and reaction. In general, Poincaré's epistemological remarks are no safe guide for the study of his contributions to physics. Poincaré usually intended his philosophical insights for specific circumstances and audiences, and never built them into a coherent system. He never quite filled the gap between the nominalism of the 'conventions' and the realism of the '*rapports vrais*'.

The most outstanding feature of Poincaré's criticism of electromagnetic theories was his overriding concern for their compatibility with the general principles of mechanics and with two more specific principles: the conservation of electric and magnetic charge, and the unity of the electric and magnetic forces. There were antecedents for these concerns in the history of electrodynamics: Helmholtz and Heaviside had given prominence to the energy principle, Maxwell and Larmor to Hamilton's principle, and Hertz to the unity of the electric force. Poincaré now considered a larger number of principles in a very systematic manner.

Energy conservation was the least problematic principle, since it played a central role in all existing theories. Poincaré, however, detected one violation of this principle in Maxwell's theory. In his opinion, Maxwell's stresses, from which all mechanical forces between electrified bodies and current carriers are

[22]Poincaré, *FS*, p. 301; *SH*, Chapter 6 (from 'Sur les principes de la mécanique', in *Bibliothèque du Congrès International de Philosophie Tenu à Paris du 6 au 12 Août 1900*, Paris, 1901), pp. 457–494, see especially p. 124.

[23]Poincaré, *SH*, p. 178.

supposed to result, contradicted the energy principle. Their work during a deformation of a volume element of the medium did not balance the variation of the energy content of the volume element. Poincaré concluded that the concept of stresses in ether was incompatible with the rest of Maxwell's system. As I have shown elsewhere, he was mistaken.[24]

Poincaré's comments on Maxwell's dynamical method were far more inspired. They led to the requirement that every fundamental theory should be cast in the Lagrangian form. However, Hertz's 1890 theory just took the fundamental equations as axioms, and provided no mechanical foundation. In 1892 Lorentz provided the missing foundation both for Hertz's theory and for his own. His complicated reasoning was based on d'Alembert's principle of virtual work. In 1899, while Lorentz had already come to doubt the usefulness of a mechanical foundation, Poincaré derived both Hertz's and Lorentz's equations by the Lagrangian method with unspecified coordinates.[25]

In a brilliant memoir of 1884, Hertz introduced the principle of the unity of electric forces: 'Electric forces which emanate from inductive actions are in every respect equivalent to equal and equally directed forces of electrostatic origin'. For instance, an electrically charged body placed near a variable current should experience a mechanical force. In Maxwell's theory this is trivially the case, since the electric field is an intrinsic modification of the medium. In Helmholtz's potential theory of 1870, however, electrostatic and electrodynamic interactions are almost completely independent: an electric current has no electrostatic effect other than that due to the accumulation of electricity at the extremities of an open current. Hertz not only noted this defect of Helmholtz's theory, but also showed that the simplest possible modification of the theory in accordance with the unity of the electric force and the energy principle led to Maxwell's equations.[26]

Poincaré was clearly impressed by Hertz's criticism, to which he devoted a whole chapter of his lectures on Helmholtz's theory.[27] While examining the unity of the electric force, he found that Helmholtz's theory violated the more fundamental principle of the equality of action and reaction. More exactly, the theory could not accommodate Rowland's result about electric convection

[24]Poincaré, *TM*, pp. 91–92; Darrigol, 'Revolution', *op. cit.*, note 3, pp. 219–220.

[25]Poincaré, *EO*, pp. 427–446 (the case of Hertz's equation is on pp. 435–436). On Lorentz's method, cf. Hirosige, 'Origins of the Lorentz theory of electrons and the concept of the electromagnetic field', *HSPS* **1** (1969), 151–209. Helmholtz also provided an action function for the Maxwell-Hertz equations in 'Das Prinzip der kleinsten Wirkung in der Elektrodynamik', *AP* **47** (1892), 1–26.

[26]H. Hertz, 'Über die Beziehungen zwischen den Maxwell'schen Elektrodynamischen Grundgleichungen und den Grundgleichungen der gegnerischen Elektrodynamik', *AP* **23** (1884), 84–103. cf. Buchwald, *The Creation of Scientific Effects: Heinrich Hertz and Electric Waves* (Chicago, 1994); Darrigol, 'Revolution', *op. cit.*, note 3.

[27]Poincaré, *Théories de Helmholtz*, *op. cit.*, note 11, pp. 114–127. This chapter is not reproduced in *EO*.

without violating this principle.[28] This shortcoming did not concern the limiting case of Helmholtz's theory that retrieved the predictions of Maxwell's theory. After Hertz's experiments on electric waves, however, no one but the inflexible Pierre Duhem could still favor the Helmholtzian view. Poincaré turned his attention to Hertz's version of Maxwell's theory.

Helmholtz himself preferred Hertz's formulation of Maxwell's theory, and used it in combination with Hamilton's principle and the ionic hypothesis to produce a new dispersion theory. In essence, this theory treated ether and the ions of transparent matter as two coupled dynamical systems. Helmholtz assumed that ether did not move except in the immediate vicinity of ions. In 1893 his disciple Richard Reiff extended this assumption to moving transparent bodies and deduced Fresnel's coefficient for the dragging of light.[29]

The concept of moving ions in a stationary ether could already be found in Lorentz's little-known Dutch memoir of 1879. The derivation of the Fresnel coefficient had appeared in Lorentz's French memoir of 1892. Yet there was at least one significant difference: Helmholtz wrote dynamical equations directly for macroscopic quantities, whereas Lorentz started with a detailed consideration of the motion of a single ion. Accordingly, Helmholtz's theory was simpler but more arbitrary. In order to retrieve Fresnel's dragging coefficient, Reiff had to drop one term in Helmholtz's action function, and Poincaré observed that Reiff's equations did not comply with the conservation of electric charge. The objection was correct, but it was not as harmful as Poincaré intended it to be: Reiff's expression for the total current could easily be corrected without leaving the framework of Helmholtz's theory. The sole interest of this episode is that it gives one more example of Poincaré's concern with general principles.[30]

2. Principles and Paradoxes

Hertz's Electrodynamics of Moving Bodies

Poincaré's use of general principles was most innovative when he discussed the electrodynamics and optics of moving bodies in accordance with the theories of Hertz, Larmor and Lorentz. In 1890 Hertz provided a complete electrodynamics of moving bodies using the fundamental equations

[28]Poincaré, *Théories de Helmholtz*, *op. cit.*, note 11, pp. 220–227. This complement is not reproduced in *EO*.

[29]Helmholtz, 'Elektromagnetische Theorie der Farbenzerstreuung', *AP* **48** (1893), 389–405; R. Reiff, 'Die Fortpflanzung des Lichtes in bewegten Medien nach der elektrischen Lichttheorie', *AP* **1** (1893), 361–367; cf. Buchwald, *op. cit.*, note 3, pp. 237–239.

[30]H. A. Lorentz, 'Over het verband tusschen de voorplantings snelheid en de dichtheid en samestelling der middenstoffen', Koninklijke Akademie van Wetenschappen, Amsterdam, *Verslagen* **18** (1879), French in *Annales de Physique et de Chimie* **9** (1880), 641–665; 'La théorie électromagnétique de Maxwell et son application aux corps mouvants', *Archives Néerlandaises* **25** (1892), 363–552; Poincaré, *EO*, pp. 592–599.

$$\nabla \times \left(\mathbf{E} - \frac{\mathbf{v}}{c} \times \mu\mathbf{H}\right) = -\frac{1}{c}\frac{\partial \mu\mathbf{H}}{\partial t},$$

$$\nabla \times \left(\mathbf{H} + \frac{\mathbf{v}}{c} \times \varepsilon\mathbf{E}\right) = \frac{1}{c}\left[\frac{\partial \varepsilon\mathbf{E}}{\partial t} + \mathbf{v}(\nabla \cdot \varepsilon\mathbf{E}) + \sigma\mathbf{E}\right]. \qquad (1)$$

The vectors **E** and **H** are the electric and magnetic forces acting on unit charges, and characterize the state of the medium at a given point. The parameters ε, μ, and σ are, respectively, the permittivity, permeability, and conductivity of the medium. The vector **v** is the velocity of ether, which is identical to the velocity of matter wherever it exists. As noted by Hertz, the form of these equations remains the same if they refer to a moving system of axes. In Poincaré's words: 'Hertz's equations have the same form in the relative motion and in the absolute motion'.[31]

When it was published, Hertz's theory was compatible with all known electrodynamic experiments. The first fundamental equation yielded induction phenomena and their dependence on relative motion only. The second included the magnetic effects of four kinds of currents: the conduction current, the displacement current, the convection current, and the Röntgen current. These last three contributions explained, repectively, Hertz's experiments of 1888 on electric waves, Rowland's experiments of 1875 on rotating electrified disks, and Röntgen's experiments of 1888 on rotating polarized dielectrics.[32]

Hertz was aware of two fundamental defects of his theory. First, it could not account for well-known results from the optics of moving bodies, like stellar aberration and Fizeau's experiment. As Poincaré noted, the invariance property of the equations created part of the trouble. It implied that the velocity of light with respect to a moving transparent body was the same as if the body did not move—a result at variance with Fizeau's experiment on the dragging of light by streaming water. Of course, the invariance was otherwise welcomed to explain that terrestrial optical experiments did not depend on the earth's motion.[33]

The other defect that Hertz deplored in his theory was that it implied non-vanishing forces acting on ether even in the complete absence of matter. British Maxwellians knew of such forces, assuming a density $\partial(\mathbf{E} \times \mathbf{H}/c)/\partial t$ in vacuum, but found them quite acceptable. In their opinion ether could well have inertia and momentum since it already carried energy. Even if such effects had not yet been observed, radiation had to exert a pressure, and a material body could receive or lose momentum without immediate compensation by other material

[31]Hertz, 'Über die Grungleichungen der Elektrodynamik für bewegte Korper', *AP* **41** (1890), 369–399; Poincaré, *EO*, p. 388.

[32]Cf. Hirosige, *op. cit.*, note 2; O. Knudsen, '19th century views on induction in moving conductors', *Centaurus* **24** (1980), 346–360; Darrigol, 'Moving bodies', *op. cit.*, note 3.

[33]Poincaré, *EO*, p. 389.

bodies. Hertz thought differently. Coming from a German tradition favoring action at a distance, he was less willing to ascribe material properties to ether. For him the equality of action and reaction had to hold when material bodies were considered alone. He therefore suggested that Maxwell's expression for stresses in ether, and perhaps even the fundamental equations, required modification.[34]

Poincaré, as we shall see below, shared Hertz's conception of the principle of reaction. Nevertheless, he judged that Hertz's theory was compatible with the principle. In his reading of Hertz's theory, there was no ether at all.[35] The vector **v** in Hertz's equations represented the velocity of matter, which was necessarily present in every portion of space, though sometimes very rarefied. With this interpretation, Hertz's unwanted force had something to grab onto, and Newton's third law was saved.[36] In 1899 Poincaré could in fact show that the energy principle together with the principle of relative motion implied the equality of action and reaction. His reasoning is shown below.[37]

Call J the total energy of the electromagnetic field, U_0 the energy communicated to the field by external sources minus the energy lost into Joule heat, and **f** the density of mechanical force. The conservation of energy gives:

$$\frac{\mathrm{d}J}{\mathrm{d}t} = \int (U_0 + \mathbf{v}\cdot\mathbf{f})\mathrm{d}\tau. \tag{2}$$

With respect to a system of axes moving with velocity **u**, the quantities just defined are invariant, since in Hertz's theory they are intrinsic properties of the medium. Consequently, the conservation of energy for the moving system reads

$$\frac{\mathrm{d}J}{\mathrm{d}t} = \int (U_0 + (\mathbf{v} + \mathbf{u})\cdot\mathbf{f})\mathrm{d}\tau. \tag{3}$$

Subtracting the two equations yields

$$\int \mathbf{u}\cdot\mathbf{f}\mathrm{d}\tau = 0. \tag{4}$$

Since **u** can take an arbitrary constant value, the total force must be zero, in harmony with the principle of reaction.

[34]Cf. Darrigol, 'Moving bodies', *op. cit.*, note 3.

[35]There were no stresses either. Consequently, Poincaré's derivation of mechanical actions in Hertz's theory differed from Hertz's (*EO*, pp. 403–420). He considered only the case of solid dielectrics immersed in air, and obtained, on the basis of the energy principle, an expression for the force density that is different from Hertz's. However, there is no observable difference, since the resultant force and moment on a given solid is the same for both expressions.

[36]In his first lectures on Hertz's theory (1892), Poincaré was not quite clear about the elimination of ether. However, he already suggested that the theory respected the reaction principle (*Oscillation Électriques*, *op. cit.*, note 11, p. 337). In 1895 he proved this result exactly, but with a cumbersome method (*PO*, Vol. 9, pp. 394–402); he made clear that he interpreted the velocity in Hertz's equations as the velocity of matter (*ibid.*, p. 397); and that 'rarefied air' could be the only matter that moved (*ibid.*, p. 392). In his 1899 presentation of Hertz's theory, Poincaré systematically avoided the word ether (*EO*, pp. 345–420).

[37]Poincaré, *EO*, pp. 420–421.

On Lorentz's Theory and the Relativity Principle

In Lorentz's theory, all electrodynamic and optical phenomena are reduced to the interactions of electrons or ions via a stationary ether. The motion of the earth with respect to the stationary ether immediately explains stellar aberration. Less obviously, the ionic scattering of light implies the Fresnel dragging effect and therefore accounts for Fizeau's result. Lorentz further explained the failure of past attempts to detect the motion of the earth through ether by terrestrial experiments. For this purpose he first wrote the field equations with respect to a moving system of axes, that is, he performed a Galilean transformation of coordinates. To first order in u/c (where $\mathbf{u}$ denotes the velocity of the earth) he could retrieve the form of the field equations with respect to ether, thanks to an additional change of variables involving the 'local time':

$$t' = t - \mathbf{u}\cdot\mathbf{r}/c^2 \tag{5}$$

and the field transformation

$$\mathbf{E}' = \mathbf{E} + \frac{1}{c}\mathbf{u} \times \mathbf{H}, \qquad \mathbf{H}' = \mathbf{H} - \frac{1}{c}\mathbf{u} \times \mathbf{E}. \tag{6}$$

To explain the negative result of second-order experiments, Lorentz assumed that the dimensions of material bodies contracted in the direction of their absolute velocity $\mathbf{v}$ in the proportion $1 - \mathbf{v}^2/2c^2$. In short, Lorentz obtained the compensation of the effects of the earth's motion, but only with certain restrictions. In the general case he expected observable effects to occur, since his stationary ether provided a privileged system of axes.[38]

Poincaré judged Lorentz's theory to be 'the one that best account[ed] for the facts' or 'the least defective', but despised the 'coup de pouce' of the Lorentz contraction. One should not, he argued, introduce a new hypothesis every time a new anomaly comes forth: 'Of hypotheses there is never lack'. Moreover, the anomalies of Lorentz's theory were all of the same nature: they concerned, at different orders of approximation, the insensitivity of optical phenomena to the earth's motion. Not to look for a common explanation was, Poincaré said, 'to concede too much to chance'.[39]

[38]Lorentz, *Versuch einer Theorie der elektrischen und optischen Erscheinungen in bewegten Körpern* (Leiden, 1895). Cf. Hirosige, *op. cit.*, note 25, and Buchwald, 'The Michelson experiment in light of electromagnetic theory before 1900', in S. Goldberg and R. Stuewer (eds), *The Michelson era in American Science*. 1870–1930 (New York, 1988), pp. 55–70.

[39]Poincaré, *EO*, pp. II, 611, 536; *SH*, p. 182, also in *FS*, p. 147; *ibid.*. Poincaré's interest in Lorentz's theory was not limited to its compatibility with general principles. In 'La théorie de Lorentz et les expériences de Zeeman', *EE* **11** (1897), 481–489, 'La théorie de Lorentz et le phénomène de Zeeman', *EE* **19** (1899), 443–460, and 'Le phénomène de Hall et la théorie de Lorentz', *CR* **128** (1899), 339–341, he discussed various magneto-optical effects in this theory. Some of his considerations were flawed, and Lorentz corrected them: cf. Buchwald, *op. cit.*, note 3, pp. 248–249.

More fundamentally, Poincaré believed that the principle of relative motion applied to matter *alone*: 'I consider it very probable that optical phenomena depend only on the relative motion of the material bodies present—light sources or optical apparatus—*and this not only to first or second order [...] but exactly*'. For the unprejudiced theorist, this conception emerged from empirical data: 'All attempts to measure the velocity of the earth in relation to the ether have led to negative results. In this case experimental physics has been more faithful to the principle than mathematical physics'. Moreover, the opposite conception pained Poincaré's epistemological conscience: 'Is it not evident that from the principle so understood [including motion with respect to imponderables] we could no longer infer anything? It could no longer tell us anything just because it would no longer fear any contradiction'. In general, principles were useful guides so long as they controlled observable quantities.[40]

Lorentz agreed that the contraction of lengths should not be purely *ad hoc*. In fact, FitzGerald, Larmor and himself believed that the contraction resulted from a more general assumption that molecular forces behaved like electric forces with respect to the earth's motion.[41] Lorentz also agreed with Poincaré that all orders in *u/c* should be treated on the same footing, and he did so in 1904. Yet, as we shall see below, he obtained his result only for certain approximations and did not exclude a future detection of the earth's motion with respect to ether. Other electron theorists agreed on this point. Poincaré was the only one (except Bucherer) to think that the principle of relative motion should apply to matter *alone*.[42]

On Lorentz's Theory and the Reaction Principle

Poincaré found another defect in Lorentz's theory, which he judged most disturbing: the theory did not comply with the principle of action and reaction. Indeed, Lorentz's microscopic force density,

$$\mathbf{f} = \rho\left(\mathbf{E} + \frac{\mathbf{v}}{c} \times \mathbf{H}\right), \tag{7}$$

and Lorentz's field equations yield the total force

[40]Poincaré, *EO*, p. 536 (his emphasis) *op. cit.*, note 21, p. 310, also in *FS*, p. 306; *ibid.*. Poincaré made a similar comment on the reaction principle in *op. cit.*, note 21, p. 312, also in *FS*, p. 310: 'The principle so understood (including a reaction of ether) will explain everything, since, whatever might be the visible movements, we always could imagine hypothetical movements that compensate them. But if it is able to explain everything, this is because it does not enable us to foresee anything; it does not enable us to decide between the different possible hypotheses, since it explains everything beforehand. It therefore becomes useless'.

[41]Cf. Hirosige, *op. cit.*, note 25; Hunt, *op. cit.*, note 3; A. Warwick, 'On the role of the FitzGerald–Lorentz contraction hypothesis in the development of Joseph Larmor's electronic theory of matter', *AHES* **43** (1991), 29–91.

[42]Cf. Paty, Miller, and Goldberg, *op. cit.*, note 1.

$$\int \mathbf{f}\mathrm{d}\tau = -\frac{\mathrm{d}}{\mathrm{d}t}\int \frac{1}{c}\mathbf{E} \times \mathbf{H}\mathrm{d}\tau. \tag{8}$$

In general the last integral is not zero. Lorentz had already noted this in his *Versuch*, but with a quieting comment: 'As far as I can see, nothing forces us to elevate the principle [of reaction] to the rank of a fundamental law of unlimited validity'.[43]

Poincaré was far more reluctant to give up the equality of action and reaction. He promptly denounced 'a grave difficulty' in Lorentz's theory, and gave concrete examples of violations of the principle. A beam of electromagnetic radiation would exert pressure on a conductor. Or a source emitting radiation in one direction should recoil in the opposite direction. Poincaré calculated that an emitted energy of three million Joules implied a considerable recoil velocity: one centimetre per second for a mass of one kilogram.[44]

As Brillouin noted, radiation pressure and the recoil effect exist 'not only in Lorentz's theory but in all electromagnetic theories that localize energy in ether as well'. Poincaré certainly knew this. His point was that Lorentz's theory had no provision for the missing momentum, whereas the Maxwell–Hertz theory could ascribe it to arbitrarily diluted dielectric matter.[45]

The only way to reconcile Lorentz's theory with the reaction principle was to abandon the stricture of an absolutely stationary ether, and to confer upon it some inertia and velocity. Against this view Poincaré objected that the velocity of ether would then be proportional to $\mathbf{E} \times \mathbf{H}$, and would therefore quadruple when the field strength doubled. He probably had in mind elastic ether theories for which the vector **H** or the vector **E** represents the velocity of ether. He noted elsewhere that Larmor's theory, in which **H** is the velocity, was absolutely incompatible with Newton's third law.[46]

Since Lorentz, unlike Larmor, was not committed to any one mechanical picture of ether processes, his ether could well carry momentum. Poincaré objected that in this case 'one could conceive of, if not perform, an experiment in which the principle of reaction would seem to be contradicted, since the

[43]Lorentz, *op. cit.*, note 37, also in *Collected Papers*, Vol. 5 (The Hague, 1937), pp. 21–28, quotation on p. 28.

[44]Poincaré, 'A propos de la théorie de M. Larmor', *EE* **3** (1895), 5–13, 289–295; **5** (1895), 385–392, also in *PO*, Vol. 9, 369–426, see pp. 391–392 (also in *EO*, part 4; the title is misleading: what Poincaré originally intended to be a discussion of Larmor's pre-electronic theory of 1893, quickly evolved into a criticism of various theories with respect to general principles and Fizeau's experiment); *EO*, p. 453. Until Lebedew's experiment of 1901, the existence of radiation pressure was not universally accepted. Cf. Stein, *op. cit.*, note 2.

[45]Brillouin, *op. cit.*, note 16. In 1900, Poincaré showed that in Lorentz's theory the fraction of the recoil which is compensated by dielectric matter is $(n^2 - 1)/(n^2 + 1)$, if n is the optical index (*PO*, Vol. 9, pp. 473–481).

[46]Poincaré, *EO*, pp. 453–454; *ibid.* pp. 631–632; on p. 632, Poincaré argues that in Larmor's theory a glass plate moving in ether should exhibit the Faraday effect, because the ether flow though the plate implies a magnetic field. In reality, for Larmor the magnetic field intensity is proportional to the absolute velocity of ether, which is zero in this case.

experimenter could only operate on ponderable bodies and had no reach on ether'. He had in mind that the negation of the reaction principle, together with the principle of relative motion, led to the possibility of perpetual motion.[47]

In 1900 Poincaré considered two bodies acting on each other, but not subject to external actions. Suppose that these two bodies are rigidly connected through a bar. Then, if the action is not equal to the reaction, the net force acting on the solid is not zero. If, moreover, the principle of relative motion applies, this force does not depend on the absolute position and velocity of the body. Consequently, the system undergoes a forever accelerating motion.[48]

In general, Poincaré saw an intimate connection between the principle of relative motion and the principle of reaction. As we saw earlier, in Hertz's theory, he could show that the former, together with energy conservation, implied the latter. More generally, he asserted: 'The impossibility of detecting the motion of matter with respect to ether, and the probable equality between action and reaction without taking into account the action of matter on ether, are two facts that seem obviously connected'. Until the early 1900s he believed that the corresponding defects of Lorentz's theory would have to be repaired at the same time.[49]

Vanishing Ether

According to Poincaré, the principle of reaction and the principle of relative motion applied to *matter* alone. In his opinion ether, if it existed, had to be a very elusive thing. In 1888, in the foreword to his lectures on the mathematical theory of light he wrote:[50]

> It matters little whether the ether really exists; that is the affair of the metaphysicians. The essential thing for us is that everything happens as if it existed, and that this hypothesis is convenient for the explanation of phenomena. After all, have we any other reason to believe in the existence of material objects? That, too, is only a convenient hypothesis; only this will never cease to do so, whereas, no doubt, some day the ether will be thrown aside as useless.

Poincaré's skepticism was partially rooted in his observation that in optics numerous, mutually contradictory theories of the elastic ether were equally fit to represent the phenomena. One may also invoke the reigning empiricism of French physics in this period. In any case, from Fresnel to Hertz, Poincaré perceived a gradual fading of the ether concept. Fresnel built his ether out of molecules with well-defined motions. Maxwell gave up mechanical pictures of ether, and just assumed a hidden motion, to be treated by Lagrange's method. What gave Maxwell's ether more body—the internal stresses and the incompressible fluid—Poincaré judged to be incompatible with the rest of the

[47]Poincaré, *op. cit.*, note 44, p. 392, also in *EO*, p. 600.
[48]Poincaré, *PO*, Vol. 9, p. 481.
[49]Poincaré, *op. cit.*, note 44, p. 413, also in *EO*, p. 613.
[50]Poincaré, *La Lumière*, *op. cit.*, note 8, p. I, also in *SH*, p. 215, and *FS*, p. 174.

system. Finally, in Poincaré's opinion Hertz's theory completely eliminated ether.

From this perspective, the theories of Larmor and Lorentz were a regression, and they could only be provisional. Larmor's was the worst, for it ascribed a small, definite velocity to the ether. This velocity was proportional to the intensity of the magnetic field, and inversely proportional to the density of ether. In Poincaré's eyes, Lodge's attempt to detect this motion could only have failed: 'To support this opinion, I only have reasons of feeling; if the result had been positive, one would have been able to measure the density of ether and, if I may put it in vulgar terms, I hate to think that ether would be so pushy [*arrivé*]'.[51]

If ether was given a new life, Poincaré believed, it could only be as a ghost. Motion with respect to ether had to remain undetectable, and ether could not carry momentum. More exactly, the principle of reaction and the principle of relative motion had to be valid for matter *alone*. As Louis de Broglie once put it:[52]

> [Poincaré] did not take the existence of ether seriously enough to believe that one would detect the relative motion of an observer with respect to this fictitious medium. Hence, he was not at all surprised by the negative results of experiments of Michelson's kind, and he followed with interest—and probably with a touch of secret irony—the efforts made by Lorentz and other theorists to conciliate this negative result with the existence of ether.

Poincaré's position was extremely original. At the turn of the century, the few physicists who wished to eliminate ether, did it because they approved Mach's criticism of mechanics; they were therefore ready to give up one or two general principles of mechanics. Emil Cohn, the only renowned electrodynamicist that would omit ether, had no qualms about violating Newton's third law.[53]

A Crisis

Despite all of these difficulties, Poincaré regarded Lorentz's theory as 'the one which illuminates the greatest number of true relations [*rapports vrais*], the one from which the most traces will be found in the final construction'.[54] But how different would be the future theory? Poincaré listed three basic conditions that it would have to meet.

(i) Account for Fizeau's experiment on the partial dragging of light waves.

(ii) Respect the principles of conservation of electricity and magnetism.

(iii) Respect the principle of reaction.

[51] Poincaré, *EO*, p. 591.
[52] De Broglie, *op. cit.*, note 5, p. 65.
[53] Cf. Miller, *Albert Einstein*, *op. cit.*, note 3, p. 181, and Darrigol, 'Emil Cohn' s electrodynamics of moving bodies', to be published in *The American Journal of Physics*.
[54] Poincaré, *SH*, p. 184, also in *FS*, p. 149.

Among the latest theories, the Helmholtz–Reiff theory violated (ii); Lorentz's and Larmor's theories satisfied (i) and (ii) but not (iii); Hertz's theory satisfied (ii) and (iii) but not (i).[55]

Poincaré went on to prove that Hertz's theory was the only theory that satisfied (ii) and (iii) for macroscopic matter. More precisely, he showed that additional linear terms in Hertz's equations necessarily vanished if both these conditions plus energy conservation were met. The flawless proof left Poincaré with a major dilemma: *the three basic conditions were not compatible*. He deplored: 'We could not hope to escape this difficulty without deeply modifying generally received ideas; one does not see, however, in which direction the modification should be done'.[56]

Intuitively, Fizeau's result contradicted the immateriality of ether. The partial dragging of light waves, Poincaré wrote, 'seems to show us two different media interpenetrating and yet moving one with respect to the other. We seem to touch the ether with the finger'—a very unpleasant feeling indeed.[57] Ether seemed to play a part, not only in the application of the reaction principle, but also as a reference for absolute motion. Indeed, the principle of relative motion, when applied to matter alone, implied a complete dragging of the light waves.[58] Quite unwittingly and with a delay of nearly half a century, Fizeau had provoked what Poincaré regarded as a major crisis of electrodynamic theory.

Poincaré's involvement in the Crémieu episode shows the extent of his doubts regarding the existing foundations of electrodynamics. The young, skilful experimenter believed he had found, in an improved version of Rowland's experiment, that rotating electrified disks produced no magnetic field. The result contradicted almost every known theory of electricity, as well as the current interpretations of electrolysis and cathode rays. Yet Poincaré took the result seriously, defended Crémieu against other physicists' criticism, and carefully reviewed past theories of electric convection. In a warning at the head of his 1901 volume of lectures, he wrote: 'Since [the manuscript] has been sent for printing, the experiments of M. Crémieu have come; perhaps they will completely modify our ideas on the electrodynamics of moving bodies'.[59]

[55]Poincaré, *op. cit.*, note 44, *PO*, Vol. 9, pp. 395–413.

[56]*EO*, p. 611.

[57]Poincaré, 'Sur les rapports de la physique expérimentale et de la physique mathématique', in *Rapports préséntés au congrès international de physique réuni à Paris en 1900*, Vol. 1 (Paris, 1900), also in *SH*, p. 181, and in *FS*, p. 149.

[58]Poincaré, *PO*, Vol. 9, p. 488.

[59]Poincaré, 'A propos des expériences de M. Crémieu', *RSPA* **12** (1901), 994–1007 (this article gives a remarkable synthesis of Poincaré's critical views on electrodynamic theories from Ampère to Lorentz); *EO*, p. II. Cf. A. Potier, 'Sur les éxpériences de M. Crémieu et une objection de M. Wilson', *EE* **31** (1902), 83–93 (includes letters by Poincaré), and the careful study by L. Indorato and G. Masotto, 'Poincaré's role in the Crémieu-Pender controversy over electric convection', *Annals of Science* **46** (1989), 117–163. According to this study, FitzGerald regarded Crémieu's negative results as a possibility. But at that time he was no longer an active contributer to electromagnetic theory, and J. J. Thomson immediately charged him with 'Irish perversity'. Other British theorists, including Larmor and Gray, took notice of Crémieu's work, but only to imagine possible flaws.

Reactions On Reaction

Poincaré was the only major theorist to take Crémieu's result seriously. But he was not quite isolated in his concern about the reaction principle. In 1898, Willy Wien formulated the problem in his report on ether motion for the *Naturforscherversammlung*. In the audience were Lorentz, as a guest of honor, and a constellation of German theoretical physicists, including Planck, Drude, Mie and Voigt. Wien listed the various optical and electrodynamic experiments concerning the motion of matter with respect to ether, and reviewed the relevant theories. He praised Lorentz's achievements, but ended his report with 'a difficulty of a fundamental nature': variable electromagnetic states imply forces that would set the ether into motion, if only it could move. He then provided a thought experiment in the style of his previous studies on thermal radiation.[60]

Consider a heated plate freely floating in ether and emitting radiation at different rates on each of its sides. The net force resulting from Maxwell's radiation pressure is non-zero, and therefore the plate should undergo an accelerated motion. If ether does not move, this constitutes a violation of the theorem of the center of mass. Wien concluded: 'This point should be kept in mind for future theoretical elaborations'.[61]

Wien's analysis differed from Poincaré's in two ways. First, Wien had no doubt about the existence of the radiation pressure, because he knew, from studies by Bartoli, Boltzmann, and himself on thermal radiation, that the second law of thermodynamics implied the existence of this pressure.[62] Second, Wien was willing to give up the principle of reaction, if the assumption of a stationary ether turned out to be the only possible one.

Lorentz agreed with Wien 'that the principle did not have to be valid for all elementary actions to which we reduce phenomena'. He nevertheless offered an alternative possibility. Ether could have, in addition to a strictly stationary component, another highly dense component that could slightly move and carry momentum. Brillouin took the alternative seriously and judged that Poincaré' s worries proceeded 'from an excessive repugnance to admit the reality of ether'.[63]

In the same year Alfred Liénard, a Professeur at the Ecole des Mines de Saint-Etienne, justified the violation of the reaction principle:

[60]W. Wien, 'Über die Fragen, welche die translatorische Bewegung des Lichtäthers betreffen', in Gesellschaft Deutscher Natursforscher und Ärtze, *Verhandlungen* (1898), 49–52, see p. 52. A more detailed version of this address is appended to AP **65** (1898).

[61]W. Wien, *op. cit.*, note 60, p. 52. Wien's example does not constiute a violation of the second law of thermodynamics, since the process is not cyclic.

[62]Cf. H. Kangro, 'Wien, Wilhelm Karl Werner Otto Franz', in C. C. Gillispie (ed.), *Dictionary of Scientific Biography* **14** (1976), pp. 337–342; B. Carazza and H. Kragh, 'Adolfo Bartoli and the problem of radiant heat', *Annals of Science* **46** (1989), 183–194.

[63]Lorentz, 'Die Fragen, welche die Bewegung des Lichtäthers betreffen (Correferat)', in Gesellschaft Deutscher Naturforscher und Ärtze, *Verhandlungen* (1898), 56–65, see pp. 64–65; Brillouin, *op. cit.*, note 16.

> This result should not surprise us: as soon as we reject the theories of action at a distance and instead assume forces that take a certain time to propagate in ether, there can no longer be equality of action and reaction at every instant; the action and the reaction do not occur at the same time. We can only require that the resultant of all forces should be zero on average.

In his 1899 lectures, Poincaré referred to Liénard's view, but only to reject it. If, for instance, electromagnetic radiation was emitted in one direction, the source permanently acquired a recoil momentum. Time averaging did not help.[64]

To summarize, at the close of the century Poincaré, and, to a lesser extent Wien, seem to have been the only experts to worry about the fate of Newton's third law. In any case, Poincaré was the only one to perceive a major crisis in electromagnetic theory and to expect major modifications.

3. Aggravating the Contradictions

For his contribution to the Lorentz jubilee volume of 1900, Poincaré provided more reflections on the reaction principle in Lorentz's theory. He first apologized for discussing a major defect of Lorentz's theory where he should have praised it. His excuse was that 'good theories were flexible' and that they contained *des rapports vrais* that would survive any assaults. To criticize them was to favor the 'development of their latent virtue'.[65]

The Fictitious Fluid

Poincaré repeated the derivation of the expression (8) of the total electromagnetic force in Lorentz's theory. If there is no other force acting on matter, this force must be equal to the rate of variation of the momentum of matter. Consequently, the following equation of conservation holds:

$$\int \zeta \mathbf{V} d\tau + \int \frac{1}{c} \mathbf{E} \times \mathbf{H} d\tau = \mathbf{C}, \qquad (9)$$

where **V** is the velocity of matter, and ζ its mass density. Poincaré further noted that the vector in the second integral was equal to Poynting's energy flux vector $c(\mathbf{E} \times \mathbf{H})$ divided by c^2.[66]

Now, the energy flux can be regarded as the flow of a fictitious fluid with the velocity **U** and the density

[64]A. Liénard, 'La théorie de Lorentz', in *EE* **14** (1898), 417–424, 456–461, see p. 457 (on p. 459 Liénard also noted that the same violation occurred in Hertz's theory, since the true Hertz admitted the existence of ether in vacuum); Poincaré, *EO*, pp. 451–453.

[65]Poincaré, 'La théorie de Lorentz et le principe de la réaction', in *Recueil des Travaux Offerts par les Auteurs à H.A. Lorentz à l'Occasion du 25ème Anniversaire de son Doctorat le 11 Décembre 1900* (The Hague, 1900), 252–278, see p. 252.

[66]*Ibid.* p. 253–256.

$$J = \frac{1}{2} E^2 + \frac{1}{2} H^2. \tag{10}$$

Then eqn (9) becomes

$$\int \zeta \mathbf{V} \mathrm{d}\tau + \int \frac{J}{c^2} \mathbf{U} \mathrm{d}\tau = \mathbf{C}. \tag{11}$$

If a mass density J/c^2 is ascribed to the fictitious fluid, the latter equation means that the total momentum of matter and fluid is conserved. Poincaré insisted that the fluid was only fictitious, since it was created or destroyed at the rate

$$\frac{\partial (J/c^2)}{\partial t} + \nabla \cdot \left(\frac{1}{c} \mathbf{E} \times \mathbf{H} \right) = - \frac{1}{c^2} \rho \mathbf{v} \cdot \mathbf{E} \tag{12}$$

predicted by Poynting's theorem, and thus did not comply with the conservation of mass.[67]

The fictitious fluid offered a formal way to save the conservation of momentum and the equality of action and reaction. However, it was not immediately clear whether the fluid could also be used to save the theorem of the center of mass, since the standard proof of this theorem relies on the conservation of mass. Using Poynting's theorem and straightforward manipulations, Poincaré derived

$$\frac{\mathrm{d}}{\mathrm{d}t} \int \frac{J}{c^2} \mathbf{r} \mathrm{d}\tau + \int \zeta \mathbf{V} \mathrm{d}\tau + \int \left(\frac{1}{c^2} \rho \mathbf{v} \cdot \mathbf{E} \right) \mathbf{r} \mathrm{d}\tau = \mathbf{C}. \tag{13}$$

When no electromagnetic energy is created or destroyed, the third integral vanishes, and the total mass of the fictitious fluid is conserved; therefore, the equation implies that the global center of mass of matter and fluid moves on a straight line with constant velocity.[68]

Poincaré continued:

> Now suppose that at certain points electromagnetic energy is destroyed and that it is transformed into non-electric energy. Then we must consider the system that includes not only matter and electromagnetic energy, but also the non-electric energy [...]. But we must agree that this non-electric energy stays at the point where the transformation has occurred and is not subsequently dragged by its usual carrier, matter. If we adopt this convention, the motion of the center of mass is still rectilinear and uniform.

In this crucial but elliptical reasoning, Poincaré meant that the fictitious fluid corresponding to the destroyed electromagnetic energy was not itself destroyed,

[67] *Ibid.* pp. 256–257.
[68] *Ibid.* pp. 257–258.

but instantaneously brought to rest and stored at the place of its destruction. Conversely, the creation of electromagnetic energy at a given point implied the liberation of latent fluid available at this point.[69]

Call K/c^2 the mass of the latent fluid. Then eqn (13) reads

$$\frac{\mathrm{d}}{\mathrm{d}t}\int \frac{J+K}{c^2}\,\mathbf{r}\,\mathrm{d}\tau + \frac{\mathrm{d}}{\mathrm{d}t}\int \zeta\mathbf{r}\,\mathrm{d}\tau = \mathbf{C}. \tag{14}$$

The total mass of the fictitious fluid, which is now represented by the integral of $(J + K)/c^2$, and the total mass of matter are both constant. Therefore, the global center of mass for matter, free fluid, and latent fluid moves uniformly in a straight line.

The Liénard Force

Poincaré drew little satisfaction from this reasoning. He repeated that the reaction principle had to apply to matter *alone*, because a violation of the principle in this sense implied, together with the principle of relative motion, the possibility of perpetual motion. Here he gave the aforementioned thought experiment with the two interacting bodies connected by a rigid bar. He also insisted that there was an intimate relationship between the principle of relative motion and the reaction principle. Since experience indicated that the former principle held for matter *alone*, the latter principle was also likely to apply to matter *alone*. Poincaré recalled that in Hertz's theory the two principles held at the same time and that their connection was transparent. Then he wondered how Lorentz's theory could violate the reaction principle and at the same time account for the negative result of ether-wind experiments.[70]

His solution for this puzzle was subtle. He first remarked that Lorentz obtained the first-order invariance of optical phenomena only if the phenomena in the moving system were described in terms of the local time (5) and in terms of the apparent field quantities (6). This condition excluded the kind of reasoning that in Hertz's theory made the law of reaction a consequence of the principle of relative motion. Moreover, the condition made room for the radiation pressure, and it implied the existence of a complementary force depending on absolute motion.[71]

To prove these last two points Poincaré considered a Hertzian oscillator placed at the focus of a parabolic mirror and emitting radiation at a constant rate. The system moves with absolute velocity $\mathbf{u}$ in the direction of emission, and is heavy enough so that this velocity does not change despite the radiation

[69] *Ibid.* pp. 258.

[70] *Ibid.* pp. 260–261, 270–272. I skip Poincaré's second section, in which he proves that in Lorentz's theory the introduction of a (diluted) material dielectric cannot save the reaction principle (see note 45).

[71] *Ibid.* pp. 272–278.

pressure. For an observer at rest with the ether and for the wave-train emitted during a unit of time, the conservation of energy reads

$$S = J + (-J/c)u, \tag{15}$$

where S is the total energy spent by the source, J is the energy of the emitted radiation, and $(-J/c)u$ is the work performed by the recoil force $-J/c$. For an observer moving with the emitter, the energy spent by the emitter should be the same as for the observer at rest, since it is measured by permanent alterations of the energy source. However, the apparent energy J' of the emitted wave train must be reckoned in terms of the apparent fields (6), and in terms of the apparent length of the wave train, referred to the local time (5). To first order in u/c, the result is

$$J' = J(1 - u/c). \tag{16}$$

This is identical to the energy spent by the oscillator. Hence the energy is also conserved for the moving observer, even though the recoil force no longer performs work.[72]

For the recoil force, Lorentz's theory gives $-J/c$ with respect to the observer at rest, and $-J'/c$ with respect to the moving observer. Poincaré called the difference, $-Ju/c^2$, 'the complementary force'. In his eyes this 'necessary consequence of the existence of the recoil' constituted a first-order violation of the principle of relative motion, when applied to matter alone:

> According to Lorentz's theory the principle of reaction does not apply to matter alone and the principle of relative motion does not apply to matter alone either. The important point is that there is an intimate and necessary connection between these two facts.[73]

The complementary force was not new. As Poincaré knew, it was first predicted by Liénard in a thorough study of Lorentz' theory. Liénard considered a system of macroscopic charges and currents at rest on earth, and applied Lorentz's local time and field transformation to the macroscopic field equations. He found that the apparent Lorentz force **f'**, calculated in terms of the transformed fields, differed from the real force **f** by the amount

$$\mathbf{f}' - \mathbf{f} = -\frac{1}{c^2}\mathbf{u}(\mathbf{j}\cdot\mathbf{E}), \tag{17}$$

where **u** is the velocity of the earth, **j** the current density, and **E** the electric field. In the case of a Hertzian oscillator, the product $\mathbf{j}\cdot\mathbf{E}$ represents the energy

[72]Poincaré's reasoning is slightly more general since it does not assume the equality of the velocities of the emitter and the moving observer.

[73]*Ibid.* p. 278. Poincaré also obtained this force by considering the energy balance when the observer moves at a velocity differing from the velocity of the emitter.

provided to the field by the oscillating current, and therefore yields Poincaré's *J* by integration. Liénard's estimate for this force, 1/600 dynes for an emitting power of 100 kW, was so small that Poincaré judged its detection 'nearly impossible'.[74]

At the end of his memoir, Poincaré provided a new proof that Fizeau's experiment violated the principle of reaction. According to Fizeau's result, the velocity of light with respect to moving water is independent of the motion only if this velocity is reckoned in terms of Lorentz's local time. In other words, the Galilean transformation must be 'corrected' to include the time transformation. This correction, Poincaré argued by analogy with his reasoning on the Hertzian oscillator, led to a violation of the reaction principle. Hence the dilemma: 'All theories respecting the principle would seem to be condemned, *unless we consent to deeply modify our ideas on electrodynamics*'. Poincaré's emphasis no doubt expressed his own inclination.[75]

The Apparent States

A less conspicuous, but most important innovation of Poincaré's memoir was his interpretation of Lorentz's theorem of corresponding states.[76] According to Lorentz, the introduction of the local time and the field transformation (6) turned the real states of the moving system into corresponding states of a *fictitious* system at rest. The properties of the states of the fictitious system did not have a direct physical meaning. They had to be translated back into properties of the states of the moving system. For example, Lorentz's proof that the laws of refraction in a moving system were the same as in a system at rest, involved two steps: first, the obvious truth that the unmodified laws hold for the fictitious system, and second, the proof that the correspondence between real and fictitious system does not change the stationary distribution of lightness and darkness.[77]

In a major conceptual leap, Poincaré realized that for observers belonging to the moving system, Lorentz's corresponding states were the 'apparent' states of the real system. If we first consider the field transformation (6) separately, it is fairly obvious that the transformed fields represent the forces acting on unit

[74]A. Liénard, 'La théorie de Lorentz et celle de Larmor, *EE* **16** (1898), 320–334, 360–365, see pp. 323–324; Poincaré, *op. cit.*, note 64, p. 278. Poincaré also derived the Liénard force in *EO*, pp. 540–543.

[75]Poincaré, *op. cit.*, note 65, p. 278.

[76]Most historians have overlooked this aspect of Poincaré's memoir. They usually state that Poincaré's physical interpretation of the local time appeared in the Saint-Louis conference of 1904 (*op. cit.*, note 21, p. 311, also in *FS*, p. 307), unless they give Einstein complete precedence on this matter. Exceptions are Scribner, *op. cit.*, note 1, Cuvaj, *op. cit.*, note 1, on 77–78, and J. Stachel *et al.* (eds), *The Collected Papers of Albert Einstein*, Vol. 2 (Princeton, 1989), p. 308n; cf. also H. Thirring, 'Elektrodynamik bewegter Körper und spezielle Relativitätstheorie', in *Handbuch der Physik*, Vol. 12 (Berlin, 1927), pp. 245–348, see p. 270.

[77]Lorentz, *op. cit.*, note 38.

electric and magnetic charges *moving together with the system.*[78] Identifying the local time with the apparent time required more genius. Poincaré *defined* the local time as follows:[79]

> I suppose that observers placed in different points set their watches by means of optical signals; that they try to correct these signals by the transmission time, but that, ignoring their translatory motion and thus believing that the signals travel at the same speed in both directions, they satisfy themselves with crossing the observations, by sending one signal from A to B, then another from B to A. The local time t' is the time indicated by watches set in this manner.

Poincaré left to the curious reader the proof that Lorentz's expression of the local time fits this definition. My version runs as follows. When B receives the signal from A, he sets his watch to zero (for example), and immediately sends back a signal to A. When A receives the latter signal, he notes the time τ that has elapsed since he sent his own signal, and sets his watch to the time $\tau/2$. By doing this he commits an error $(\tau/2) - t_-$, if t_- is the time that light really takes to travel from B to A. This time, and the time of the reciprocal travel are given by

$$t_- = \frac{AB}{c+u}, \qquad t_+ = \frac{AB}{c-u}, \tag{18}$$

where u is the common velocity of the two observers with respect to ether. The time τ is clearly the sum of these two travelling times. Therefore, to first order in u/c the error committed in setting the watch A is

$$\frac{\tau}{2} - t_- = \frac{t_+ - t_-}{2} \sim \frac{uAB}{c^2}. \tag{19}$$

In other words, the setting of A's watch differs from the setting of B's watch by the amount uAB/c^2, in conformity with Lorentz's expression of the local time.[80]

[78]As obvious as it is, the remark is not to be found in the writings of Lorentz and Larmor, presumably because the field transformation was always used together with the time transformation, and thus seemed to lose physical meaning.

[79]Poincaré, *op. cit.*, note 65, p. 272. In the case of observers at rest, Poincaré first gave this method of synchronization in 'La mesure du temps', *Revue de Métaphysique et de Morale* **6** (1898), 1–13, also in *La Valeur de la Science* (Paris, 1905), Chapter 5, see pp. 34–35; double signaling is required if the distance between the observers is a priori unknown. In his lectures of 1899 (*EO*, p. 532), Poincaré was not yet aware of the relation between local time and apparent time; his proof that according to Lorentz's theory the absolute motion of the earth could not be detected by optical means relied on the smallness of the difference between local time and absolute time.

[80]Poincaré provided a similar reasoning in 'Les limites de la loi de Newton' (Sorbonne lectures of 1906–1907, H. Vergne (ed.)), *Bulletin Astronomique Publié par l'Observatoire de Paris* **17** (1953), 121–365, Chapter 11: 'Dynamique de l'électron', see pp. 217–218.

Since the local time is that measured by moving observers who do as if the propagation of light were isotropic, free electromagnetic disturbances should propagate at the velocity *c* with respect to moving axes, if only the field evolution is expressed in terms of the local time. As Poincaré noted, Lorentz's theory complies with this condition, since the transformed field equations, governing the evolution of the apparent field with respect to the local time, have the same form as for a system at rest.[81]

Poincaré's definition of the local time involved two basic components. The first is the condition that the moving observers 'ignore their motion'. It derives from the relativity principle, according to which, motion with respect to ether cannot be detected. The second component is the assumption that absolute light propagation is isotropic, which implies, together with the first component, that the apparent propagation is also isotropic. This assumption is a consequence of Lorentz's theory. However, in Poincaré's eyes it had a much wider significance. In certain situations it was the only available basis for timing events:[82]

> When an astronomer tells me that some stellar phenomenon, which his telescope reveals to him at this moment, happened, nevertheless, fifty years ago, I seek his meaning, and to that end I shall ask him how he knows it, that is, how he has measured the velocity of light. He has begun by *supposing* that light has a constant velocity, and in particular that its velocity is the same in all directions. That is a postulate without which no measurement of this velocity could be attempted. This postulate could never be verified directly by experiment; it might be contradicted by it if the results of different measurements were not concordant. We should think ourselves fortunate that this contradiction has not happened and that the slight discordances which may happen can be readily explained. At all events, the postulate conforms to the principle of sufficient reason, and has been accepted by everybody; what I wish to emphasize is that it furnishes us with a new rule for the investigation of simultaneity.

These reflexions on the measurement of time, which Poincaré gave in 1898, no doubt inspired his new interpretation of Lorentz's local time.

The Electromagnetic Momentum

Poincaré's objections failed to disturb the addressee. In his letter of acknowledgement, Lorentz agreed that the reaction principle should be violated in any theory that accounts for Fizeau's result, but wondered: 'Must we, in truth, worry ourselves about it?'. In his opinion, the principle of reaction 'should not be regarded as a fundamental principle of physics'. Regarding the conservation of momentum, he congratulated Poincaré for his 'beautiful formula' (9) and suggested that the quantity $(\mathbf{E} \times \mathbf{H})/c$ be regarded as 'equivalent to a momentum'. He even speculated that the equivalence

[81]Poincaré, *op. cit.*, note 65, p. 274.
[82]Poincaré, *op. cit.*, note 79, *FS*, pp. 232–233.

could some day become an identity, 'if we manage to consider ponderable matter as a modification of ether'. As we shall soon see, Abraham soon accomplished this prophecy.[83]

Lorentz said nothing about the Liénard force, although it invalidated his own earlier proof that electromagnetic forces did not depend on the earth's motion to first order in u/c. He very likely accepted Liénard's and Poincaré's derivations of this force, since he later reproduced them in his article for Klein's *Encyclopädie der mathematischen Wissenschaften.* However, a first-order violation of the relativity principle being no surprise to him, he had no urge to comment on it.[84]

The first beneficiary of Poincaré's reflections on the reaction principle was a bright newcomer to electron theory, Max Abraham. Two pioneering electron theorists, Larmor and Wiechert, had speculated that mechanics could be reduced to electromagnetism. They believed that the electromagnetic inertia discovered by J. J. Thomson in 1881 was the only inertia of the electron, and that the equations of motion of electrons derived from the more fundamental field dynamics. In his contribution to the Lorentz jubilee of 1900, Wien reactivated this programme and calculated that the mass of an electron, if it was electromagnetic in nature, should depend on its velocity. Walter Kaufmann, a Göttingen experimentalist, soon found the expected dependence in refined measurements of the magnetic deflection of fast electrons from a radium source. His colleague Abraham then imagined a purely electromagnetic spherical electron, and derived its mechanical properties with rigor and elegance.[85]

In this context Abraham introduced 'the electromagnetic momentum' $(\mathbf{E} \times \mathbf{H})/c$, for which he referred to Poincaré. In Abraham's opinion, the notion not only saved the law of conservation of momentum, but it became, in the electromagnetic view of nature, the only kind of momentum. The momentum of any piece of matter was just the sum of the momenta of the fields of its electrons; and the inertial mass of an electron derived from its electromagnetic momentum. After Kaufmann's subsequent verification of his mass formulas, Abraham drew the moral: 'Whereas H. Poincaré had used the violation of the momentum law in the narrow sense against electron theory, the generalized momentum law has now become the most solid basis of this theory'. Ironically, Poincaré is often regarded as the inventor of the notion of electromagnetic momentum. This is a double historical mistake: the notion already existed in

[83]Lorentz to Poincaré, 20 Jan 1901, pub. in Miller, 'Other approaches', *op. cit.*, note 1, pp. 70–71.

[84]Lorentz, 'Weiterbildung der Maxwellschen Theorie. Elektronentheorie', in *Encyclopädie der Mathematischen Wissenschaften mit Einschluss ihrer Anwendungen*, Vol. V:2, pp. 145–288, see p. 264.

[85]Cf. R. McCormmach, 'Einstein, Lorentz, and the electromagnetic view of nature', *HSPS* **2** (1970), 41–87; Miller, Albert Einstein, *op. cit.*, note 1; J. Cushing, Electromagnetic mass, relativity, and the Kaufmann experiments', *AJP* **49** (1981), 1133–1149.

Maxwell's theory, and it was only a fiction in Poincaré's analysis of Lorentz's theory.[86]

By 1903 physicists commonly accepted violations of the momentum law 'in the narrow sense'. For example, at the *Naturforscherversammlung* of that year, Ostwald acknowledged Lebedev's then recent measurement of the radiation pressure, and speculated on related corrections to the motion of celestial bodies. In the ensuing discussion, Drude commented that violations of the momentum law were already well-known in electrodynamics, and that the best theory was Lorentz's, precisely because it made the violation explicit. The more conservative and less informed von Oettingen was certainly worried: 'Unfortunately, I do not know the works of Lorentz mentioned by Drude; but I would leave this room with a very strange [*unheimlich*] feeling if Newton's third law of motion was overthrown and the facts of light pressure violated the law of the center of mass'. Greater authorities contributing to the discussion, namely Wien, Boltzmann, Klein and Schwarzschild, felt quite at ease with the violation.[87]

A Generalized Crisis

In his lecture for the Saint-Louis congress of 1904, Poincaré acknowledged the experimental confirmation of the radiation pressure, as well as Kaufmann's measurements of the velocity dependence of the mass of the electron. The latter effect, he admitted, 'seemed rather to confirm [...] the consequences of the theory contrary to Newton's principle [of reaction]'. He explained that Kaufmann's results seemed to confirm Abraham's idea that the electron's mass was entirely due to the inertia of the surrounding ether. Then mass would no longer be conserved. 'Need I point out', Poincaré asked, 'that the fall of Lavoisier's principle involves that of Newton's? The latter signifies that the center of gravity of an isolated system moves in a straight line; but if there is no longer a constant mass, there is no longer a center of gravity, we no longer know even what it is'. The resulting mechanics would be 'entirely new'. Not only would it violate older dynamical principles, but, as a consequence of the electromagnetic nature of mass, it would not admit velocities higher than that of light.[88]

In Poincaré's opinion, this threatening of Newton's and Lavoisier's principles were symptoms of a more general crisis of physics. In the first part of his lecture, Poincaré described the historical transition from the physics of central forces to the physics of principles. In the second part he asked: 'Are

[86]M. Abraham, 'Dynamik des Elektrons' Göttinger Nachrichten (1902), 20–41; 'Prinzipien der Dynamik des Elektrons', *PZ* 4 (1902), 57–63; 'Zur Theorie der Strahlung und des Strahlungsdruckes', *AP* **14** (1904), 236–287, see p. 237; the historical mistake is, e.g., in Langevin, *op. cit.*, note 5, pp. 164–165.

[87]'75. Versammlung der Naturforscher und Ärtze zu Kassel von 20 bis 26 September 1903. Drei Referaten über den gegenwärtigen Stand der Mechanik. Diskussion', *PZ* 4 (1903), 788–793, see p. 792.

[88]Poincaré, *op. cit.*, note 21, *FS*, pp. 309, 310, 312. Cf. also Poincaré, 'La fin de la matiére', *The Athenaeum* (1906), in *SH*, 2nd edn (1907), Chapter 14.

we on the eve of a second crisis?'. He now saw that all great principles of physics were in danger, save for the 'more vague and more general' principle of least action.[89]

'Carnot's principle', or the second law of thermodynamics, would only be 'an imperfect principle, a concession to the infirmity of our senses'. Such was the increasingly popular conception of Maxwell, Boltzmann and Gibbs, now confirmed by Gouy's experiments on Brownian motion. The other principle of thermodynamics was also in danger, as a result of Curie's caloric experiments on radium. Poincaré conceded, however, that Ramsay had found a plausible counterpart to the emitted energy: a slow, steady transmutation of the radium.[90]

Finally, the relativity principle had survived many attacks, but only at the price of accumulating hypotheses. Poincaré certainly admired Lorentz's 'ingenious idea' of the local time. He repeated, as if it were Lorentz's, his definition of the local time by means of cross-signalling. He noted, without proof, that the relative velocities, if measured according to the local time, could never exceed the velocity of light if absolute velocities could not. However, he deplored the need for 'complementary hypotheses', like the FitzGerald contraction, and the hypothesis that all forces were affected by motion in the same manner as electromagnetic forces.[91]

Poincaré warned his audience against giving up the general principles too quickly. Yes, they were in danger, but physicists had not yet the right to declare their final collapse. A more satisfactory theory of the electrodynamics of moving bodies, Poincaré hoped, would be more respectful of principles: 'Let us take, then, the theory of Lorentz, turn it in all senses, modify it little by little, and perhaps everything will arrange itself'. As a first possible modification, he suggested the replacement of the Lorentz contraction with 'a more simple and natural hypothesis [...], that it is the ether which is modified when it moves relative to the material medium which penetrates it'.[92]

4. Solving the Crisis

Soon after the Saint-Louis conference, Poincaré no longer contemplated such drastic modification of Lorentz's theory. He studied Lorentz's memoir of 1904 more carefully, judged it 'extremely important', and decided to perfect it.[93]

Lorentz's essential advance was the ability to treat all orders in u/c on the same footing. He maintained his earlier field equations for the stationary ether

[89]Poincaré, *FS*, p. 314.
[90]*Ibid.* pp. 304, 313.
[91]*Ibid.* pp. 306–307, 312, 307–308.
[92]*Ibid.* p. 315.
[93]Lorentz, 'Electromagnetic phenomena in a system with any velocity less than that of light', Royal Academy, Amsterdam, *Proceedings* 6 (1904), 809–831; Poincaré to Lorentz, undated (between Sept. 1904 and May 1905), pub. in Miller, 'Other approaches', *op. cit.*, note 1, pp. 76–78. Cf. Hirosige, *op. cit.*, note 25; Miller, *op. cit.*, note 1; Paty, *Einstein Philosophe*, *op cit.*, note 1, pp. 45–48; Darrigol, 'The electron theories', *op cit.*, note 3.

and the Lorentz force formula. He further assumed that a global translatory motion of the system altered all forces as if they were of electromagnetic origin, that the longitudinal extension of a moving electron was $(1 - u^2/c^2)^{1/2}$ times its transverse extension, and finally that all mass was of electromagnetic origin. With these elements he could prove that optical phenomena did not depend on a global translatory motion of the system to any order in u/c.

As he had done in the *Versuch*, he wrote the microscopic field equations with respect to a moving system of axes, that is, in modern terminology, he performed a Galilean transformation of coordinates. Then he provided a coordinate and field transformation that turned the field equations *exactly* into those for a system at rest, except for the source terms. Combining this transformation with the previous Galilean transformation gives the modern Lorentz transformation, except for a global scale factor ε.

Lorentz did not have the transformations for velocity, current and charge which we now consider to be the correct ones. Nevertheless, he could show that for an oscillating electric dipole in the moving system, his transformation turned the emitted radiation into that of a dipole at rest. Then he derived (in the quasi-stationary approximation, and for small relative velocities) the equation of motion of a purely electromagnetic electron, and showed that it was invariant under his transformations if and only if the scale factor ε was equal to unity. This removed an embarrassing ambiguity in the behavior of moving solids. Finally, combining the transformation properties for dipole radiation and electronic motion, Lorentz applied the technique of corresponding states to optical phenomena, and obtained their invariance at any order in u/c.

In sum, Lorentz, in this complex memoir, met Poincaré's requirement that a common explanation should be given to the invariance of optical phenomena at any order in u/c. But he did this only with certain approximations, for instance with the dipole approximation. In his mind, motion with respect to ether remained detectable in principle. Also, Lorentz still regarded his 'corresponding states' as states of a fictitious system, despite Poincaré's earlier remark that these states were the ones apparent to moving observers. This made the derivation of physical invariance properties impractical. Finally, the deformations of Lorentz's electron implied the existence of non-electromagnetic cohesive forces. Such forces contradicted the electromagnetic world-view, as Abraham noted.[94] At any rate, their role in electron dynamics was unclear.

The Dynamics of the Electron (1905–1906)

Poincaré's faith in the relativity principle was such that he saw in Lorentz's memoir an opportunity to achieve a complete, exact invariance of

[94]Abraham, 'Die Grundhypothesen der Elektronentheorie', *PZ* 5 (1904), 576–579. Cf. Miller, *Albert Einstein, op. cit.*, note 1, pp. 75–77.

electrodynamic phenomena with respect to uniform translatory motion. His first improvement upon Lorentz's considerations concerned the new field and coordinate transformations. With an eye for symmetry, Poincaré focused on what he called 'the Lorentz group', that is, the group of linear transformations that left the field equations invariant; whereas Lorentz followed a two-step procedure, applying first a Galilean transformation of coordinates, and then restoring the form of the field equations with respect to ether. Poincaré noted that the coordinate transformations were, up to a global scale factor, those that left the quadratic form $x^2 + y^2 + z^2 - t^2$ invariant (taking $c = 1$), and further that they could be regarded as rotations in a four-dimensional space with an imaginary fourth coordinate. He rectified Lorentz's transformation equations for charge and velocity, and thus achieved a complete invariance of the field equations, including the source terms.[95]

Next, Poincaré discovered a short cut for deriving the value of the global scale factor (Lorentz's ε). In order for the Lorentz transformations to be relevant to the electrodynamics of moving bodies, he argued, they had to belong to a sub-group for which ε is a function of the velocity in of the system only.[96] Then, the inverse of a transformation corresponding to the velocity u and to the scale factor ε is another transformation of the sub-group, with the velocity $-u$ and the scale factor $1/\varepsilon$. Combining the latter transformation with a rotation of 180° around an axis perpendicular to u yields a transformation with the original velocity u, but with the scale factor $1/\varepsilon$. Since, by assumption, the scale factor depends only on the velocity, we have $\varepsilon = 1/\varepsilon$, or $\varepsilon = 1$.[97]

The argument could not have been made in Lorentz's original theory, for at least two reasons: Lorentz did not reason in terms of transformation groups, and he did not regard his 'corresponding states' as states of the moving system. But what was Poincaré's interpretation of the transformed variables? On this essential point Poincaré was nearly silent. He gave his version of 'Lorentz's idea' succinctly:[98]

> The reason we can, without modifying any apparent phenomenon, confer to the whole system a common translation, is that the equations of an electromagnetic medium are not changed under certain transformations which I shall call *the Lorentz transformations*; two systems, one at rest, the other in translation, thus become exact images of one another.

[95]Poincaré to Lorentz, cf. note 93; 'Sur la dynamique de l'électron', *CR* **140** (5 June 1905), 1504–1508, also in *PO*, Vol. 9, pp. 489–493; 'Sur la dynamique de l'électron', Circolo matematico di Palermo, *Rendiconti* **21** (1906), 129–176, also in *PO*, Vol. 9, 494–550, see pp. 498–503, 542, 513–515. Cf. Paty's perceptive reading of this memoir in *Einstein Philosophe*, *op. cit.*, note 1, pp. 49–21.

[96]Poincaré implictly excluded mirror reflections.

[97]Poincaré to Lorentz, second letter written between October 1904 and May 1905, in Miller, 'Other approaches', *op. cit.*, note 1, pp. 79–80; Poincaré, *PO*, Vol. 9, p. 515. Poincaré implicitly excluded the inversion $\varepsilon = -1$.

[98]*Ibid.* p. 495.

In other words, for Poincaré, the mathematical symmetry implied a physical symmetry. This implication was only stated, rather than proved. Very likely, Poincaré regarded the transformed variables as corresponding to the 'apparent phenomena' recorded by moving observers. This was the view he earlier proposed in the memoir for the Lorentz jubilee, and later extended in lectures on electron dynamics, as we shall see below.

Poincaré applied the improved Lorentz transformation to the Lorentz force $\mathbf{f} = \rho[\mathrm{E} + (\mathbf{v}/c) \times \mathbf{H}]$ and obtained

$$\mathbf{f}' = (\gamma,1)\left[\mathbf{f} - \frac{1}{c^2}\,\rho(\mathbf{v}\cdot\mathbf{E})\mathbf{u}\right], \tag{20}$$

where $(\gamma, 1)$ denotes the multiplication by $\gamma = (1 - v^2/c^2)^{-1/2}$ of the component along $\mathbf{u}$. The second term 'recalls a result once obtained by Liénard'. Indeed, when summed over the conduction electrons of a metal and approximated to first order, this term gives Liénard's complementary force. Poincaré no longer regarded this force as violating the relativity principle. If all forces, including inertial forces, were of fundamentally electromagnetic origin, the condition of (statical or dynamical) equilibrium of any material body could be reduced to the vanishing of the electromagnetic force acting on the volume elements of the various electrons; and this condition was left invariant by the transformation (20). Not every force needed to be of electromagnetic origin, however, but one could assume with Lorentz that motion affected all forces exactly in the same manner. Then the principle of relativity held without exception.[99]

As Poincaré showed, non-electromagnetic forces were necessary to keep the parts of an electron together. In general, a purely electromagnetic equilibrium condition would imply, for an electron at rest, the vanishing of the electric field, and therefore the vanishing of its charge.[100] One way to avoid the paradox was to introduce a purely kinematical constraint on the electron's shape. For example, Abraham's electron was absolutely rigid, and Langevin's was incompressible. However, Lorentz had shown that the equations of motion for these models implied violations of the relativity principle.

More generally, Poincaré argued that the longitudinal contraction imagined by Lorentz was the only deformation of the electron that was compatible with the relativity principle, because it resulted from the Lorentz transformation, with the proper value (one) of the scale factor.[101] But this contraction could not be produced by purely kinematical means; it required an additional potential energy. Poincaré tried an energy proportional to the electron's volume, in other words, a uniform, cohesive pressure. He derived the resulting dynamics, and

[99]Poincaré to Lorentz, cf. note 93, p. 77; *PO*, Vol. 9, pp. 490, 503.
[100]*Ibid.* p. 503.
[101]*Ibid.* p. 536.

proved its compatibility with the relativity principle. The latter proof was based on the Lorentz invariance of the electromagnetic Lagrangian $(1/2)(E^2 - H^2)$ and of the additional pressure terms.[102]

Finally, Poincaré tried to generalize Newton's theory of gravitation in a Lorentz-invariant manner. Here he used the four-dimensional pseudo-Euclidian space and the corresponding invariants. He also required that the generalized gravitational force should transform according to formula (20) derived for electromagnetic forces, and should propagate with the velocity of light.[103]

The New Mechanics (1906–1908)

Although he had achieved complete relativistic invariance for electrodynamics and gravitation, Poincaré was not completely satisfied. In his opinion, the assumption that all forces behaved like electromagnetic forces with respect to the system's motion, and the assumption that gravitation propagated with the velocity of light, required further justification:[104]

> We cannot content ourselves with simply juxtaposed formulas that would agree only by some happy coincidence; the formulas should, so to say, penetrate each other. Our mind will be satisfied only when we believe that we perceive the reason for this agreement, in order that we may fancy that we have predicted it.

Poincaré compared the present theory with Ptolemaic astronomy, and predicted some drastic simplification. In the new physics the omnipresence of the velocity of light would no longer be a coincidence. Perhaps everything in the world would have an electromagnetic origin. More likely:[105]

> The common part of all physical phenomena would only be an appearance, something that would pertain to our methods of measurement. How do we perform our measurements? By superposing objects that are regarded as rigid bodies, would one first answer; but this is no longer true in the present theory, if one assumes the Lorentz contraction. In this theory, two equal lengths are, by definition, two lengths which light takes equal time to travel through. Perhaps it would be sufficient to renounce this definition so that Lorentz's theory would be as completely overturned [*bouleversée*] as Ptolemy's system was through Copernicus' intervention.

From this extract we may infer that Poincaré detected an implicit convention of Lorentz's theory: two lengths (sharing the same motion) are equal if and only if light takes the same time to travel through them, from the point of view of an observer at rest. From this perspective, the Michelson–Morley experiment, as analyzed by an observer at rest with the ether, provides a measurement of the relative length of the perpendicular and parallel arms of the interferometer, the

[102]Poincaré to Lorentz, third letter written between October 1904 and May 1905, in Miller, 'Other approaches', *op. cit.*, note 1, pp. 81–82; *PO*, Vol. 9, pp. 491, 521–538.
[103]*Ibid.* pp. 492–493, 538–550.
[104]*Ibid.* p. 497.
[105]*Ibid.* p. 498.

result being the Lorentz contraction. However, it is not quite clear what Poincaré intended by dropping this convention. Was he just repeating the earlier suggestion he made at the Saint-Louis conference, that the absolute velocity of light could depend on the proximity of moving matter? Or did he imply a revision of the concept of time?

The first alternative seems more plausible. In any case, Poincaré maintained Lorentz's convention for the rest of his life. What changed was the status he gave to the Lorentz transformation. As we saw earlier in the 1906 memoir, Poincaré introduced the Lorentz transformations as those leaving Lorentz's field equations invariant. In his Sorbonne lectures of 1906–1907 entitled 'The limits of Newton's law', he introduced Lorentz's coordinate transformations as those giving the 'apparent space and time coordinates'. The proof involved a generalization of his earlier argument about the synchronization of moving clocks.[106]

Poincaré first repeated this argument, and introduced a basic condition of transitivity: if A is synchronized with B, and if B is synchronized with C, then A must be synchronized with C, for any given choice of the relative positions of the moving clocks A, B, and C. Otherwise, the earth's absolute motion could be detected. To first order in u/c the condition is met, because the error committed in synchronizing two clocks A and B by cross-signaling is proportional to $\mathbf{u}\cdot\mathbf{AB}$. Poincaré further proved that the condition held *exactly*, if the Lorentz contraction was assumed. His reasoning went as follows.[107]

Consider an observer moving with the constant velocity $\mathbf{u}$ and emitting a flash of light at time zero. At later time t, the source has travelled the distance ut, and the emitted light is located on a sphere centered at the emission point and with radius ct. From the point of view of moving observers, this light shell has an ellipsoidal shape, because the rods used to measure the shell are contracted in the direction parallel to $\mathbf{u}$ in the proportion

$$\gamma^{-1} = \sqrt{1 - u^2/c^2}. \tag{21}$$

The half-axes of the generating ellipse are

$$a = \gamma ct, \qquad b = ct. \tag{22}$$

[106]Poincaré, *op. cit.*, note 80, p. 221.

[107]*Ibid.* pp. 218–220. Poincaré repeated the reasoning in 'La dynamique de l'électron', *RSPA* **19** (1908), 386–402, also in *PO*, Vol. 9, pp. 551–586, see pp. 564–567; a less explicit version is in 'La mécanique nouvelle' (Conférence pour l'Association Française pour le Progrès des Sciences, Lille, 1909), *Revue d'électricité* **13** (1910), 23–28, also in *La Mécanique Nouvelle. Conférence. Mémoire et Note sur la Théorie de la Relativité* (Paris, 1924), pp. 1–17, German translation in the Wolfskehl lectures (Göttingen, 1909): *Sechs Vorträge über ausgewählte Gegenstände aus der reinen Mathematik und mathematischen Physik* (Leipzig, 1910), and in the Berlin lecture (14 Oct. 1910): 'Die neue Mathematik', *Himmel und Erde* **23** (1911), 97–116.

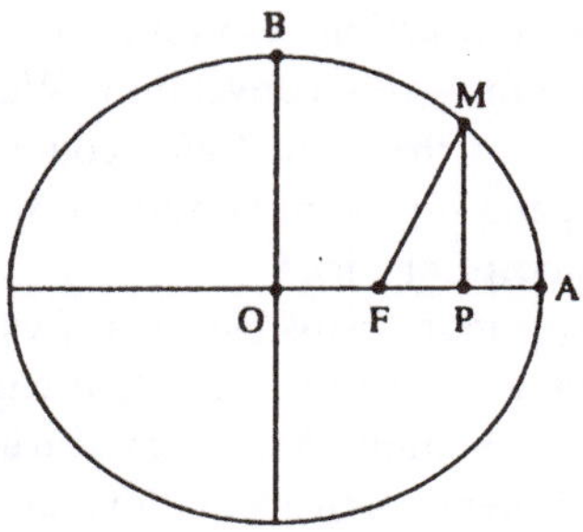

Fig. 1. Poincaré's light ellipsoid (a = OA, b = OB, f = OF).

The eccentricity is

$$e = \sqrt{1 - b^2/a^2} = u/c. \tag{23}$$

The focal distance,

$$f = ea = \gamma ut, \tag{24}$$

is equal to the apparent distance between the center of the ellipse and the position of the source. Therefore, the Lorentz contraction is the contraction for which the source is at the focus F of the light ellipsoid (see Fig. 1).

Now consider a second observer travelling with the same velocity **u** and receiving the flash of light at the time t. The position M of this observer belongs to the ellipsoid, and the distance FM represents the apparent distance between the two observers, which is invariable. The apparent time t' that light takes to travel between the two observers must be such that the apparent velocity of light is c; it is therefore equal to FM/c. According to a well-known property of ellipses (that gives their equation in polar coordinates), we have

$$\mathrm{FM} + e\mathrm{FP} = b^2/a, \tag{25}$$

where P denotes the projection of M on the larger axis of the ellipse. The length FP being identical to the difference x' of the apparent abscissas of the two observers, we reach

$$t' = \gamma^{-1}t - x'u/c^2. \tag{26}$$

Since this expression is a linear function of x', the synchronization of moving clocks is transitive, as was to be proved.

Call x the true abscissa at time t of the second observer with respect to the point of emission of the light flash. With respect to the same origin, and at the same time t, the true abscissa of the first observer is ut. Hence,

$$x' = \gamma(x - ut). \tag{27}$$

Inserting this in the formula (26) for the apparent time yields

$$t' = \gamma(t - ux/c^2), \tag{28}$$

in conformity with the Lorentz transformation.

The Lorentz contraction plays a central role in this derivation. In the last analysis, the contraction is necessary to save the relativity principle, provided that light propagates with the constant velocity c with respect to the stationary ether. Admittedly, Poincaré did not make this point very clear. He still called the contraction a hypothesis, and made a confusing comment:[108] 'At first glance this hypothesis [...] seems to be quite extraordinary; so far we can only say in its favor that it is the immediate translation of Michelson's experimental result, if we *define* the lengths by the times which light takes to travel through them'. Then the contraction would depend not only on a convention on length measurement, but also on a specific experiment. In fact, Michelson's result is a consequence of the relativity principle, and therefore the contraction of lengths results from the relativity principle.

Despite some lingering obscurity, Poincaré's new way of introducing the Lorentz transformation had an essential quality: it provided, independently of the Maxwell–Lorentz equations, a physical interpretation of the transformed coordinates as those measured by moving observers. With this interpretation, the Lorentz invariance of a theory implied its compatibility with the relativity principle.

In his Sorbonne lectures of 1906, Poincaré derived the relativistic dynamics of the electron from this point of view. Like Einstein, he no longer assumed a specific structure of the electron, and no longer relied on the electron's self-field. He simply required that the new dynamical equation be invariant under infinitesimal Lorentz transformations.[109]

Comparison with Einstein

Poincaré and Einstein shared a belief in the general validity of the relativity principle, and both gave an organizing purpose to the principle. As is well known, Einstein was little disturbed by Kaufmann's 1905 data, which seemed to confirm Abraham's electron dynamics and to exclude Lorentz's. Despite the experimenter's 'admirable care', Einstein judged that Abraham's theory had 'little probability'. At the other extreme, Lorentz was 'au bout de son latin' and judged that he was 'forced to abandon' his own theory.[110]

[108]Poincaré, 'La dynamique de l'électron', *op. cit.*, note 107, *PO*, Vol. 9, p. 567.

[109]Poincaré, *op. cit.*, note 80, pp. 222–228.

[110]A. Einstein, 'Über das Relativitätsprinzip und die aus demselben gezogenen Folgerungen', *Jahrbuch der Radioaktiviät* 4 (1907), 411–462, see pp. 437–439; Lorentz to Poincaré, 8 Mar 1906, in Miller, 'Other approaches', *op. cit.*, note 1, pp. 83–84. Cf. Stachel *et al.*, *op. cit.*, note 76, pp. 270–273.

Poincaré's reaction was more moderate: 'The principle of relativity would not have the rigorous value that one was tempted to give to it [...]. However, before we definitively adopt this conclusion, some reflection is necessary'. Poincaré recommended that someone repeat Kaufmann's experiment, and suggested that Kaufmann's determination of the electrostatic field in his tube could be flawed. A few minutes earlier in the same address, he had reasserted his faith in the relativity principle: 'It is impossible to escape the feeling that the principle of relativity is a general law of Nature'.[111]

Another similarity between Poincaré and Einstein is that they both regarded the Lorentz-transformed coordinates and fields as those measured by moving observers. Poincaré had this insight in 1900, to first order in u/c. By 1907 he introduced the Lorentz transformation on this basis, as Einstein had done in 1905. Both physicists regarded the Lorentz invariance as a constraint to be imposed on every kind of force, including gravitation and the cohesive forces in an electron. Finally, they agreed that the invariance was sufficient to determine electron dynamics.

Noting this resemblance, one may wonder whether Poincaré benefitted from Einstein's views when he prepared his 1906–1907 lectures. This seems unlikely, because both Poincaré's introduction to the Lorentz transformation and his derivation of electron dynamics differed substantially from their Einsteinian counterparts. In Einstein's reasonings, there was no light ellipsoid and the Lorentz contraction was not a hypothesis; the relativistic electron dynamics resulted from applying a Lorentz transformation to the dynamics of slow motion, not from requiring invariance under infinitesimal Lorentz transformations.

We should therefore conclude that Poincaré's views on electrodynamics independently led to a theory which, from a practical point of view, hardly differed from Einstein's. Not only were the two theories empirically equivalent, but they led to similar procedures for the derivation and application of relativistic invariant theories. The only difference concerned the role of ether, which Poincaré maintained, and Einstein completely denied.

As we saw earlier, Poincaré's critiques of electrodynamics gradually stripped ether of all of its virtues. We may then wonder why he did not take the final step and eliminate ether completely. The most plausible answer is that he wished to maintain the notions of true space and time. These notions were so adequate in describing the course of current mechanical phenomena that it was foolish to give them up for the sake of the fourth or fifth decimal. Moreover, these notions were irrefutable, since they were no more than conventions, which did not by themselves determine the global content of the theory.[112]

[111]Poincaré, 'La dynamique de l'électron', *op. cit.*, note 107, *PO*, Vol. 9, pp. 572, 567. At the end of his introduction to the Palermo memoir (*PO*, Vol. 9, p. 498), Poincaré wrote: 'The discovery of magnetocathodic rays might seem to endanger the whole theory'.

[112]Very likely Poincaré's attitude paralleled his earlier considerations on the convenience of Euclidean geometry. Cf. Paty, *Einstein Philosophe*, *op. cit.*, note 1, pp. 264–271.

By maintaining a privileged frame of reference, Poincaré favored intuitiveness over unambiguity. Unlike Einstein, he was not disturbed by the inherent ambiguity in the choice of a privileged frame. And he did not hesitate to introduce notions and distinctions that had no observable counterpart. A good example is the light ellipsoid which he used in his discussion of the apparent time. The ellipsoid represents the location of a light pulse at a given value of the absolute time and for geometers belonging to a moving system. Such a notion is an absurdity from the Einsteinian view-point, for it mixes up kinematic attributes belonging to different frames of reference. Some commentators have inferred from this that Poincaré's views on space and time measurement were not consistent.[113] In reality, they were only incommensurable with Einstein's.

The Mass–Energy Relation

By 1905 Poincaré had achieved a complete reconciliation of electrodynamics with the relativity principle. In the past he had expected that this achievement would bring with it compatibility with the reaction principle. In fact, it did not, since the general framework of Lorentz's theory was still maintained. In 1908 Poincaré did not hesitate to conclude: 'We must *renounce the principle of reaction*'.[114]

What became of the earlier arguments linking the violation of the reaction principle with a violation of the relativity principle? Poincaré gave a partial answer to this question when he argued that Liénard's complementary force did not contradict the invariance of an electron's equation of motion, because all forces, *including inertial forces*, were complemented in the same manner. In other words, the velocity-dependence of the electron's mass exactly compensated for the Liénard force.

The argument concerned the microscopic evolution of the system at the electronic and sub-electronic level. It did not directly solve the macroscopic paradoxes which Poincaré had raised in his contribution to the Lorentz jubilee. This was left to Einstein's and Langevin's ingenuity.

Poincaré's main concern was that a violation of the reaction principle would lead to a violation of the theorem of the center of mass, and to the possibility of perpetual motion. As we saw earlier, he demonstrated this paradox in two different manners: analytically, and with a thought experiment. One of Einstein's proofs of the equivalence between mass and energy was based on an improvement of the thought experiment. Poincaré's example was based on uniting two bodies whose mutual action was not balanced. Einstein made this more specific by considering a source of radiation and an

[113]For instance Cuvaj, *op. cit.*, note 1.
[114]Poincaré, *PO*, Vol. 9, p. 570 (his emphasis).

absorber that face each other and belong to the same solid. He imagined the following cycle.[115]

The source emits a radiation pulse with energy E in the direction of the absorber, which implies a recoil momentum $-E/c$ for the solid. When the pulse reaches the absorber, the solid is brought back to rest. A massless carrier then brings the energy E back to the absorber, completing the cycle.

During this whole operation the solid shifts by the amount $-(E/Mc)(L/c)$ (in a first approximation), where M is the mass of the solid, and L the distance between the emitter and the absorber. Consequently, a system initially at rest can move without compensation.

In order to avoid this paradox, Einstein proposed that the return of the energy E to the emitter involved a transfer of mass E/c^2. During the transfer the center of mass does not move. Therefore, the geometrical center of the solid moves by the amount $(E/c^2)L/M$ (in a first approximation), which exactly compensates the displacement in the first phase of the cycle.

Einstein's argument is slightly defective, because the absorbtion of the radiation normally implies an irreversible dissipation. However, it remains true that the mass–energy equivalence removes the violation of the theorem of the center of mass in the first part of the cycle: the transfer of luminous energy now involves a transfer of mass from one side of the body to the other, so that the body can globally move and still have the same center of mass.

In order to prove, quite generally, that the mass–energy relation saved the theorem of the center of mass, Einstein rederived eqn (13) of Poincaré's memoir:

$$\frac{\mathrm{d}}{\mathrm{d}t}\int \frac{J}{c^2}\,\mathbf{r}\,\mathrm{d}\tau + \int \zeta \mathbf{V}\,\mathrm{d}\tau + \int \left(\frac{1}{c^2}\rho\mathbf{v}\cdot\mathbf{E}\right)\mathbf{r}\,\mathrm{d}\tau = \mathbf{C}, \tag{29}$$

and realized that it could be rewritten as

$$\frac{\mathrm{d}}{\mathrm{d}t}\left(\int \frac{J}{c^2}\,\mathbf{r}\,\mathrm{d}\tau + \int \zeta\mathbf{r}\,\mathrm{d}\tau\right) = \mathbf{C}, \tag{30}$$

provided that an increase $\rho\mathbf{v}\cdot\mathbf{E}$ of the energy content of matter implied an increase $\rho\mathbf{v}\cdot\mathbf{E}/c^2$ of its mass:

$$\frac{\partial \zeta}{\partial t} + \nabla\cdot(\zeta\mathbf{V}) = \frac{1}{c^2}\rho\mathbf{v}\cdot\mathbf{E}. \tag{31}$$

[115] Einstein, 'Das Prinzip von der Erhaltung der Schwerpunktsbewegung und die Trägheit der Energie', *AP* **20** (1906), 627–633, par. 1.

With this assumption and that of a mass density J/c^2 for the electromagnetic field, the total mass of matter and field is conserved:

$$\frac{\mathrm{d}}{\mathrm{d}t}\int\left(\frac{J}{c^2}+\zeta\right)\mathrm{d}\tau=0, \tag{32}$$

and eqn (30) means that the motion of the center of mass is uniform.[116]

Einstein published these two arguments in 1906, with a passing reference to Poincaré's memoir: 'Although the relevant [...] elementary formal considerations can be found in a memoir by H. Poincaré, I shall not, for the sake of clarity, base my reasonings upon this memoir'.[117] Einstein's first published derivation of the mass–energy equivalence (1905) has no exact counterpart in Poincaré's memoir, but it has some similarities with Poincaré's discussion of Liénard's complementary force. Poincaré considered the same process of emission of radiation from the point of view of two observers, one at rest, the other moving with the absolute velocity **u**. He wrote the equation for the conservation of energy for each observer, taking into account the effect of motion on the apparent energy J emitted in unit time. He found that the two conservation equations were compatible if and only if the recoil forces perceived by the two observers differed by Liénard's amount Ju/c^2.[118] We may note with the benefit of hindsight that a decrease in the mass of the emitter by J/c^2 exactly compensates the Liénard force, since it implies an additional difference $(J/c^2)u$ in the variations of the emitter's momentum as measured by the two observers.

In his argument of 1905 Einstein also considered an emission process and its energy balance from two different points of view. However, instead of Poincaré's unidirectional process, he took a symmetrical process, in which the same quantity of light $J/2$ is emitted in two opposite directions. This simplifies the argument because the source does not recoil. To second order in u/c, the energy balances provided by the two observers agree if and only if the mass of the source decreases by J/c^2 during the emission process. Indeed, for an observer attached to the source, the energy S provided by the source is simply equal to the energy J of the emitted radiation, whereas with respect to an observer moving with the velocity u on the emission line, the energy of the emitted radiation is[119]

$$J'=\gamma\left(1+\frac{u}{c}\right)\frac{J}{2}+\gamma\left(1-\frac{u}{c}\right)\frac{J}{2}\sim J+\frac{1}{2}\frac{J}{c^2}u^2. \tag{33}$$

[116]*Ibid.*, par. 2.

[117]*Ibid.* p. 627.

[118]As mentioned in note 73, Poincaré's full reasoning involved two different absolute velocities for the emitter and the moving observer.

[119]Einstein, 'Ist die Trägheit eines Körpers von seinem Energieinhalt abhängig?' *AP* **18** (1905), 639–641. The reasoning requires the transformation properties of the energy of a light complex to second order, which Poincaré did not have.

Langevin apparently gave a similar argument in 1906, independently of Einstein. He discussed exactly the same emission process, but in terms of momentum exchange. For the moving observer, the total momentum of the emitted radiation is, to first order,

$$\left(1-\frac{u}{c}\right)\frac{J}{2c}-\left(1+\frac{u}{c}\right)\frac{J}{2c}=-\frac{Ju}{c^2}. \tag{34}$$

The conservation of momentum implies that the momentum of the source changes by the opposite amount, uJ/c^2. Since the velocity $(-u)$ of the source does not change, its mass must change by the amount $-J/c^2$.[120]

There is no strong evidence that Einstein and Langevin conceived their arguments as simple adaptations of Poincaré's earlier reasoning. They were probably inspired, however, by Poincaré's way of combining principles and thought experiments. In that case, why did not Poincaré himself reach the mass–energy equivalence in 1900?[121] A plausible answer is that a new particle dynamics had not yet been developed. There was as yet no reason to doubt the universal validity of the conservation of mass. Having reached eqn (13), which strongly suggests a relation between mass and energy content, Poincaré could only deplore the paradox. However, the ability to detect genuine paradoxes seems as essential to the progress of science as the ability to solve them.

Acknowledgements—I would like to thank Françoise Balibar, Matthias Dörries, Vena Kostroun, David Stump and Scott Walter for their comments on the first draft of this paper.

[120]Langevin, 'L'inertie de l'énergie et ses conséquences', *Journal de Physique* 3 (1913), 553–582, also in *Oeuvres Scientifiques* (Paris, 1950), 397–426, see pp. 418–419. Langevin also proved that the mass of a cavity containing electromagnetic radiation depended on the energy of this radiation (*ibid.* pp. 416–418). According to Langevin, *ibid.* on 414, these considerations were independent of Einstein's, and were included in his course at the Collège de France of 1906. In his *Paul Langevin, Mon Père* (Paris, 1971), pp. 58–59, André Langevin gives the testimony of E. Bauer, who attended the lectures.

[121]I have not found any direct evidence that Poincaré accepted the equivalence between mass and energy content after 1905. Nor have I found evidence for the contrary. His statement of 1908 (*PO*, Vol. 9, p. 568) that 'l'énergie n'a pas de masse' occurred in a discussion of the violation of the reaction principle in radiation processes; it only meant that energy did not have a *mechanical* mass. In the following paragraph (*ibid.*, pp. 571, 573) Poincaré recalled that according to the new mechanics there would be no mechanical mass at all; every mass would be of electromagnetic origin.

Phys. perspect. 4 (2002) 421–446
1422–6944/02/010001–26

Physics in Perspective

Reconsidering a Scientific Revolution: The Case of Einstein *versus* Lorentz

Michel Janssen*

The relationship between Albert Einstein's special theory of relativity and Hendrik A. Lorentz's ether theory is best understood in terms of competing interpretations of Lorentz invariance. In the 1890s, Lorentz proved and exploited the Lorentz invariance of Maxwell's equations, the laws governing electromagnetic fields in the ether, with what he called the *theorem of corresponding states*. To account for the negative results of attempts to detect the earth's motion through the ether, Lorentz, in effect, had to assume that the laws governing the matter interacting with the fields are Lorentz invariant as well. This additional assumption can be seen as a generalization of the well-known contraction hypothesis. In Lorentz's theory, it remained an unexplained coincidence that both the laws governing fields and the laws governing matter should be Lorentz invariant. In special relativity, by contrast, the Lorentz invariance of all physical laws directly reflects the Minkowski space-time structure posited by the theory. One can use this observation to produce a common-cause argument to show that the relativistic interpretation of Lorentz invariance is preferable to Lorentz's interpretation.

Key words: Special relativity; ether theory; Albert Einstein; Hendrik A. Lorentz; Lorentz invariance; theorem of corresponding states; contraction hypothesis; Trouton-Noble experiment; common-cause argument; ad hoc hypothesis; empirical equivalence.

Einstein's *versus* Lorentz's Interpretation of Lorentz Invariance

Both Albert Einstein's introduction of the special theory of relativity in 1905 and Nicolas Copernicus's introduction of his heliocentric model of the world in 1543 would make many people's list of the great revolutions in the history of science. A comparison of the case of Einstein *versus* Hendrik Antoon Lorentz to the case of Copernicus *versus* Claudius Ptolemy, however, suggests that accepting Einstein's introduction of special relativity uncritically as the epitome of a scientific revolution may seriously distort our perception and assessment of this fascinating episode in the history of science.

To focus on one main protagonist, be it Lorentz or Ptolemy, is obviously an oversimplification, but a useful one in that it allows me to highlight three important parallels that I want to draw between the two cases. First, in both cases the central innovation was a reinterpretation of a well-known feature of the old formalism.

* Michel Janssen is Assistant Professor in the Program in History of Science and Technology at the University of Minnesota.

Second, in both cases the strongest argument in favor of the new and against the old interpretation was essentially a common-cause argument, that is, the new interpretation traced to a common cause what in the old interpretation were unexplained coincidences. Third, in both cases the central innovation in and of itself did not constitute a wholesale replacement of one tradition or worldview by another, but is best understood as one of several factors that brought about this transition over time. In short, the central innovation is best thought of as an important catalyst of a revolution rather than as a revolution itself.

In the case of Copernicus *versus* Ptolemy, this picture, which significantly qualifies the revolutionary character of Copernicus's contribution, is well established.[1] In the case of Einstein *versus* Lorentz, however, one still tends to think in terms of a battle of paradigms, research programs, and the like. The time has come to reject this picture. In this section, I will present a new picture, fleshing out the above analogy with the Copernican revolution. In the next section, I will use this new picture to resolve some long-standing controversies in the philosophy of science concerning the case of Einstein *versus* Lorentz and indicate directions for further research. In particular, I will argue for the importance of analyzing experiments such as those of Frederick Trouton and Henry Noble on charged capacitors in torsion balances for understanding the development of special relativity.

Let me first briefly consider the case of Copernicus *versus* Ptolemy. It suffices for my purposes to use the crudest versions of their models of the motion of the sun and the planets. In the simplified Ptolemaic model, the motion of the sun is represented by a uniform circular motion with the earth at its center, while the motion of a planet is represented by the composite of two uniform circular motions: the planet moving on one circle (the epicycle) whose center is moving on another circle (the deferent) with the earth at its center. Ptolemaic astronomers knew that the motion of the planets is correlated with the motion of the sun: either the epicycle motion or the deferent motion is in phase with the motion of the sun (see the diagram on the left in figure 1). They also knew that the motion of a planet around the earth can be replaced by the motion of a planet around the sun and either the sun around the earth or the earth around the sun.

Before Copernicus, these alternative representations of the motion of a planet were not accorded any physical significance. One of Copernicus's predecessors, for instance, had the sun moving around the earth for one planet and the earth moving around the sun for another. Copernicus was the first to have the earth and all of the other planets moving around the sun, thus introducing his heliocentric model of the world. One of the main virtues of this model, as Copernicus himself realized, was that it explains the striking correlations between the motions of the planets and the motion of the sun, which in the Ptolemaic model were purely coincidental. In the Copernican model these correlations arise because one component of a planet's motion as seen from the earth reflects the earth's own motion around the sun (see the diagram on the right in figure 1).

The case of Copernicus *versus* Ptolemy thus exhibits the first two characteristics I mentioned above. The central innovation was a reinterpretation of a well-known feature of the existing formalism, and the basic argument in its favor can be cast in

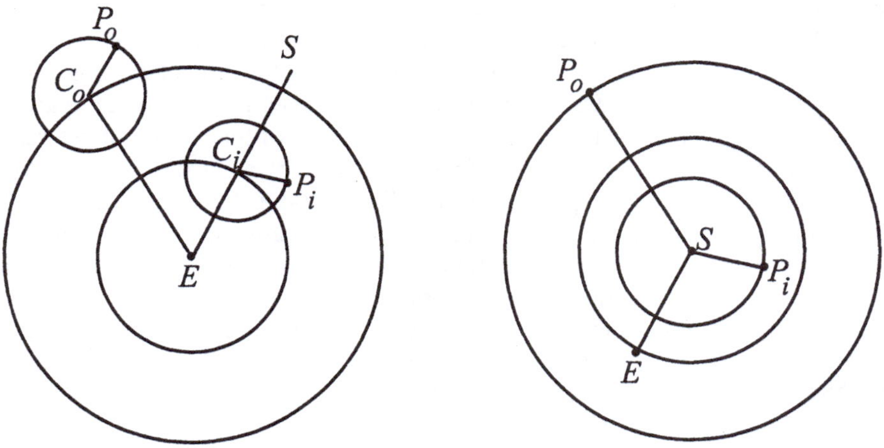

Fig. 1. The diagram on the left shows the motion of an inner planet P_i and an outer planet P_o around the earth E in a simplified Ptolemaic model. P_i and P_o move on epicycles with centers C_i and C_o, respectively. C_i and C_o, in turn, move on deferents with center E. (All circular motions indicated in the figure are counterclockwise.) The epicycle motion of the outer planet and the deferent motion of the inner planet are perfectly correlated to the motion of the sun: the line segments C_oP_o and EC_i both point in the direction of the sun (ES). In the Ptolemaic model this is an unexplained coincidence. The diagram on the right shows the corresponding simplified Copernican model. It is obtained from the Ptolemaic diagram on the left by: (*i*) placing the sun S at the center C_i of the inner planet's epicycle; (*ii*) setting the radius C_oP_o of the outer planet's epicycle equal to the radius $EC_i = ES$ of the inner planet's deferent; and (*iii*) referring the relative motion of the earth, the sun, and the two planets to the sun rather than to the earth. The Copernican model offers a simple explanation for the correlation between the motion of the sun, the outer planet's epicycle motion and the inner planet's deferent motion: these motions just reflect the earth's motion around the sun.

the form of a common-cause argument. What about the third characteristic? Here we have to bear in mind that the Ptolemaic astronomical tradition was embedded in an Aristotelian geocentric worldview. Despite the above arguments in favor of a heliocentric model, a moving earth is simply incompatible with Aristotelian physics. It took the work of Galileo Galilei, Johannes Kepler, and Isaac Newton to make physical sense of the heliocentric model. Copernicus's contribution thus is best understood as a catalyst of a larger revolution in which the Aristotelian worldview was overthrown.

I now turn to the case of Einstein *versus* Lorentz. Lorentz's theory, the most promising fundamental theory of physics around the turn of the last century, was a combination of Newtonian mechanics and Maxwellian electrodynamics. Newton provided Lorentz with the space-time arena for his theory and with the laws—Newton's laws of motion—governing matter in that arena. James Clerk Maxwell provided the laws—Maxwell's equations—governing electric and magnetic fields conceived as states of a medium called the ether. In Lorentz's theory, there is a strict separation of ether and matter. The only way in which the two can interact is through tiny charged particles first called ions, later called electrons. The electrons generate electric and magnetic fields in the ether, and these fields in turn exert forces on material bodies by exerting electromagnetic forces on the electrons contained in these bodies.

In principle, Maxwell's equations can be used to compute the field configuration in the ether that will be produced by the charge distribution carried by a given matter configuration. There is one problem though. If Newton's laws hold in terms of the space-time coordinates of one frame of reference in Newtonian space-time, they also hold in terms of the space-time coordinates of any frame of reference in uniform motion with respect to the first. Such frames are called *inertial frames*. Transformations from the space-time coordinates (and other quantities) of one inertial frame to another are called Galilean transformations. Laws that hold in all inertial frames in Newtonian space-time if they hold in one of them are called invariant under Galilean transformations or *Galilean invariant* for short. Maxwell's equations do not have this property. They hold only in those inertial frames in which the ether is at rest. In Lorentz's theory, the earth is moving through an immobile ether. Thus, Maxwell's equations do not hold in the frames of reference actually used to describe terrestrial experiments.

Lorentz found a way around this problem. He replaced the real space-time coordinates of an arbitrary inertial frame in Newtonian space-time by cleverly chosen fictive space-time coordinates that depend on the frame's velocity with respect to the ether. If the frame's velocity happens to be zero, the fictive coordinates are identical to the real coordinates. He likewise replaced the real electric and magnetic fields by fictive fields. In terms of these fictive fields as functions of the fictive space-time coordinates, Maxwell's equations hold in every frame in which the ether is either at rest or in uniform motion. Lorentz also had to replace electric charge and current densities by appropriate fictive quantities. At first, the equations he obtained by just introducing fictive space-time coordinates and fictive fields were close enough to Maxwell's equations for his purposes. Later he also introduced fictive charge and current densities. Only after 1905, however, did he choose these particular fictive quantities such that the equations for the fictive quantities were precisely Maxwell's equations.

Thus, Maxwell's equations hold in every inertial frame, provided that the appropriate space-time coordinates, fields, and charge and current densities for that frame are used. They are therefore called invariant under Lorentz transformations or *Lorentz invariant* for short.

Suppose Maxwell's equations allow a certain configuration of the real fields in terms of the real space-time coordinates in an inertial frame at rest in the ether. It follows from their Lorentz invariance that the equations allow the same configuration in an inertial frame moving through the ether, but now of the fictive rather than the real fields, and in terms of the fictive rather than the real space-time coordinates. This configuration of fictive fields translates into a configuration of real fields in the moving frame that is different from the configuration in the initial frame at rest in the ether. Lorentz referred to two such configurations of real fields as corresponding states, and to the mathematical result that if one is allowed the other also is as the *theorem of corresponding states*. In 1895 he proved an approximate first-order version of this theorem, one that was valid as long as he neglected quantities smaller than the ratio of the velocity of the moving frame to the velocity of light.[2] A few years later, in 1899, he proved an exact version of this theorem.[3]

Lorentz was particularly interested in configurations of electric and magnetic fields constituting light waves and producing interference patterns. Most optical experiments eventually boil down to the observation of such interference patterns. According to the first-order version of Lorentz's theorem, there is no difference between the interference patterns associated with the field configurations of two corresponding states. Lorentz's result thus explained in one fell swoop the negative result of a broad class of first-order ether-drift experiments (that is, experiments aimed at detecting the earth's motion through the ether and having the same accuracy as the first-order version of Lorentz's theorem). According to the later exact version of Lorentz's theorem, there is one small but important difference between the interference patterns associated with two corresponding states: the pattern in the moving frame is obtained by taking the pattern of the corresponding state in the frame at rest in the ether and contracting it ever so slightly in the direction of motion.

It takes a second-order ether-drift experiment (one that can detect quantities as small as the square of the ratio of the velocity of the moving frame to the velocity of light) to distinguish the contracted pattern expected in a moving frame from the uncontracted pattern expected in a frame at rest in the ether. The famous Michelson-Morley experiment of 1887 had been that accurate, yet no difference had been found. Lorentz already had tried to explain this result in 1892 before he had found the exact version of his theorem.[4] He—and, independently of him, the Irish physicist George Francis FitzGerald—had recognized that the null result of the Michelson-Morley experiment could be explained by assuming that matter moving through the ether experiences a contraction—the same contraction, it turns out, that Lorentz's exact theorem of corresponding states predicts for interference patterns.

Once he had found the exact theorem of corresponding states, Lorentz immediately saw how to ensure that his theory would predict in full generality that no observations of interference patterns, no matter how accurate, would ever reveal the earth's motion through the ether. He only had to add the following physical assumption to his mathematical theorem: *a matter configuration producing a certain field configuration in a frame at rest in the ether will, when the system is set in motion, change into the matter configuration producing the corresponding state of that field configuration in the frame moving with the system*. One of the changes in the matter configuration will be the contraction familiar from the original Lorentz-FitzGerald contraction hypothesis. Because of this contraction, interference patterns observed on the earth, that is, in a frame moving through the ether, will be indistinguishable from the corresponding pattern in a frame at rest. Lorentz's new assumption, however, which I will call the generalized contraction hypothesis, entails many more changes. Consider for instance the electrons emitting the light waves that produce the interference patterns. As Lorentz himself realized, it follows from the generalized contraction hypothesis that the frequency with which these electrons oscillate and even their masses will depend on their velocity with respect to the ether. The generalized contraction hypothesis thus is much stronger than the former contraction hypothesis. In modern terms, it constitutes, in effect, the assumption that the laws governing matter are Lorentz invariant, just as are the laws governing the fields in the ether.

Instead of simply adding the generalized contraction hypothesis to his theory, Lorentz assumed, in a more piecemeal fashion, that all shapes, masses, and forces, contrary to what one would expect on the basis of the Galilean-invariant Newtonian laws, depend on a system's velocity with respect to the ether. In this way he tried to ensure that matter configurations that produce certain field configurations at rest in the ether would, when set in motion, change into the matter configurations that produce the corresponding states of those field configurations in motion through the ether. In other words, Lorentz decreed a number of important exceptions to the Galilean-invariant Newtonian laws that are supposed to govern matter, so that the laws effectively governing matter are Lorentz invariant.

Why, one can legitimately ask, would the laws governing matter have the property of Lorentz invariance, which so far appeared to be nothing but a peculiar property of Maxwell's equations? Is it perhaps because all physical phenomena ultimately might be reducible to electrodynamics? This possibility was taken very seriously at the beginning of the last century, but turned out to be problematic.[5] Lorentz himself flirted with the idea of a purely electromagnetic physics, but soon abandoned it. In the final analysis, it is thus left as an unexplained coincidence in Lorentz's theory that both matter and fields are governed by laws that are Lorentz invariant. In Einstein's special theory of relativity the Lorentz invariance of these different laws is traced to a common origin.

The title of the paper in which Einstein introduced his special theory of relativity—"On the electrodynamics of moving bodies"[6]—shows that at least part of Einstein's concern was the same as Lorentz's. Einstein's approach, however, was completely different. What Lorentz was working on is what Einstein (figure 2) later would call a *constructive theory*, a type of theory in which one gives a detailed model of certain features of the physical world. Einstein, by contrast, developed a *theory of principle*, a type of theory in which one only gives constraints on such modeling, constraints that are based on empirically well-confirmed regularities.[7]

The two regularities on which Einstein built his theory are expressed in two postulates. The first, the *relativity postulate*, says that all physical laws hold in every frame in which the laws of mechanics hold. I will continue to call such frames inertial frames. The second, the *light postulate*, says that the velocity of light in a vacuum is independent of the motion of its source. On the assumption that the world is such that these two postulates are satisfied, and carefully examining the concept of simultaneity of events occurring at a distance from one another, Einstein derived the relationship between the space-time coordinates that two observers in different inertial frames read off identical sets of ideal rods and clocks. He found that the space-time coordinates of two such observers are related by a Lorentz transformation and not by a Galilean transformation. In other words, from the point of view of the observer in one frame, the space-time coordinates of the observer in the other frame are just Lorentz's fictive space-time coordinates for that frame.

The expressions for the fictive space-time coordinates of Lorentz's exact theorem of corresponding states encode three effects that one would not expect in a Newtonian space-time but that necessarily occur in a universe satisfying Einstein's

Fig. 2. Albert Einstein and Hendrik A. Lorentz in front of Paul Ehrenfest's house in Leiden in 1921. Picture taken by Ehrenfest. Reproduced by permission of the Boerhave Museum, Leiden, The Netherlands.

two postulates: relativity of simultaneity, length contraction, and time dilation. Consider again the two observers above. Appealing to the light postulate, both observers can synchronize clocks at different locations with the help of light signals sent back and forth between them. However, two clocks synchronized in this manner by an observer at rest with respect to both clocks will not be properly synchronized according to an observer moving with respect to these clocks. A direct consequence is that the two observers will disagree on the length of rods and other objects and the rate of clocks and other processes. Each observer will find that the rods of the other will be shorter and that the clocks of the other will tick at a slower rate.

Because the space-time coordinates that Einstein defines in different inertial frames are related by Lorentz transformations rather than Galilean transformations, the apparent incompatibility between the relativity postulate and Maxwell's equations is removed. The physical laws in a universe in which Einstein's two postulates hold must be Lorentz invariant rather than Galilean invariant. And, as Einstein showed in the second part of his 1905 paper, Maxwell's equations are Lorentz invariant.

In 1905 Einstein was familiar only with the first-order version of Lorentz's theorem of corresponding states. He had to derive the expressions for the space-time coordinates, the fields, and the charge and current densities in the moving frame required to ensure the exact Lorentz invariance of Maxwell's equations himself. The physical interpretation he gave of all these quantities was something Lorentz had never even considered. Possibly, however, Einstein got the basic idea from the French mathematician Henri Poincaré. Poincaré considered the situation of an observer in uniform motion through the ether synchronizing clocks with the help of light signals while pretending to be at rest in ether and thus assuming the velocity of light to be the same in all directions. He found, within the limits of accuracy of the first-order theorem of corresponding states, that moving clocks synchronized in this manner will not indicate the real time but the fictive or, as Lorentz called it, the local time of the first-order version of Lorentz's theorem.[8] Einstein, in effect, gave an exact version of Poincaré's calculation that showed that the time indicated by moving clocks synchronized with the help of light signals is the local time of the exact version of Lorentz's theorem.

Einstein also found part of another result that had been found by Poincaré, namely, that the Lorentz transformations form what mathematicians call a *group*.[9] Roughly, what this means is that the inverse of a Lorentz transformation as well as the combination of any Lorentz transformation followed by another are themselves Lorentz transformations. This group property is important to guarantee that the situation for two observers in relative motion to one another is completely symmetric, so that, for instance, each observer will find that the rods of the other are shorter.

These results tremendously simplified the task that Lorentz had set himself of accounting for the absence of any signs of the earth's motion through the ether. If an observer on earth, that is, in a frame moving through the ether, always measures the fictive or Lorentz-transformed quantities for that frame, which satisfy the same equations as the real quantities in a frame at rest in the ether, it is immediately clear that that observer never can detect the earth's motion. It is no longer necessary to show that two corresponding states that are different in terms of the real quantities, are indistinguishable empirically (for instance, that the interference patterns are indistinguishable). The group property of Lorentz transformations furthermore ensures that the situation of an observer at rest in the ether as seen from the point of view of an observer in motion through the ether is the same as the situation of the moving observer as seen from the point of view of the observer at rest. Thus, there is no way to tell which of the two observers is truly at rest in the ether. Lorentz gave Einstein all of the credit for these insights, which led to such a dramatic improvement in his own theory, and curiously failed to acknowledge Poincaré at all for them.[10]

Lorentz admired the way in which Einstein had set up his theory as a theory of principle, perhaps in part because it allowed him to look upon his own theory as providing a concrete model of a world in which, for all intents and purposes, Einstein's two postulates are satisfied. Needless to say, however, Lorentz's picture of such a world was not what Einstein had in mind. For one thing, there is no room for Lorentz's ether in Einstein's theory. The singling out of one inertial frame as truly at rest in the ether is incompatible with Einstein's notion that all inertial frames are physically equivalent. For Einstein, the Lorentz-transformed quantities are the correctly measured real quantities. In the final version of Lorentz's theory, measurements never reveal the real quantities because all measurements are made with devices that are systematically distorted and improperly synchronized as a result of their and the observer's motion through the ether. Lorentz continued to look upon Lorentz invariance as a property shared by all laws governing systems in Newtonian space-time. Einstein recognized that Lorentz invariance reflects a new space-time structure.

In 1905 Einstein talked about space and time rather than about space-time. The notion of space-time was introduced by the German mathematician Hermann Minkowski, who developed the geometry of the new space-time implicit in Einstein's theory, now known as Minkowski space-time.[11] In this way, and despite Einstein's initial reservations about Minkowski's work, Minkowski provided the space-time part of any specific model of a world satisfying Einstein's two postulates.

There are important parallels between the geometry of Minkowski space-time and the standard Euclidean geometry of ordinary space. The choice of a particular inertial frame in Minkowski space-time is similar to the choice of a particular set of orthogonal axes to serve as a Cartesian coordinate system in Euclidean space. Using the space-time coordinates of such inertial frames, one can compute lengths and distances in Minkowski space-time with a formula similar to the Pythagorean theorem used when computing lengths and distances in terms of Cartesian coordinates in Euclidean space. A Lorentz transformation to get from one inertial frame in Minkowski space-time to another is similar to the rotation of a set of orthogonal axes in Euclidean space to get from one Cartesian coordinate system to another. The freedom to pick any inertial frame in Minkowski space-time is reflected in the invariance under Lorentz transformations of all laws governing physical systems in Minkowski space-time just as the freedom to pick any Cartesian coordinate system in Euclidean space is reflected in the invariance under spatial rotations of all laws governing physical systems in Euclidean space.

Minkowski also gave a straightforward geometrical interpretation of the three basic effects Einstein discussed in his 1905 paper, the relativity of simultaneity, length contraction, and time dilation (figure 3). Two observers in relative motion to one another will have different ways of slicing Minkowski space-time into three-dimensional hyperplanes of simultaneous events. Events that lie on the same simultaneity slice for one observer will not lie on the same simultaneity slice for another. Length contraction and time dilation arise because different observers will look upon the length of different space-time cross sections of a system extended in space and time (say a rod or a clock persisting through time) as the length of the system or the duration of a process in that system.

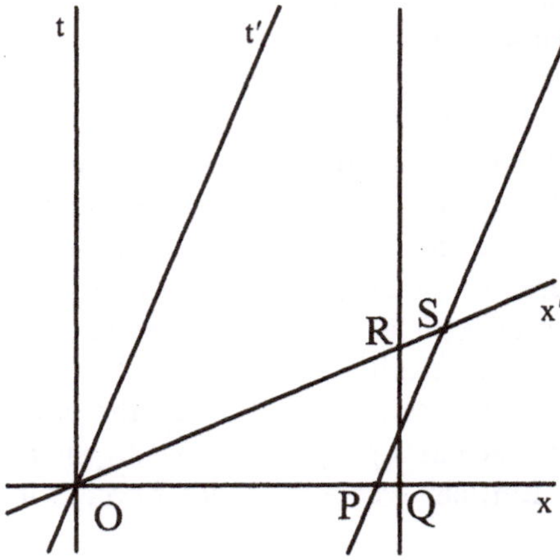

Fig. 3. This diagram shows the space and time axes of two inertial frames in a two-dimensional Minkowski space-time (two spatial dimensions of our actual four-dimensional space-time have been suppressed). Because the geometry of Minkowski space-time is different from the geometry of ordinary Euclidean space, not all features of the space-time geometry can be captured in a drawing on a Euclidean piece of paper. The space and time axes, for instance, are in fact perpendicular to one another in both frames, even though they are only drawn perpendicularly for the unprimed frame. The primed frame is moving to the right with respect to the unprimed frame, which is reflected in the t'-axis being tilted to the right with respect to the t-axis. The corresponding tilting of the x'-axis with respect to the x-axis reflects the disagreement between observers in the two frames over the simultaneity of events. For the observer in the unprimed frame, horizontal slices of space-time (in other words, slices parallel to the x-axis) consist of simultaneous events. For the observer in the primed frame, slices parallel to the x'-axis consist of simultaneous events. The diagram also illustrates the phenomenon of length contraction. Let the two vertical lines be the worldlines of the end points of a rod at rest in the unprimed frame (a worldline traces an object through space-time). Let the t'-axis and the line parallel to it through points P and S be the worldlines of the end points of an identical rod at rest in the primed frame. The length of a rod, at rest or in motion, can be defined as the distance between its end points at a given time. In the diagram, the left ends of the two rods meet at O, the origin of the two space-time coordinate systems. Observers in the two frames will not agree about the distance between the end points of the rods at the time of event O, because they will pick different events on the worldlines of the right ends of the rods as simultaneous with O. The observer in the unprimed frame will say that P and Q are simultaneous with O and will conclude that his rod is longer than the other one ($OQ > OP$). The observer in the primed frame will say that R and S are simultaneous with O and will conclude that her rod is longer than the other one ($OS > OR$).

As Minkowski's work vividly illustrates, the case of Einstein *versus* Lorentz does indeed exhibit the first two characteristics I identified in making a comparison with the case of Copernicus *versus* Ptolemy. First, from a modern point of view, the central innovation of Einstein's 1905 paper is the reinterpretation of Lorentz invariance, a well-known feature of the formalism of Lorentz's theory, as reflecting a new space-time structure. Second, the strongest argument for this reinterpretation is that it traces the Lorentz invariance of Maxwell's equations and the Lorentz invariance of other physical laws to a common cause, namely this new space-time structure.

The case also exhibits the third characteristic I emphasized. Special relativity overthrew Lorentz's interpretation of Lorentz invariance, but it did not replace—nor was it ever intended to replace—Lorentz's theory with a theory of comparable scope. What eventually superseded Lorentz's theory was quantum theory, not relativity theory. Einstein's reinterpretation of Lorentz invariance, however, did strike at the heart of Lorentz's theory. Special relativity left no room for Lorentz's ether, one of the two basic constituents of his universe. As with Copernicus who set the earth in motion even though this flew in the face of Aristotelian physics, Einstein in his relativity paper abolished the ether without explaining how to conceive of electromagnetic fields in the absence of a substratum carrying such fields. Paul Ehrenfest probably expressed a sentiment felt by many of his colleagues when in 1913 he suggested adding a third postulate to Einstein's theory explicitly declaring the combination of the first two to be satisfactory.[12]

Ehrenfest's quip nicely illustrates that it was not clear immediately after 1905 what the physics of a world in accordance with the postulates of special relativity would look like. Ehrenfest was thinking mainly about the problem of how to make sense of electromagnetic fields in the absence of an ether. But it also was far from straightforward to find all of the new physical laws that special relativity called for. Even the formulation of a satisfactory relativistic mechanics was not simply a matter of making a few adjustments to the laws of Newtonian mechanics to make them Lorentz invariant. In the cases of thermodynamics and gravitation theory, the situation was notoriously more complicated, to say nothing of the development of a satisfactory relativistic quantum theory a few decades later. These brief remarks must suffice here to support my claim that Einstein's introduction of the special theory of relativity in 1905 is best understood as one of the catalysts of a much broader revolution in early twentieth-century physics.

Philosophers of Science on the Case of Einstein *versus* Lorentz

Various philosophers of science have tried to articulate why Einstein's theory is preferable to the parts of Lorentz's theory it replaced. In this section, I will examine a few of these attempts, discuss their shortcomings, and then return to the common-cause argument advocated in the preceding section as a promising alternative.

Until about thirty years ago, comparisons between the theories of Einstein and Lorentz focused almost without exception on the Michelson-Morley experiment. Einstein's own popular expositions of special relativity, in which this experiment is given a prominent role, are partly to blame for this. But whatever its exact origins, the following myth was part of physics folklore for several generations and still can be found in the popular literature.[13] Einstein, the story went, developed special relativity out of dissatisfaction with the way in which Lorentz had tried to explain away the result of the Michelson-Morley experiment by introducing, in a completely ad hoc fashion, his infamous contraction hypothesis. Einstein, by contrast, accepted at face value what the Michelson-Morley experiment was telling us, namely, that there are no preferred inertial frames and that light has the same velocity in all directions in all of them. He elevated these experimental dicta to the status of postulates and accepted whatever these postulates entailed.

The Einstein part of this story has been thoroughly discredited. Gerald Holton showed convincingly that the Michelson-Morley experiment played no role of any significance in the considerations that led Einstein to special relativity.[14] For a long time, it was not even clear whether Einstein knew about the Michelson-Morley experiment when he formulated the theory. Equally damning, as John Stachel has pointed out, is that after 1905 Einstein consistently and exclusively cited the Michelson-Morley experiment in support of the relativity postulate, never in support of the light postulate.[15] And in Einstein's 1905 paper itself, the light postulate says that the velocity of light is independent of the velocity of its source, not, as the folklore story wants to have it, that it is independent of the velocity of the inertial frame in which it is measured.

Even if all of this is granted, one could still embrace the Lorentz part of the folklore story and argue that Einstein's theory is better than the relevant part of Lorentz's simply because Lorentz had to resort to an ad hoc hypothesis to account for the Michelson-Morley experiment, whereas Einstein did not. On the basis of my reconstruction of the development of Lorentz's theory in the preceding section, it can be shown conclusively that this part of the folklore story also is false. Earlier attempts to show this in the extensive literature on the subject all have been unsatisfactory.

What exactly is it that makes a hypothesis—or, more appropriately perhaps, a theory—ad hoc? When Einstein used this charge to put down the contraction hypothesis, he probably meant two things: first, that Lorentz introduced the contraction hypothesis for the sole purpose of accounting for one particular experimental result; second, that the contraction hypothesis was highly contrived by the standards of Lorentz's own theory. In the draft of a letter to Einstein of 1915 (discovered by A. J. Kox),[16] Lorentz (figure 4) conceded the first point, but took exception to the second. I want to emphasize that Einstein's objections clearly are directed toward the original contraction hypothesis. To the best of my knowledge, Einstein never commented on what I called the generalized contraction hypothesis. Sadly, this same observation can be made for all subsequent discussions about whether or not the contraction hypothesis is ad hoc.

An influential definition of the term "ad hoc"—which can be seen as amplifying the first part of what I suggested Einstein meant by the term—was given by Karl R. Popper.[17] For Popper, "ad hoc" is synonymous with "non-falsifiable." The contraction hypothesis, according to Popper, is unsatisfactory because it did not give any new predictions that could be used to put it to an experimental test. As Lorentz himself noted, one cannot detect a moving object's contraction directly because any device used to measure its length would experience the same contraction as the object itself. This does not mean, however, that the hypothesis cannot be tested. In response to Popper, Adolf Grünbaum drew attention to the Kennedy-Thorndike experiment of 1932,[18] a variant of the Michelson-Morley experiment, which from an ether-theoretic point of view should have detected the earth's motion through the ether, even if one factors in the Lorentz-FitzGerald contraction.* In accordance with special relativity and later versions of Lorentz's theory, no such motion had been detected. This falsification of the contraction hypothesis shows convincingly that it is not ad hoc in the Popperian sense of the term.

Popper accepted Grünbaum's criticism, but insisted that Lorentz's theory was more ad hoc, that is, less testable than Einstein's. If we bear in mind that it is the original contraction hypothesis that is under discussion here, Popper does have a point. First note that to examine the testability of a hypothesis, we also need to examine the theory in which the hypothesis is embedded, in this case, the version of Lorentz's theory based on the first-order theorem of corresponding states. This theory was set up to deal with first-order experiments. The contraction hypothesis was added to deal with one particular second-order experiment. The proposed contraction, a second-order effect, in principle can affect a wide range of phenomena, providing an equally wide range of conceivable tests. The problem is that a first-order theory cannot be used to predict the consequences of second-order effects. Thus, the most we can expect of the theory is that it will give definite predictions for the outcome of experiments that are but minor variants of the Michelson-Morley experiment, such as the Kennedy-Thorndike experiment. The possibilities of testing the contraction hypothesis in its original form are thus very limited. This is not true of the generalized contraction hypothesis, which is part of Lorentz's mature theory based on the exact theorem of corresponding states. This theory gives definite predictions for any experiment used to test it. There is, of course, something peculiar about these predictions. The prediction is always that no effect of the earth's presumed motion through the ether will be detected. This is a problematic feature of the theory. But it has nothing to do with lack of testability. All that matters on that score is that the theory predicts a definite result. The generalized contraction hypothesis therefore passes Popper's falsifiability test with flying colors, whereas the original contraction hypothesis earned only a marginally passing grade courtesy of Roy Kennedy and Edward Thorndike.

---

* Consider light traveling back and forth along a rod of length l with mirrors attached to both ends. If the rod were at rest in the ether, the amount of time required for a round trip would be equal to $2l/c$, where c is the velocity of light. If the rod is at rest in some laboratory on earth and thus moving through the ether along with the earth with some velocity v, the ether theory (without any additional hypotheses) predicts that the travel time will be longer. How much longer depends on the orientation of the rod with respect to the ether wind. If the rod is parallel to the ether wind, the expected travel time increases to $l/(c+v) + l/(c-v) = (2l/c)/(1 - v^2/c^2)$. If the rod is perpendicular to the ether wind, the expected travel time only increases to $(2l/c)/\sqrt{1 - v^2/c^2}$. The basic idea behind the Michelson-Morley experiment was to measure the expected changes in travel time as the orientation of the moving rod with respect to the ether wind varies. No such changes were found. By adding the contraction hypothesis to the theory, one can account for this null result. According to the contraction hypothesis, a rod moving with velocity v through the ether will contract by a factor $\sqrt{1 - v^2/c^2}$ in the direction of motion. As a result, the expected travel time when the rod is parallel to the ether wind becomes $(2l/c)/\sqrt{1 - v^2/c^2}$, which is the same as the expected travel time when the rod is perpendicular to the ether wind. In other words, the expected travel time no longer depends on the orientation of the rod with respect to the ether wind, in accordance with the null result of the Michelson-Morley experiment. The expected travel time, however, still depends on the velocity v and should therefore vary as a result of the earth's diurnal and annual rotation. This is the effect sought in vain in the Kennedy-Thorndike experiment.

As Grünbaum pointed out,[19] Lorentz's theory can account for both the Michelson-Morley and the Kennedy-Thorndike experiment if a second hypothesis is added to the contraction hypothesis, postulating another second-order effect. This is the so-called clock-retardation hypothesis, which says that any process in a system moving through the ether takes longer than that same process in the system at rest in the ether.* Before continuing Grünbaum's analysis, I will make two comments.

First, the clock-retardation hypothesis follows from the generalized contraction hypothesis. This explains how it can be that Lorentz's first-order theory plus contraction hypothesis gives the wrong result for the Kennedy-Thorndike experiment, whereas the mature version of the theory gives the correct result. Second, the role of the retardation effect in the Kennedy-Thorndike experiment underscores just how limited the testability of the original contraction hypothesis is. It shows that even the case of the Kennedy-Thorndike experiment is problematic. To predict a definite outcome for this experiment, some assumption must be made about how the contraction of a moving source will affect the frequency of the light it emits, even if it is just the assumption that the frequency will not be affected at all.

Grünbaum showed that what he called the "doubly-amended theory"—Lorentz's core theory plus contraction and clock-retardation hypotheses—is still a testable theory. The clock-retardation hypothesis ensures that the theory gives the relativistic formula for the Doppler effect, which was verified experimentally by Herbert Ives and G. R. Stillwell in 1938.

Thus without becoming ad hoc in the falsificationist sense, the doubly-amended theory can account for three optical experiments that usually are thought to provide strong empirical support for the special theory of relativity: the Michelson-Morley experiment, the Kennedy-Thorndike experiment, and the Ives-Stillwell experiment.[20] Can it account for all experiments that special relativity can account for? In the usual parlance, is the doubly-amended theory empirically equivalent to special relativity? It is not. Can the theory be further amended so that it becomes empirically equivalent to special relativity without becoming ad hoc in the falsificationist sense? It can.

It is sometimes suggested that the doubly-amended theory itself is already empirically equivalent to special relativity. This is because the contraction hypothesis, the clock-retardation hypothesis, and the stipulation that the light-signaling method of Poincaré and Einstein be used to synchronize clocks suffice to show that the space-time coordinates of any two inertial observers are related by a Lorentz transformation. This, however, does not imply empirical equivalence. At best, it

* What is actually measured in both the Michelson-Morley experiment and the Kennedy-Thorndike experiment are phase shifts rather than travel times. To obtain the phase shift the travel time must be divided by the period of the light used. According to the clock-retardation hypothesis, a light source moving through the ether with velocity v emits light with a period $T/\sqrt{1-v^2/c^2}$, where T is the period of the light that would be emitted by that same light source if it were at rest in the ether. If one only factors in the contraction hypothesis, the ether theory predicts that it will take an amount of time $(2l/c)/\sqrt{1-v^2/c^2}$ for light to travel back and forth along a rod of length l moving through the ether with velocity v (see the preceding footnote). The corresponding phase shift, $2l/cT$, is exactly equal to the expected phase shift for a round trip if the rod were at rest in the ether.

Fig. 4. Albert Einstein and Hendrik A. Lorentz in Leiden in the 1920s. Courtesy of American Institute of Physics Emilio Segrè Visual Archives.

shows that two observers will use the same space-time coordinates according to the two theories. What must be shown to prove empirical equivalence is that the doubly-amended theory only allows systems and processes described in terms of such coordinates that also are allowed by special relativity. In other words, it must be shown that all physical laws of the doubly-amended theory are Lorentz invariant.

How should the doubly-amended theory be further amended to become empirically equivalent to special relativity? Here the exact theorem of corresponding states, which tells us how motion through the ether affects field configurations, and the generalized contraction hypothesis, which says that motion through the ether affects matter configurations the same way, provide all the guidance we need. For instance, the theorem of corresponding states—or, in modern terms, the Lorentz invariance of Maxwell's equations—allows us to see, by comparing field configurations in systems in motion through the ether to their corresponding states in systems at rest in the ether, that interference patterns experience a contraction and that frequencies of electromagnetic waves experience a retardation as a result of a system's motion through the ether. The two amendments of the doubly-amended theory postulate that matter is subject to these same two effects. The theorem of corresponding states likewise tells us that two things happening simultaneously at different points in a field configuration in a system at rest in the ether will happen one after the other at the corresponding points in the corresponding state of that field configuration in a system moving through the ether. To obtain full empirical equivalence with special relativity, it thus will be necessary to add a third amendment to the doubly-amended theory saying that matter is subject to this same effect. Every amendment added to the theory in this manner for the purpose of explaining new experiments will bring the theory a step closer to being Lorentz invariant and thus a step closer to being empirically equivalent to special relativity. There is no need to worry about testability. The theory can be subjected to as many additional tests as special relativity itself.

Despite its limitations, the doubly-amended theory has become a popular toy model for Lorentz's mature theory. It is used for instance by Elie Zahar in his study of the case of Einstein *versus* Lorentz.[21] I now turn to Zahar's study, which is cast in the mold of the methodology of scientific research programmes of his teacher Imre Lakatos.[22]

Zahar takes it upon himself to exonerate Lorentz's theory of all charges of ad hoc-ness. He essentially rehearses Grünbaum's argument to show that the theory is not ad hoc in Popper's falsificationist sense. He then tries to establish that neither Lorentz's core theory amended by the contraction hypothesis, nor the doubly-amended theory, are ad hoc in the sense that these theories were "obtained from [their] predecessor[s] through a modification of the auxiliary hypotheses which does not accord with the spirit of the heuristic of the programme."[23] For my purposes, this can be thought of simply as a more precise statement of the second part of what I suggested Einstein had in mind when he denounced the contraction hypothesis as ad hoc, namely, that judged by the standards of Lorentz's own theory the contraction hypothesis is a highly contrived way of accounting for the result of the Michelson-Morley experiment.

Zahar offers the following argument for his claim that the contraction hypothesis is not ad hoc in this sense. Lorentz, he claims, derived the contraction hypothesis from what Zahar dubs the molecular-force hypothesis. In view of the Lorentz invariance of Maxwell's equations, Zahar continues, it was natural to add the molecular-force hypothesis to the theory. It was not something cooked up to deal with the Michelson-Morley experiment. Zahar's argument is in error on several counts.[24] First, in the letter to Einstein mentioned earlier, Lorentz admitted somewhat sheepishly that he had found his supporting molecular-force argument for the contraction hypothesis only after he had recognized that such a contraction could explain the result of the Michelson-Morley experiment. More seriously, Zahar mistakes a plausibility argument for a derivation, and misunderstands the role of Lorentz invariance in the first-order version of Lorentz's theory.

Suppose that motion through the ether affects inter-molecular forces the same way it affects electrostatic forces. For electrostatic forces, Lorentz could derive the following result: if a charge configuration is in electrostatic equilibrium at rest in the ether, then that same charge configuration contracted in just the way postulated by the contraction hypothesis is in electrostatic equilibrium in motion through the ether.[25] The problem is that there is no such thing as electrostatic equilibrium. Other forces are needed to maintain the stability of any static charge distribution. Clearly then, the forces that hold matter together must be very different from electrostatic forces. Lorentz's result, to be sure, is suggestive, and Lorentz used it as the basis for a plausibility argument for the contraction hypothesis that he still defended when he wrote to Einstein years later; but he did not intend the argument to be a derivation of the contraction hypothesis.

Lorentz's treatment of electrostatics in inertial frames moving through the ether is similar to his treatment of optics in such frames. It involves the introduction of fictive quantities in the moving frame satisfying the same equations as the real quantities in a frame at rest in the ether. This is why Zahar claims that the molecular-force hypothesis is suggested by the Lorentz invariance of Maxwell's equations. However, in this early version of Lorentz's theory, no connection is made between the treatment of electrostatics and the treatment of optics based on the first-order theorem of corresponding states. It is only in the mature version of Lorentz's theory that both optics and electrostatics are treated on the basis of the theorem of corresponding states. Lorentz actually found the exact version of the theorem by combining the fictive quantities he had introduced earlier and separately in his discussions of optics and electrostatics in moving frames.

Although Zahar's argument fails to clear the original contraction hypothesis of the charge of ad hoc-ness under consideration, a variant of his argument does clear the generalized contraction hypothesis on this score. Given the exact theorem of corresponding states, it is natural to add the generalized contraction hypothesis to the theory. This move certainly "accords with the spirit of the heuristic of the programme," to use Zahar's phrase. Moreover, the generalized contraction hypothesis, unlike Zahar's molecular-force hypothesis, allows Lorentz to derive from his theory, rather than just make plausible, that the earth's motion through the ether will not be revealed in experiments such as the Michelson-Morley experiment. Both of these observations remain true if instead of adding the generalized contraction hypothesis lock, stock, and barrel, we add more specific hypotheses that ensure that in a given type of experiment motion through the ether affects matter configurations the same way it affects field configurations.

In summary: if Grünbaum's doubly-amended theory is replaced by a more realistic model of Lorentz's mature theory of 1899 and 1904, a solid case can be made for the claim that that theory is not ad hoc by any of the criteria considered here. In other words, with the help of this more realistic model, the Lorentz part of the myth of the Michelson-Morley experiment can be debunked as thoroughly as the Einstein part.

If Lorentz's theory is not ad hoc and empirically equivalent to special relativity, can we still say that Einstein's theory is superior to the part of Lorentz's theory it replaced and that it was rational on the part of scientists in the first decades of this century to prefer Einstein's theory over Lorentz's? Like Zahar—and Grünbaum, for that matter—I want to answer these questions affirmatively.

In arguing for this position, Zahar again takes his cue from Lakatos, who pointed out that of two competing research programs one may be stronger than the other in terms of "the power of the positive heuristic."[26] The central heuristic that Zahar claims gives Einstein's research program the edge over Lorentz's is the strategy it recommends to look for Lorentz-invariant modifications of classical laws.[27] One of Zahar's favorite examples is Max Planck's modification of Newton's second law to make it Lorentz invariant.[28] This modification had, in effect, already been given by Lorentz when he first derived the exact theorem of corresponding states and introduced the generalized contraction hypothesis.[29] This illustrates what my analysis of the generalized contraction hypothesis shows more generally, namely, that there is no problem exploiting in the context of Lorentz's research program the heuristic power that comes with the insight that physical laws must be Lorentz invariant.

There are additional problems with Zahar's argument. As Lakatos and Zahar concede, greater heuristic power must translate at some point into empirical successes not matched by the competition. Since Lorentz's mature theory and special relativity are empirically equivalent, Zahar has to turn to general relativity to find such empirical successes. This means that he has to assume that there is sufficient continuity between special and general relativity to consider them as belonging to one research program with a core theoretical commitment and a set of strategies for the development of the theory that stayed more or less fixed. As Zahar realizes,[30] this assumption is problematic. There is an even more serious problem. The acceptance of special relativity by the physics community in the early decades of this century had nothing to do with general relativity. Consequently, an account of what makes special relativity preferable to Lorentz's theory that depends on virtues of general relativity cannot shed any light on the rationality of the choices that were made. But the point of Zahar's study was precisely to capture that rationality.[31]

As part of his analysis of the doubly-amended theory, Grünbaum offers a more accurate diagnosis of the trouble with Lorentz's theory.[32] What makes the theory unsatisfactory are not the elements that are added, but some of the original elements, notably the ether and Newtonian space-time, that are rendered more and more invisible with every amendment. This suggests that Ockham's razor is all that is needed to settle the case of Einstein *versus* Lorentz. The problem with this type of argument is that it derives its force from a blanket rejection of unobservable elements in scientific theories, whereas it is widely accepted that such elements should not be banned automatically. Rather than condemning unobservables in general, I think it is wiser to demand that arguments be put forward on a case-by-case basis to show why a particular unobservable is otiose.

The common-cause argument I advocated in the first section of this essay provides such an argument. All that is needed for the argument to go through is one of the strongest principles of inductive inference, namely, the principle that tells us to infer to a common cause whenever one can be found. The argument basically turns on the observation that the Lorentz invariance of all physical laws, on which Einstein and Lorentz agree, points to a Minkowskian rather than a Newtonian space-time structure.

Lorentz invariance manifests itself in many different phenomena. Ultimately, these phenomena form the input of the common-cause argument. The most obvious examples are length contraction and time dilation. The length of a system in uniform motion is less than that of an identical system at rest by a factor that depends only on the velocity of the moving system. A process in the moving system takes longer than the corresponding process in the system at rest by that same factor. In the Newtonian space-time of Lorentz's theory this is a consequence of the unexplained coincidence that all of these systems are governed by Lorentz-invariant laws. In the Minkowski space-time of special relativity, it is a consequence of the way in which particular space-time slices are used to define the length of a system or the duration of a process.

One of the morals of my discussion of Grünbaum's doubly-amended theory was that there is more to Lorentz invariance than the length-contraction and time-dilation effects encoded in the basic Lorentz transformation equations for the space-time coordinates. It would strengthen my common-cause argument if additional phenomena can be found that point to the Minkowski space-time structure in the way that length contraction and time dilation do, but whose analysis requires other Lorentz transformation equations.

The Trouton-Noble experiment of 1903 provides a striking example of the sort of phenomenon I have in mind.[33] Trouton and Noble tried to detect the earth's motion through the ether with the help of a charged and carefully insulated capacitor suspended on a torsion wire (figure 5; see the appendix for a

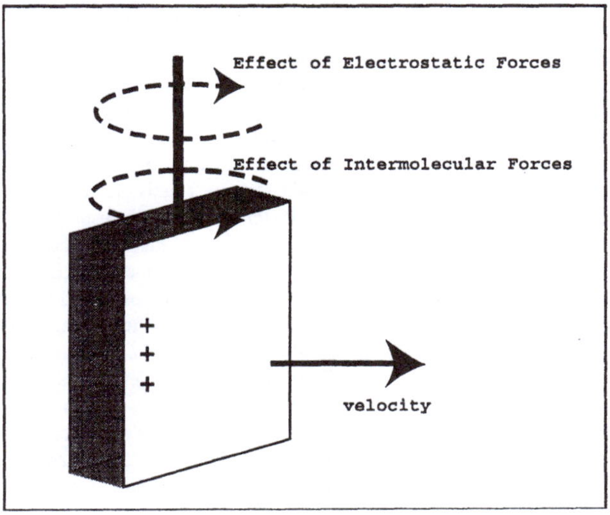

Fig. 5. Turning couples on a charged moving capacitor hanging from a torsion wire. The Coulomb forces between the plates exert a turning couple that tries to align the plates with the velocity of the capacitor. This turning couple is exactly compensated by a turning couple coming from the intermolecular forces that prevent the plates from collapsing onto one another.

discussion of a related earlier experiment). The electrostatic forces between the plates of the moving capacitor produce a turning couple that tries to align these plates with the direction of motion. The capacitor thus was expected to act as a kind of weather vane for ether drift. No turning couple was found, however, neither by Trouton and Noble, nor by later investigators repeating the experiment with more sensitive equipment.[34]

To account for the absence of the turning couple, proponents of both Lorentz's theory and special relativity can point out that the Lorentz transformation equations for forces, used to derive the turning couple produced by the electrostatic forces, also apply to the intermolecular forces that prevent the plates of the capacitor from collapsing onto one another under the influence of their electrostatic attraction. These intermolecular forces therefore also will produce a turning couple that will compensate the one coming from the electrostatic forces.[35] In the final analysis, the existence of these turning couples exactly canceling one another is an unexplained coincidence in Lorentz's theory. In special relativity, however, these turning couples simply come from the standard convention for choosing space-time slices of the moving system in the definition of the momentum or the angular momentum contained in both its electromagnetic field and its material parts.[36] If an alternative convention is adopted, there will be no turning couples at all.[37] In other words, in special relativity the delicately balanced turning couples in the standard analysis of the Trouton-Noble experiment can simply be defined away.*

* The momentum of spatially extended systems such as a capacitor and its electromagnetic field is obtained by integrating the corresponding momentum *density* over the volume of the system. In space-time terms, the volume of a system is some three-dimensional slice of that system in four-dimensional space-time. The definition of momentum thus depends on what convention is adopted for choosing such three-dimensional slices. Under the standard convention we find two delicately balanced turning couples acting on a moving charged capacitor, but under an alternative convention we find no turning couples at all.The difference between these two conventions can be illustrated with the help of figure 3, used earlier to explain the phenomenon of length contraction. In this figure two spatial dimensions were suppressed, leaving only one space dimension and time. Let the region between the t'-axis and the line parallel to the t'-axis through P and S represent the volume between the plates of the capacitor as the system moves with respect to the unprimed frame. It is only in this region that the capacitor's electromagnetic field has a significant momentum density. The same is true for the momentum density of the material part of the capacitor. Integrating the momentum density expressed in the space-time coordinates of the unprimed frame over some spatial slice of this region, we find the momentum of (different parts of) the system in the unprimed frame. The standard convention is to use the set of slices parallel to OP, that is, slices consisting of simultaneous events in the unprimed frame. The alternative convention is to use the set of slices parallel to OS, that is, slices consisting of simultaneous events in the primed frame in which the capacitor is at rest. With the standard convention, the integrals are over different sets of slices depending on the frame in which the system's momentum is evaluated; with the alternative convention, the integrals are always over the same set of slices no matter in which frame the system's momentum is evaluated.Both conventions (and the resulting definitions of momentum) have advantages and disadvantages. The standard convention is probably the more natural choice for a definition of ordinary volume integrals in a space-time setting. The advantage of the alternative convention is that it gives a definition of momentum such that the momentum of a system in one inertial frame is always related to that system's momentum in another inertial frame by a Lorentz transformation. This guarantees, for instance, that momentum will always be in the direction of

It is not just for the purpose of adding grist to the mill of my common-cause argument that I think the analysis of experiments such as the Trouton-Noble experiment is important. If one accepts the basic picture of the case of Einstein *versus* Lorentz outlined in this paper, it will be clear that the emergence and development of our understanding of Lorentz invariance should be the focus of work in the history and philosophy of special relativity. There is only so much one can learn from studying the optical ether-drift experiments that did or did not influence Einstein in formulating his theory and from studying the state of affairs in the electrodynamics of moving bodies around the turn of the last century. What my analysis suggests is that one should look for situations where the Lorentz invariance of Maxwell's equations calls for modifications of the Galilean-invariant laws that are supposed to govern the material systems interacting with the electromagnetic fields. The contraction hypothesis and the clock-retardation hypothesis are just two examples of such modifications. Lorentz had already introduced several other modifications, notably the velocity dependence of mass and forces, by the time Einstein published his 1905 paper. These other modifications can be ignored in the context of optical ether-drift experiments, but they play a central role in the analysis of both the experiments of Trouton and Noble and the more famous experiments of Walter Kaufmann and others on the velocity-dependence of the electron mass.[38] The analysis of these experiments is thus of great importance for a history of special relativity conceived as a history of the discovery and interpretation of Lorentz invariance.

Appendix: The Trouton Experiment and $E = mc^2$

The Trouton-Noble experiment grew out of another experiment aimed at detecting the earth's motion through the ether with the help of a capacitor.[39] The original experiment, suggested by FitzGerald to his assistant Trouton,[40] beautifully illustrates relativity's most famous equation, $E = mc^2$.

Consider figure 6. If the power is switched on, an electromagnetic field is built up between the plates of the capacitor. If the system is at rest in the ether, the charges

---

motion. Given the definition of momentum based on the standard convention, this will only be true for systems with special symmetries and for *closed* systems, that is, for systems that do not exchange energy or momentum with their surroundings. For closed systems, it does not matter over which spatial slice the integration is done and the two definitions of momentum considered here coincide. A charged capacitor and its electromagnetic field form a closed system, but its components considered separately do not. The total momentum of the system will therefore be in the direction of motion under both definitions, but under the standard definition the momentum of the electromagnetic field and the momentum of the material part of the capacitor will have equal and opposite components that are not in the direction of motion. Under the alternative definition, there are no such components. This is why there will be delicately balanced turning couples in one case and not in the other. Whenever there is momentum that is not in the direction of motion there will be a turning couple that tries to align the momentum with the system's velocity. The upshot of the analysis outlined in this footnote is that these turning couples are simply artifacts of the standard convention for choosing spatial slices in the definition of momentum for spatially extended systems.

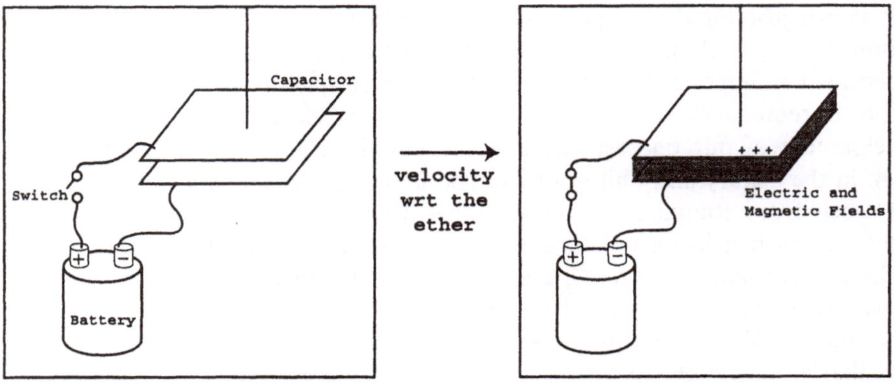

Fig. 6. The basic principle behind the Trouton experiment.

will produce only an electric field; if it is moving, the charges also will produce a magnetic field.

FitzGerald thought that the energy for this magnetic field would come from a decrease in the kinetic energy of the capacitor, which according to elementary Newtonian mechanics would result in a decrease in its velocity. A capacitor on a wire suspended from the ceiling of the laboratory therefore should experience a sudden jolt opposite to the direction of motion the moment the power is switched on. Trouton did not find such jolts.[41]

FitzGerald died before the final results were in. It was thus left to other theorists to interpret Trouton's findings. Joseph Larmor argued as follows.[42] Suppose the capacitor is held fixed in the laboratory. According to FitzGerald the energy for the magnetic field then would have to come from a tiny decrease in the kinetic energy of the earth as a whole. But this would violate the so-called center-of-mass theorem, according to which nothing happening inside an isolated system can change the state of motion of the center of mass of the system. One might object that the earth-capacitor system is not fully isolated because it interacts with the ether. However, since Larmor, like Lorentz, assumed the ether to be immobile, the center-of-mass theorem is also violated if we consider the ether-earth-capacitor system. Larmor's simple rebuttal of FitzGerald's suggestion thus seems perfectly adequate.

The problem is that it was unclear whether the center-of-mass theorem actually holds in the ether theories of Lorentz and Larmor. The theorem is closely related to Newton's third law, which says that action equals reaction. This law cannot hold in a theory in which the ether exerts forces on matter but matter never exerts forces on the ether. Poincaré saw this as a very serious problem for Lorentz's theory, especially because it would lead to violations of the center-of-mass theorem.[43] One of the examples Poincaré considered was that of a mirror recoiling upon reflecting light because of radiation pressure. Since light did not have mass or momentum in Lorentz's theory, this recoil effect violates both the center-of-mass theorem and another closely related law, conservation of momentum. Max Abraham suggested ascribing momentum to the electromagnetic field,[44] which at least restores momentum conservation in such experiments.

Electromagnetic momentum plays a central role in Lorentz's subsequent analysis of the Trouton experiment.[45] Lorentz reasoned as follows. When the moving capacitor is charged, electromagnetic momentum is created pointing in the direction of motion. For the total momentum to be conserved, this gain in electromagnetic momentum must be compensated by a loss of mechanical momentum of the capacitor. This means, according to elementary Newtonian mechanics, that the capacitor's velocity must decrease. In other words, Lorentz agreed with FitzGerald that if a moving capacitor is charged, it should experience a sudden jolt. Lorentz also showed, however, that Trouton's experiment had not been sensitive enough to detect this effect.

Notice the peculiar situation we have found. Larmor argued that if the effect predicted by FitzGerald does occur, the center-of-mass theorem is violated. Lorentz argued that if the effect does *not* occur, conservation of momentum is violated. It seems that momentum conservation and the center-of-mass theorem, two laws that are essentially equivalent in Newtonian mechanics, cannot both hold.

This is where the most famous equation of all of modern physics comes in. Using a thought experiment that is a variant on Poincaré's recoiling mirror example, Einstein showed that the equivalence of mass and energy expressed in $E = mc^2$ is a necessary and sufficient condition for the center-of-mass theorem to hold in addition to momentum conservation.[46] If the mass-energy equivalence is taken into account in Lorentz's analysis of the Trouton experiment, the contradiction between the conclusions of Lorentz and Larmor disappears.

When the moving capacitor is charged, energy is transferred from the battery to the capacitor. Energy is equivalent to mass, moving mass means momentum, so with the energy transfer there is also momentum transfer from the battery to the capacitor. The arrows in figure 7 represent the momentum of the battery and the capacitor before and after the power is switched on. When the power is

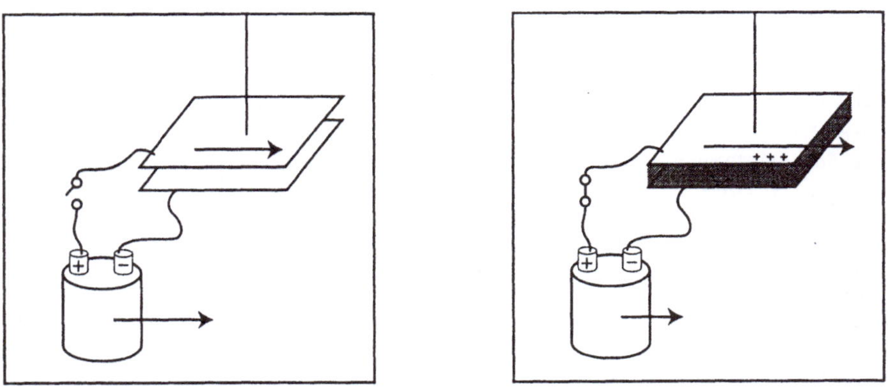

Fig. 7. Momentum transfer in the Trouton experiment.

switched on, the battery's momentum decreases and the capacitor's momentum increases by the same amount. The total momentum of the system is conserved.

In relativistic mechanics, as in Newtonian mechanics, momentum is equal to the product of mass and velocity. The difference is that in relativistic mechanics a change in momentum can be due to a change in energy, and hence mass, as well as to a change in velocity. The changes in momentum in the Trouton experiment are of the former variety. Hence, there will be no jolt upon charging the capacitor. There is no violation of the center-of-mass theorem.

Acknowledgments

I am grateful to Jon Dorling, Tony Duncan, Gordon Fleming, Gerd Graßhoff, Adolf Grünbaum, A. J. Kox, John Norton, Jürgen Renn, Merrilee Salmon, Wesley Salmon, and John Stachel for helpful discussions. I also want to thank Roger H. Stuewer for his extensive comments on an earlier version of this paper. Finally, I want to thank Aaron Oppenheimer and Nickdokht Torkzadeh for drawing figures 5–7.

References

1 See, for instance, Noel M. Swerdlow and Otto Neugebauer, *Mathematical Astronomy in Copernicus's De Revolutionibus*, 2 Vols. (New York: Springer, 1984).

2 Hendrik Antoon Lorentz, *Versuch einer Theorie der electrischen und optischen Erscheinungen in bewegten Körpern* (Leiden: Brill, 1895).

3 H. A. Lorentz, "Vereenvoudigde theorie der electrische en optische verschijnselen in lichamen die zich bewegen," *Verslagen van de gewone vergaderingen der wis- en natuurkundige afdeeling, Koninklijke Akademie van Wetenschappen te Amsterdam* **7** (1899), 507–522; "Electromagnetic Phenomena in Systems Moving with Any Velocity Less Than That of Light," *Proceedings of the Section of Sciences, Koninklijke Akademie van Wetenschappen te Amsterdam* **6** (1904), 809–831.

4 H. A. Lorentz, "De relatieve beweging van de aarde en den aether," *Verslagen van de gewone vergaderingen der wis- en natuurkundige afdeeling, Koninklijke Akademie van Wetenschappen te Amsterdam* **1** (1892), 74–79; *Versuch* (ref. 2).

5 Russell McCormmach, "H. A. Lorentz and the Electromagnetic View of Nature," *Isis* **61** (1970), 459–497.

6 Albert Einstein, "Zur Elektrodynamik bewegter Körper," *Annalen der Physik* **17** (1905), 891-921.

7 Albert Einstein, "Time, Space, and Gravitation," in *The Times* (London), November 28, 1919, pp. 13–14.

8 Henri Poincaré, "La théorie de Lorentz et le principe de réaction," *Archives Néerlandais des sciences exactes et naturelles* **5** (1900), 252–278.

9 H. Poincaré, "Sur la dynamique d'électron," *Rendiconti del Circolo Matematico di Palermo* **21** (1906), 129–175.

10 H. A. Lorentz, *The Theory of Electrons*, 2nd ed. (Leipzig: Teubner, 1915).

11 Hermann Minkowski, "Raum und Zeit," *Physikalische Zeitschrift* **10** (1909), 104–111.

12 Paul Ehrenfest, *Zur Krise der Lichtaether-hypothese* (Leiden: Eduard Ijdo and Berlin: Julius Springer, 1913).

13 See, for instance, Tony Hey and Patrick Walters, *Einstein's Mirror* (Cambridge: Cambridge University Press, 1997), pp. 39–43.

14 Gerald Holton, "Einstein, Michelson and the "Crucial" Experiment," *Isis* **60** (1969), 133–197.

15 John Stachel, "Einstein and Michelson: the Context of Discovery and the Context of Justification," *Astronomische Nachrichten* **303** I (1982), 47–53.

16 Robert Schulmann, *et al.*, ed., *The Collected Papers of Albert Einstein.* Vol. 8. *The Berlin Years. Correspondence: 1914–1918* (Princeton: Princeton University Press, 1998), Doc. 43.
17 Karl R. Popper, *The Logic of Scientific Discovery* (New York: Basic books and London: Hutchinson, 1959).
18 Adolf Grünbaum, "The Falsifiability of the Lorentz-FitzGerald Contraction Hypothesis," *British Journal for the Philosophy of Science* **37** (1959), 48–50.
19 Adolf Grünbaum, *Philosophical Problems of Space and Time*, 2nd ed. (Dordrecht: D. Reidel, 1973); "Ad Hoc Auxiliary Hypotheses and Falsificationism," *British Journal for the Philosophy of Science* **27** (1976), 329–362.
20 See, for example, Howard P. Robertson, "Postulate *versus* Observation in the Special Theory of Relativity," *Reviews of Modern Physics* **21** (1949), 378–382.
21 Elie Zahar, "Why Did Einstein's Programme Supersede Lorentz's?" *British Journal for the Philosophy of Science* **24** (1973), 95-123, 223-262; *Einstein's Revolution: a Study in Heuristic* (La Salle, IL: Open Court, 1989).
22 Imre Lakatos, "Falsification and the Methodology of Scientific Research Programs," in Imre Lakatos and Alan Musgrave, ed., *Criticism and the Growth of Knowledge* (Cambridge: Cambridge University Press, 1970), pp. 91–196.
23 Zahar, "Why Did Einstein's Programme Supersede Lorentz's?" (ref. 21), p. 101.
24 See also Arthur I. Miller, "On Lorentz's Methodology," *British Journal for the Philosophy of Science* **25** (1974), 29–45; Kenneth F. Schaffner, "Einstein versus Lorentz: Research Programmes and the Logic of Comparative Theory Evaluation," *ibid.*, pp. 45–78.
25 Lorentz, "De relatieve beweging" (ref. 4); *Versuch* (ref. 2).
26 Lakatos, "Methodology of Scientific Research Programs" (ref. 22), p. 164.
27 Zahar, "Why Did Einstein's Programme Supersede Lorentz's?" (ref. 21), p. 243.
28 *Ibid.*, pp. 246–247.
29 Lorentz, "Vereenvoudigde theorie" (ref. 3).
30 Zahar, "Why Did Einstein's Programme Supersede Lorentz's?" (ref. 21), pp. 249–251.
31 For similar criticism of Zahar's study, see R. M. Nugayev, "The History of Quantum Mechanics as a Decisive Argument Favoring Einstein over Lorentz," *Philosophy of Science* **52** (1985), 44–63. As the title of his paper suggests, Nugayev argues that to understand the rationality of the acceptance of special relativity in the physics community by the early 1910s one should take into account concurrent developments in quantum theory.
32 Grünbaum, "Ad Hoc Auxiliary Hypotheses" (ref. 19), p. 360.
33 Frederick T. Trouton and Henry R. Noble, "The Mechanical Forces Acting on a Charged Electric Condenser Moving through Space," *Philosophical Transactions of the Royal Society of London* **202** (1903), 165–181.
34 See, for instance, Howard C. Hayden, "High Sensitivity Trouton-Noble Experiment," *Review of Scientific Instruments* **65** (1994), 788–792.
35 Max Laue, "Zur Theorie des Versuches von Trouton und Noble," *Annalen der Physik* **38** (1912), 370–384.
36 Max Laue, "Zur Dynamik der Relativitätstheorie," *Annalen der Physik* **35** (1911), 524–542; Wolfgang Pauli, "Relativitätstheorie," in Arnold Sommerfeld, ed., *Encyclopädie der mathematischen Wissenschaften, mit Einschluss ihrer Anwendungen.* Vol. 5. *Physik*, Part. 2 (Leipzig: Teubner, 1921); translated as *Theory of Relativity* (London: Pergamon, 1958).
37 J. W. Butler, "On the Trouton-Noble Experiment," *American Journal of Physics* **36** (1968), 936–941; Michel Janssen, "A Comparison Between Lorentz's Ether Theory and Special Relativity in the Light of the Experiments of Trouton and Noble," Ph.D. Dissertation, University of Pittsburgh, 1995; Saul A. Teukolsky, "The Explanation of the Trouton-Noble Experiment Revisited," *American Journal of Physics* **64** (1996), 1104–1109. All three accounts were inspired by Fritz Rohrlich, "Self-energy and the Stability of the Classical Electron," *American Journal of Physics* **28** (1960), 639–643; *Classical Charged Particles: Foundations of their Theory* (Reading, Mass.: Addison-Wesley, 1965).
38 For a discussion of these experiments, see, for instance, Arthur I. Miller, *Albert Einstein's Special Theory of Relativity. Emergence (1905) and Early Interpretation (1905–1911)* (Reading, Mass.: Addison-Wesley, 1981).

39 For a more detailed version of the analysis given in this appendix, see Michel Janssen, "The Trouton Experiment, $E = mc^2$, and a Slice of Minkowski Space-Time," to be published in a *Festschrift* for John Stachel.

40 The historical context is discussed by Andrew Warwick, "The Sturdy Protestants of Science: Larmor, Trouton, and the Earth's Motion Through the Ether," in Jed Z. Buchwald, ed., *Scientific Practice. Theories and Stories of Doing Physics* (Chicago: University of Chicago Press, 1995), pp. 300–343.

41 Frederick T. Trouton, "The Results of an Electrical Experiment, Involving the Relative Motion of the Earth and Ether, Suggested by the Late Professor FitzGerald," *Transactions of the Royal Dublin Society* **7** (1902), 379–384.

42 Joseph Larmor, "Can Convection Through the Æther Be Detected Electrically?" in Joseph Larmor, ed., *The Scientific Writings of the Late George Francis FitzGerald* (Dublin: Hodges, Figgis, & Co. and London: Longmans, Green, & Co., 1902), pp. 566–569.

43 Poincaré, "La théorie de Lorentz" (ref. 8).

44 Max Abraham, "Prinzipien der Dynamik des Elektrons," *Annalen der Physik* **10** (1903), 105–179.

45 Lorentz, "Electromagnetic Phenomena" (ref. 3).

46 Albert Einstein, "Das Prinzip von der Erhaltung der Schwerpunktsbewegung und die Trägheit der Energie," *Annalen der Physik* **20** (1906), 627–633.

Program in History of Science and Technology
Tate Laboratory of Physics
University of Minnesota
116 Church Street SE
Minneapolis, MN 55455 USA
e-mail: janss011@tc.umn.edu

WISDOM IS -

Wisdom is
the booby prize
given when you've been
unwise.

Piet Hein

Arthur I. Miller

THE SPECIAL RELATIVITY THEORY: EINSTEIN'S RESPONSE TO THE PHYSICS OF 1905

IMAGINE THAT YOU are on the editorial board of a prestigious physics journal and that you receive a paper that is unorthodox in style and format. Its title has little to do with most of its content; it has no citations to current literature; a significant portion of its first half seems to be philosophical banter on the nature of certain basic physical concepts taken for granted by everyone; the only experiment explicitly discussed could be explained adequately using current physical theory and is not considered to be of fundamental importance. Yet, with a minimum of mathematics, the little-known author deduces exactly a result that has heretofore required several drastic approximations. Furthermore, you are struck by certain of the author's general principles, and you feel that they promise additional simplifications. So you decide to publish the paper. This could well have been the frame of mind of the most eminent theoretical physicist on the Curatorium of the *Annalen der Physik*, Max Planck, when he received from the editor's office Albert Einstein's 1905 paper "On the Electrodynamics of Moving Bodies"—the relativity paper.[1]

The kind of title Einstein had given his paper customarily signaled a discussion of the properties of moving bulk matter, either magnetic or dielectric. Einstein analyzed neither of these topics. In fact, the paper's first quarter contains a philosophical analysis of the notions of time and length. The paper's second half dispatches quickly certain problems of such fundamental importance that they generally rate separate papers—for example, the characteristics of radiation reflected from a moving mirror—and he concludes with certain results from the dynamics of electrons that generally appear at the beginning of papers in which electrons are discussed. The only experiment developed in detail is at the paper's beginning and concerns the generation of current in a closed circuit as a result of the circuit's motion relative to a magnet, that is, electromagnetic induction.

The phenomenon of electromagnetic induction had ushered the Western world into the age of technology because it is fundamental to electrical dynamos. Everyone knew dynamos worked, but there remained fundamental problems concerning their operation. This essay discusses the connection that Einstein realized in 1905 between problems concerning huge electrical dynamos, radiation, moving electrical bulk media, the dynamics of electrons, and the nature of space and time. In order to set the stage for Einstein's bold approach to the physics of 1905, let us review the treatment of these topics by scientists and philosophers of whom Einstein has acknowledged he was aware before 1905.[2]

THE NATURE OF SPACE AND TIME

In his *Science of Mechanics* (1883) the philosopher-scientist Ernst Mach leveled a devastating critique at the Newtonian notions of absolute space and time. According to Newton, absolute space was the ultimate receptacle in which all phenomena occurred, and absolute time flowed independent of the motion of clocks. Mach considered these notions to be "metaphysical obscurities" because they were unavailable to our sense perceptions. Consequently, Mach disagreed with Immanuel Kant, who by 1781 had elevated Newton's notions of absolute space and time to knowledge that we possessed before all else, that is, a priori intuitions. According to Kant, these intuitions serve as basic organizing principles that enable our minds to construct knowledge from the potpourri of sense perceptions. Thus, for example, we are driven irresistibly toward a three-dimensional Euclidean geometry and the law of causality, and then to such higher-order organizing principles as Newton's physics. Although the discovery of non-Euclidean geometries in 1827 had dealt the Kantian view a serious blow, Kant's emphasis on the role of a priori organizing principles was nevertheless considered important to an understanding of how exact laws of nature are possible. A priori organizing principles played an important role in the neo-Kantian frameworks of such influential philosopher-scientists as Hermann von Helmholtz, Heinrich Hertz, and Henri Poincaré, whose writings impressed Einstein no less than did Mach's "incorruptible skepticism."[3]

Although Mach and Poincaré probed the relation between time and sense perceptions, in their work time remained absolute because there was no reason for it to depend on motion. But Mach and Poincaré insisted on replacing motion relative to Newton's absolute space with motion relative to the distant stars or, even better, motion relative to the substance that electrodynamicists assumed to fill Newton's cosmic receptacle—the ether.[4] This brings us to electromagnetism.

ELECTROMAGNETIC THEORY

Newton's mechanics of 1687 had unified terrestrial and extraterrestrial phenomena. The next great synthesis occurred not quite two hundred years later, when James Clerk Maxwell unified electromagnetism and optics. Whereas in the Newtonian mechanics disturbances propagated instantaneously through empty space, in Maxwell's theory disturbances propagated at a large but finite velocity through an ether, like ripples in a pond. In 1892 there appeared the result of over two decades of elaborations and purifications of Maxwell's theory—the electromagnetic theory of that master of theoretical physics, Hendrik Antoon Lorentz.

Lorentz assumed that the sources of the electromagnetic fields were as yet undiscovered electrons, which moved about in an all-pervasive, absolutely resting ether.[5] The five fundamental equations of Lorentz's theory are:[6]

Maxwell-Lorentz Equations:

$$\vec{\nabla} \times \vec{E} = -\frac{1}{c}\frac{\partial \vec{B}}{\partial t} \qquad (1)$$

$$\vec{\nabla} \times \vec{B} = \frac{1}{c}\frac{\partial \vec{E}}{\partial t} + \frac{4\pi}{c}\rho\vec{V} \qquad (2)$$

$$\vec{\nabla} \cdot \vec{E} = 4\pi\rho \qquad (3)$$

$$\vec{\nabla} \cdot \vec{B} = 0 \qquad (4)$$

Lorentz Force Equation:

$$\vec{F} = \rho\vec{E} + \rho\frac{\vec{V}}{c} \times \vec{B} \qquad (5)$$

where $\vec{E}$ and $\vec{B}$ are the electric and magnetic fields, respectively, and ρ is the electron's volume density of charge. Since Lorentz's fundamental equations are written relative to a reference system at rest in the ether, which we shall call S, then c is the velocity of light measured in S, and $\vec{v}$ is the electron's velocity relative to S. The Maxwell-Lorentz equations possess the property expected of a wave theory of light, namely, that relative to S the velocity of light is independent of the source's motion and is always c. But this may not necessarily be the result of measuring the velocity of light in a reference system moving with a uniform linear velocity relative to the ether, that is, in an inertial reference system. Therefore, the reference systems in the ether are preferred reference systems. To be sure, despite much effort, experiments had not revealed that the earth's motion through the ether had any effect on optical or electromagnetic phenomena.

Concerning the velocity of light, Newtonian mechanics predicted that the velocity of light emitted from a moving source should differ from the velocity of the light emitted from a source at rest by the amount of the source's velocity; consequently, the velocity of light c' from a source moving with velocity v is given by Newton's law for the addition of velocities,

$$\vec{c}\,' = \vec{c} + \vec{v}. \tag{6}$$

On the other hand, according to the wave theory of light, the quantity c' measured by an observer at rest in the ether is

$$\vec{c}\,' = \vec{c}, \tag{7}$$

and Lorentz's equations agreed with this requirement. But the effect of the ether on the measuring apparatus was expected to yield a result in agreement with (6), where c' is the velocity of the light relative to the earth and v is the ether's velocity relative to the earth. However, experiments accurate to second-order in the ratio v/c, where v is the velocity of a body that is moving relative to the ether and c is the velocity of light, led to (7). To this order of accuracy, optical and electromagnetic phenomena occurred on the moving earth as if the earth were at rest in the ether. Therefore, to second-order accuracy in v/c, Newtonian mechanics and electromagnetism are inconsistent with optical phenomena occurring in inertial reference systems.

In an 1895 monograph titled *Treatise on a Theory of Electrical and Optical Phenomena in Moving Bodies*,[7] Lorentz responded fully to the failure of the first-order experiments to detect any effects of the earth's motion on optical and electromagnetic phenomena; these experiments were called ether-drift experiments. For regions of the ether that are free of matter, or within neutral matter that is neither magnetic nor dielectric, the Lorentz equations in the ether-fixed reference system are the set of equations (S) (see Fig. 1). Applying the modified space and time transformations to (S), Lorentz obtained their analogues in the inertial reference system S_r.

We can appreciate Lorentz's achievement at a glance because to first-order accuracy in the quantity v/c, the Maxwell-Lorentz equations have the same form in the inertial system S_r as in the ether-fixed system S, and thus the same physical laws pertain to both these reference systems; in other words, to this order of accuracy neither optical nor electromagnetic experiments could reveal the motion of the system S_r. Lorentz called this stunning and desirable result the "theorem of corresponding states." It rested on the hypothesis of the mathematical "local time coordinate" t_L; the real or physical time was still Newton's absolute time. Hence, to order v/c, the velocity of light in S_r was the same as in S, that is, $c' = c$. To this order of accuracy, then, Lorentz's theorem of corresponding states removed the inconsistency between Newton's prediction and that of electromagnetic theory in favor of electromagnetic theory.

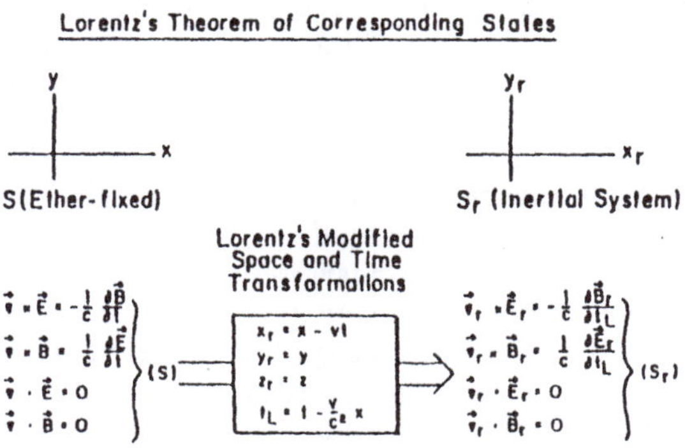

FIG. 1
Lorentz's modified space and time transformations contain the "local time coordinate" t_L and Lorentz referred to the electromagnetic field quantities $\vec{E}_r = \vec{E} + \vec{v}/c \times \vec{B}$ and $\vec{B}_r = \vec{B} - \vec{v}/c \times \vec{E}$ as "new" vectors.

But Lorentz had not yet explained the only reliable experiment accurate to second-order accuracy in v/c, namely, the 1887 experiment of Albert A. Michelson and Edward Williams Morley, in which light had been found to take the same time to race back and forth along each of two orthogonal rods of equal length that were at rest on the moving earth. He discussed this experiment in the final chapter of the 1895 treatise, which contained two other experiments that, as the chapter's title indicated, could "not be explained without further ado."[8] In order to explain the Michelson-Morley experiment, Lorentz indulged in a physics of desperation. From Newton's law for the addition of velocities, which the theorem of corresponding states was supposed to have obviated, he proposed the hypothesis that the dimensions of the rod in the direction of the earth's motion contracted by an amount $\sqrt{(1 - v^2/c^2)}$. In short, Lorentz's contraction hypothesis was admittedly ad hoc. This blemish on Lorentz's theory was emphasized in the philosophic-scientific criticism of that titan of international science, Henri Poincaré. Nevertheless, Poincaré was impressed with Lorentz's theorem of corresponding states because he eschewed absolute motion.[9]

So successful had been Newton's physics that many scientists had attempted to reduce all of physical theory to it; that is, they pursued a mechanical world-picture. For example, they attempted to simulate the contiguous actions of the ether with increasingly complex mechanical models. But these attempts paled before the successes of Lorentz's theory. Thus, in 1900 Wilhelm Wien suggested the "possibility of an electromagnetic foundation for mechanics," that is, pursuance of an electromagnetic world-picture based on Lorentz's electromagnetic theory.[10] A far-reaching implication of this program was that the electron's mass originated in its own electromagnetic field and should therefore be a velocity-dependent quantity. From studying the behavior of fast electrons that had been injected transversely into parallel electric and magnetic fields, Walter Kaufmann gave data for a dependence of the electron's mass on its velocity that increased without limit as the electron's velocity approached that of light.[11] Kaufmann's colleague at Göttingen, Max Abraham, developed the first field-theoretical description of an elementary particle.[12] Depending on whether his rigid-sphere electron experienced a force transverse or parallel to its motion, and with certain severe restrictions placed on the electron's acceleration, Abraham predicted that it had transverse (m_T) and longitudinal (m_L) masses:

$$m_T = \frac{m_o^e}{2\beta^3}\left[(1+\beta^2)\log\left(\frac{1+\beta}{1-\beta}\right) - 2\beta\right] \tag{8}$$

$$m_L = \frac{m_o^e}{\beta^3}\left[\frac{2\beta}{1-\beta^2} - \log\left(\frac{1+\beta}{1-\beta}\right)\right] \tag{9}$$

where $m_o^e = e^2/2Rc^2$ is the electron's electrostatic (that is, rest) mass and $\beta = v/c$. The transverse mass of Abraham's theory agreed with Kaufmann's data, and the goal of an electromagnetic world-picture appeared to be within reach. However, Abraham's theory offered no explanation for the Michelson-Morley experiment, and by 1904 it was in violent disagreement with the new optical experiments of Lord Rayleigh and D. B. Brace, which were accurate to second-order in v/c.[13]

Prompted by the new second-order data and by Kaufmann's measurements, as well as by Poincaré's criticisms, Lorentz proposed his own theory of the electron in which the contraction hypothesis was deemed no longer to be ad hoc, because it became one of several hypotheses that could explain more than one experiment.[14] Lorentz's electron can be likened to a balloon smeared with a uniform distribution of charge. While at rest, Lorentz's electron is assumed to be a sphere; but moving, it undergoes a Lorentz contraction, and its mass becomes a two-component quantity:

$$m_T = \frac{4}{3}\frac{m_o^e}{\sqrt{1-\beta^2}} \tag{10}$$

$$m_L = \frac{4}{3}\frac{m_o^e}{(1-\beta^2)^{3/2}} \tag{11}$$

where $m_o^e = e^2/2Rc^2$ is the electron's electrostatic (that is, rest) mass and $\beta = v/c$.

Lorentz's m_T agreed with Kaufmann's data as well as did Abraham's. But Abraham immediately leveled a severe fundamental criticism at Lorentz's theory: Lorentz's deformable electron was unstable because it could explode under the enormous repulsive forces among its constituent parts.[15] From newly discovered Lorentz-Poincaré correspondence we know that Poincaré had recognized this problem independently and then cracked it with his unmatched arsenal of mathematics. Poincaré's resulting papers were the penultimate effort toward an electromagnetic world-picture based on Lorentz's electromagnetic theory.[16] They included such advanced notions of mathematics as group-theoretical methods and four-dimensional spaces. Using a term familiar from fundamental studies in geometry, Poincaré renamed Lorentz's 1904 theorem of corresponding states that embraced all extant data—and, it was *hoped*, future ones as well—the "principle of relativity."[17] Einstein, we know, had not encountered Poincaré's 1905 version of Lorentz's theory of the electron when he wrote the relativity paper.

To summarize, by 1905 physicists believed that electromagnetic theory was proceeding in the correct direction. Many of them felt sure that with a little more tinkering, Lorentz's theory of the electron could serve as the cornerstone for a unified field-theoretical view of nature. Lorentz's was a dynamical theory that *explained* such effects as the presumed contraction of length, the observed variation of mass with velocity, and the fact that the measured velocity of light always turned out to be the same—all explained as resulting from the interaction of electrons with the ether. The stage was set for a great new era in science to emerge from what everyone considered to be the cutting edge of scientific research. But, as we shall see, this turned out not to be the case. We move next to an area of science and engineering whose basic problems were deemed unimportant for progress in basic physical theory: the area of electromagnetic induction. In German-speaking countries problems in this area combined technology and basic research. They received a particularly interesting treatment because, as the intellectual historian J. T. Merz has written, the "German man of science was a philosopher."[18]

ELECTRICAL DYNAMOS

In 1831 Michael Faraday discovered that relative motion between a wire loop and a magnet produces a current in the wire. Faraday interpreted this result as follows: the magnet affects the wire loop through its lines of force, which emanate from the magnet's north pole and enter through its south pole; consequently, relative motion between the loop and the magnet results in the loop's cutting the lines of force. Faraday's law states that the rate at which the lines of force are cut determines the strength of the current induced; furthermore, the direction and magnitude of the induced current depend on only the relative velocity between the loop and magnet.

But Faraday's interpretation of electromagnetic induction differed when circuit and magnet were rotating relative to each other. An apparatus of the sort in Fig. 2 was important to Faraday because by 1851 he had convinced himself that lines of force participated in the magnet's linear motion but not in its rotation. In Fig. 2 the wire loop makes sliding contact with the rotating magnet's periphery and touches one of the magnet's poles; hence, this sort of electromagnetic dynamo became known as a unipolar dynamo. Faraday's interpretation of unipolar induction is: if the loop rotates counterclockwise, a current appears in it owing to its cutting the lines of force. If the loop remains at rest, and the magnet rotates clockwise, a current of the same magnitude and direction appears in the loop; the current was thought to originate in the magnet owing to the magnet's rotating through its lines of force. When magnet and loop turn together in the same direction, there is no net current in the loop because the loop's current is canceled by the magnet's internal current. Thus, two different explanations were required for the current induced in the loop, depending on whether the loop or the magnet rotated.[19] Clearly, the experimental data were more easily understood in terms of relative motion between the wire loop and the magnet with its co-moving lines of force.

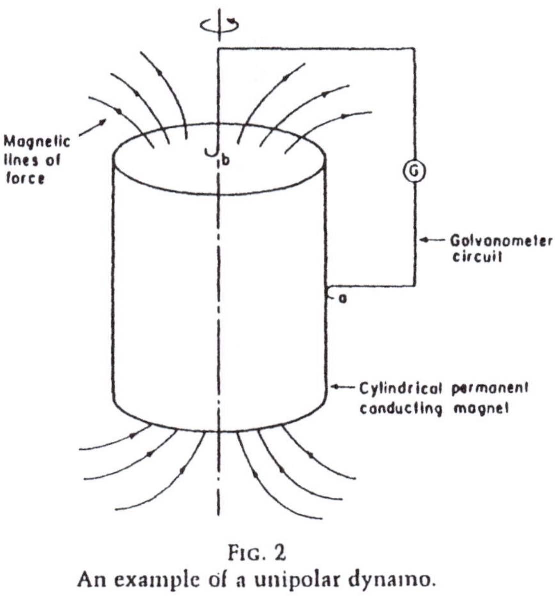

FIG. 2
An example of a unipolar dynamo.

Representing the magnetic field by lines of force found fertile ground in the German-speaking countries. Although by the end of the nineteenth century most British scientists considered lines of force to be useful chiefly for pedagogy, the Kantian philosophical position of scientists and engineers in German-speaking countries led them to consider lines of force as a fundamental *Anschauung*. In this context *Anschauung* refers to the intuition through pictures formed in the mind's eye from previous visualizations of physical processes in the world of perceptions; *Anschauung* is superior to viewing merely with senses.[20] In short, lines of force were seen everywhere. There ensued a controversy in the German-speaking scientific-engineering community on the merits of the *Anschauungen* of Faraday versus the rotating-line view, and experiments were offered to distinguish between them. One engineer emphasized that this controversy was "not an academic moot point" for two chief reasons. First, there was intense research and development toward unipolar direct-current and alternating-current generators owing to the relative simplicity of these machines vis-à-vis multipolar machines. Second, calculating the effect of a rotating magnet on stationary armature coils required knowing whether the magnetic field lines remained stationary; this problem appeared also in multipolar machines with stationary field magnets.[21]

On the theoretical side, the problem of how current arises in unipolar induction led some physicists to question the universal validity of Faraday's law of induction. Faraday's law was taken to be valid for closed wire circuits. The problem was how Faraday's law could explain the current that arose in the open wire in Fig. 2 as a result of slipping between the magnet's surface and the wire. This was a complicated boundary-value problem because it concerned assumptions on how electromagnetic quantities varied in the transition layer (that is, the surface of slip) between the spinning magnet and the ether. The widely read texts of August Föppl (1894) and the 1904 edition rewritten by Max Abraham emphasized the relation between the laws of mechanics and electromagnetism in the explanation of electromagnetic induction: the laws of mechanics are used to discuss the motion of circuit and magnet; the current in the loop depends on only the relative velocity between loop and magnet, and this result agrees with the principle of relative motion from mechanics.[22] This principle asserts that physical phenomena occur as a result of the relative motion of material bodies and that the laws of mechanics are the same for all inertial systems. Abraham stressed that experiments supported the principle of relative motion for Lorentz's theory only to first-order accuracy in v/c. Whereas Föppl's use of complicated notions of surfaces of slip allowed him to express only faith in the validity of Faraday's law, Abraham avoided any problems of unipolar induction by adhering strictly to Hertz's electromagnetic theory, in which the principle of relative motion held exactly. Then he calculated the current for the one case of the loop turning where Faraday's law could be applied with its line-cutting interpretation; the inverse case followed by replacing v with $-v$ in Faraday's law.

In Abraham's view, only the induced current, and not the lines of force, was measurable; that is, the equations of electromagnetic theory dictated what was to be observed, and lines of force were merely auxiliary quantities. In 1904 Lorentz enlarged his electromagnetic theory to describe moving bulk magnetic and dielectric matter, and his results agreed with empirical data better than did Hertz's. Lorentz's extension required his postulating three different sorts of electrons: conduction electrons responsible for electric current; polarization electrons that produce dielectric properties; and magnetization electrons to explain magnetic properties of matter. Using the local time coordinate, Lorentz showed that the effect of a moving magnet on a stationary current loop—that is, the electric field due to the moving magnet—had its origin in the moving magnet developing dielectric properties. But this explanation for electromagnetic induction involved a complicated intermingling of two different kinds of electrons in a substance that was assumed to possess only conduction and magnetization electrons. Lorentz himself found this explanation puzzling.[23]

Besides appending the hypotheses of additional electrons to an already overburdened superstructure, Lorentz by the end of 1904 was still unable to explain the stability of his electron. Then there was Max Planck's 1900 explanation for the radiation from hot substances, which brings us to the final ingredient necessary for describing Einstein's unique view of the physics of 1905—problems concerning radiation.

RADIATION

Since the late nineteenth century, the light emitted from a cavity within a hot substance had fascinated scientists because its characteristics are independent of the substance's constitution.

To his horror, Planck found that the successful empirical formula he offered to describe cavity radiation required that energy be exchanged in only discrete amounts between the cavity radiation and the hot substance's constituent electrons.[24] Yet, according to electromagnetic theory, processes involving radiation should be continuous. Furthermore, if only certain discontinuous exchanges of energy were allowed, then the constituent electrons' states of motion were similarly restricted. Consequently, Planck's formula violated both mechanics and electromagnetism. Einstein recalled that in 1904 this situation was a "second fundamental crisis" (in addition to the instability of Lorentz's electron).[25] He was alone in this assessment.

EINSTEIN'S VIEW OF PHYSICAL THEORY

Einstein's opinion of Planck's theory of cavity radiation received confirmation from the application of his 1904 results on the behavior of atoms in gases—that is, statistical mechanics—to Planck's radiation formula. He found that light exhibited particulate properties in regions of space whose size was that of the electron's volume.[26] Thus, current electromagnetic theory was insufficient for discussing the nature of the electron. In addition, his research on particles in solution (Brownian motion) convinced him of the insufficiency of thermodynamics and mechanics in microscopic volumes. These two results led Einstein to conclude that scientists of 1905 were "out of [their] depth":[27] neither the mechanical nor the electromagnetic world-picture could succeed. Einstein's first two papers in the now famous volume 17 of *Annalen der Physik* explored the consequences of particles of light, that is, light quanta[28] and Brownian motion.[29] In the third paper of that 1905 trilogy—the relativity paper—Einstein again dealt with the characteristics of light.

Einstein's studies of the great philosopher-scientists von Helmholtz, Hertz, and Poincaré had shown him the power of a fundamental analysis within a neo-Kantian framework. But when it came to what to analyze and where to begin anew, he was in a quandary. Theories based on assumptions concerning the constitution of matter were inadequate for a fundamental analysis. At this point, Einstein was aided by his predilection for visual thinking, which was reinforced when he encountered the writings of Ludwig Boltzmann and von Helmholtz, in particular, at the

Eidgenössische Technische Hochschule in Zurich. Before Einstein, however, no one had combined visual thinking so effectively with the thought-experiment, that is, an experiment capable of being performed in the mind.

In 1895, after ten years of obstinate pondering, Einstein had conceived of a thought-experiment that finally revealed the "germ of the special relativity theory." The essence of this experiment is: (1) Current physics asserts that an observer who is moving alongside a light wave whose source is in the ether should be able to discern the effects of his motion by, for example, measuring the velocity of light. (2) But to Einstein it was "intuitively clear" that the laws of optics could not depend on the state of the observer's motion. Statements (1) and (2) are mutually contradictory, and to Einstein this thought-experiment contained a paradox.[30]

In their own ways, Lorentz and Poincaré were also attempting to resolve this paradox—by one degree of accuracy at a time, that is—by proposing, as we saw, hypotheses such as the local time and the ad hoc contraction of moving bodies. But Einstein's 1904 results on the nature of light had convinced him that theories of matter could lead neither to exact explanations for ether-drift experiments nor to a consistent world-picture.

I turn next to the basic scientific-philosophic considerations that led Einstein to discover that the key to the paradox lay, as he said later, in the "axiom of the absolute character of time, viz., of simultaneity, [which] unrecognizedly was anchored in the unconscious."[31] My research, based on Einstein's writings, has resulted in the following reconstruction of Einstein's thinking toward this momentous discovery.[32] Of all the ether-drift experiments

Galilean Transformations	*Lorentz's Modified Space and Time Transformations*
$x_r = x - vt$	$x_r = x - vt$
$y_r = y$	$y_r = y$
$z_r = z$	$z_r = z$
$t_r = t$	$t_L = t - \frac{v}{c^2}x$
(a)	(b)

FIG. 3

(a) The coordinates (x_r, y_r, z_r, t_r) and (x, y, z, t) refer to the two *inertial* reference systems S_r and S.

(b) The coordinates (x_r, y_r, z_r, $t_r = t$) and (x, y, z, t) refer to the inertial reference system S_r and to the ether-fixed reference system S; and t_L is the mathematical "local time coordinate."

that had been performed, those accurate to first order in v/c were explainable by Lorentz's theorem of corresponding states, which was based on the set of modified space and time transformations that included the mathematical local time coordinate (see Fig. 3b). Although the Maxwell-Lorentz equations remained unchanged under the modified space and time transformations of Fig. 3b, the same was not true of the equations of mechanics that were transformed between inertial reference systems according to the Galilean transformations (Fig. 3a). Thus, according to Lorentz's modified transformations, the laws of mechanics were not the same in every inertial reference system. But this result violated Newton's exact principle of relative motion, which states that no mechanical experiment could reveal an inertial system's motion. The mathematical statement of Newton's principle of relative motion lies in the Galilean transformations of Fig. 3a, where for mechanics the reference systems S_r and S are *both* inertial reference systems. Consequently, the transformation rules for the laws of mechanics and of electromagnetism depended on two different notions of time—one physical and the other mathematical—contrary to Lorentz's goal of unification. Thus, whereas most scientists in 1905 considered the tension between mechanics and electromagnetism to be rooted in the inability of mechanics to explain the measured velocity of light, Einstein delved deeper and found that current physics rendered mechanics and electromagnetism *incompatible*.

From the texts of Föppl and Abraham, Einstein could have learned well the intimate connection between mechanics and electromagnetism for interpreting electromagnetic induction. Faraday's law was basic also to Lorentz's electromagnetic theory, where the local time made it possible to calculate the moving magnet's effect on either a resting conductor or the open circuit in unipolar induction. But, as we recall, this calculation rested on certain special assumptions on the constitution of matter. Moreover, electromagnetic theory explained electromagnetic induction in two different ways, depending on whether the conductor or the magnet was moving, even though the physically measurable effect was a function of their relative velocity alone: when the conductor moved relative to the magnet, a current flowed owing to a force on the conductor's electrons; in the inverse case, a current flowed owing to an electric field at the conductor's site. For Einstein, two explanations for an effect that depended on only the relative velocity between magnet and conductor was more than a shortcoming of Lorentz's theory; it was, as he wrote in the relativity paper, an asymmetry that was "not inherent in the phenomena."[33] Einstein found the theoretical situation in electromagnetic induction so "unbearable," as he recalled in 1919,[34] that he focused on the necessity for an equivalence of viewpoints between observers on the wire loop and on the magnet, rather than on the source of the moving magnet's effect on the wire. He turned the fashionable programmatic research efforts of physics sideways by enlarging Newton's principle of relative motion to treat mechanics and electromagnetism on an equal footing, instead of attempting to reduce one to the other. In the relativity paper, he referred to this widened version of Newton's principle of relative motion as the "principle of relativity." Since, as we saw, the Galilean transformations could not explain the velocity of light, the modified transformations of Lorentz, with their apparently ubiquitous local time coordinate, would have to play a role in relating phenomena between reference systems. From Lorentz's modified transformations Einstein could have deduced a new result for the addition of velocities:

$$w_r = \frac{w - v}{1 - vw/c^2} \qquad (12)$$

where w_r is the velocity of a moving point relative to S_r and w is its velocity relative to S. For the case of $w = c$, then $w_r = c' = c$, instead of Newton's addition law of velocities, $c' = c - v$. Thus Einstein could have realized that, to first order in v/c, the addition law for velocities from Lorentz's modified transformations produced a result that agreed with the intuition of his thought-experimenter. Since Lorentz's modified transformation differed from the Galilean transformation only in the local time coordinate, Einstein asked himself whether the local time might be the physical "time"? But this step required asserting that the times in inertial reference systems differed because the local time coordinate depends on their relative velocity. Yet the absoluteness of time had always been accepted. Furthermore, the thought-experimenter's intuition demanded an examination of the mathematical relation between Newton's principle of relative motion and Lorentz's theorem of corresponding states. After all, the spatial coordinates of Lorentz's modified transformation equations of 1895 were mathematically the same as the ones in the Galilean transformations, and the local time coordinate had been invented for use in electromagnetic theory. Einstein's imposing a Newtonian unity upon Lorentz's modified transformation equations meant also his asserting the equivalence of the reference systems S and S_r. This was a big step, for it meant rejecting Lorentz's ether, and with it the dynamical interpretations of an enormously successful and, for the most part, satisfying theory. To his surprise, Einstein had found that the notion of time was both the central point and the Achilles' heel of the electrodynamics of moving bodies.

For aid in analyzing the nature of time, Einstein benefited from the critical reasoning "especially . . . of David Hume's and Ernst Mach's philosophical writings." Their analyses of sense perceptions offered strong evidence that exact laws of nature could not be induced from empirical data. And their analyses of the limits imposed by sense perceptions on notions of causality and of time enabled Einstein to realize that the high value of the velocity of light, compared with the other velocities we encounter daily, had prevented our appreciating that "the absolute character of time, viz., of simultaneity, unrecognizedly was anchored in the unconscious."[35] Einstein's fundamental analysis of physical theory went far beyond science as it is normally conceived: from an analysis of electromagnetic induction into an analysis of sensations, and then into an analysis of thinking itself. He concluded that the customary sensation-based notions of time and simultaneity resulted in a physics burdened with asymmetries, unobservable quantities, and ad hoc hypotheses. Thus prevented from lapsing into a dogmatic slumber, Einstein took recourse in the neo-Kantian view that was predicated on the usefulness of organizing principles such as the second law of thermodynamics. He enlarged Newton's principle of relativity to include Lorentz's theory, and then he raised to axioms this principle and the basic principle of every wave theory of light: in the ether-fixed system S the velocity of light is independent of the source's motion and is always c.

Einstein's two principles do not attempt to explain anything—for example, why the measured velocity of light always turns out to be c. Whereas Lorentz's and Poincaré's principle of relativity was based on a theory of matter that was heavily dependent on experimental data and was contingent on negative results of future ether-drift experiments, Einstein's was independent of assumptions concerning electrons and was an axiom. As Einstein wrote in 1907,[36] the principle of relativity was the basis of a theory that specified the form that laws of physics should assume in order to be used again to investigate the constitution of matter;

it was a theory of principle. In short, Einstein moved boldly counter to the prevailing currents of theoretical physics by resolving problems in a Gordian manner, that is, by formulating a view of physics in which certain problems do not occur, a view in which the 1895 paradox becomes a mere fiction. Consequently, Einstein's relativity paper looked insignificant alongside the papers of Abraham, Lorentz, and Poincaré, which used the most *au courant* methods of mathematical physics in order to derive what Einstein took to be axiomatic.

Straightway in the relativity paper, Einstein emphasized that the basic problems confronting physical theory concerned not the constitution of matter but understanding the equivalence of viewpoints between moving observers. For this purpose, he began with the simplest thought-experiment illustrating the problems of electromagnetic induction: a magnet and conducting loop in relative inertial motion. No mathematics was necessary to demonstrate that Maxwell's electrodynamics led observers on the wire loop and the magnet to different interpretations for the physically measurable effect, the current induced in the conductor. Thus, Faraday's law was universally valid, though it had been misinterpreted.

In the second paragraph of the relativity paper Einstein mentioned a facet of the results of ether-drift experiments accurate to the first order in *v/c* (which he left unnamed) that may have been glimpsed only by Abraham,[37] but within the context of the electromagnetic world-picture: namely, the laws of electrodynamics and mechanics were valid in inertial reference systems to first-order accuracy in *v/c*. Einstein's masterstroke was to link the experiment of magnet and conductor, which he had just explained, to the ether-drift experiments. He reasoned that electromagnetic induction depends on the laws of mechanics and of electromagnetism, which cover optics also; he then "conjectured" that Newton's principle of relative motion covers these three disciplines to order *v/c*. He boldly continued: "We will raise this conjecture (whose intent will from now on be referred to as the 'Principle of Relativity') to the status of a postulate."[38] Employing the principle of relativity and the principle governing the velocity of light, Einstein went on to develop a view of physics so powerful that the only mention he made of problems involving unipolar induction was to dismiss them as "meaningless."[39] In the relativity paper the Lorentz contraction and most of the other hypotheses of Lorentz's theory surfaced only as "secondary consequences,"[40] and it was unnecessary for Einstein to review every extant ether-drift experiment, because in his view their results were *ab initio* a foregone conclusion.

With further simple thought-experiments the first part of the relativity paper demonstrated how imprecise were current notions of time and length, and the relative nature of these notions emerged. Most readers of 1905, however, emphasized the end of the paper, where Einstein deduced exactly a formula for the electron's transverse mass that was almost Lorentz's, and so they interpreted Einstein's work as a valuable generalization of Lorentz's theory of the electron. Soon Kaufmann himself pointed out that these two equations should be the same, and he was one of the first to use the name "Lorentz-Einstein theory."[41] But what did Einstein have to say in the relativity paper about the data in Kaufmann's 1902 and 1903 papers? Nothing, because it is reasonable to conjecture that Einstein knew well that they disagreed with his prediction for the electron's transverse mass.[42] Einstein considered this prediction to be of secondary importance anyway. In a 1907 paper reviewing the status of the principle of relativity,[43] Einstein discussed Kaufmann's data from late 1905, which, as Kaufmann had written in no uncertain terms, were "*not consistent with the Lorentz-Einstein fundamental assumption.*"[44] Einstein acutely emphasized that the "systematic deviation" between the Lorentz-Einstein predictions and the data could indicate a hitherto "unnoticed source of error"; consequently, Einstein called for further experiments. Then he dismissed Kaufmann's data, because in his "opinion" they supported theories that did not "embrace a greater complex of phenomena," that is, the electron theories of Abraham and A. H. Bucherer, which could not explain optical experiments accurate to second order in v/c.[45]

A letter among Poincaré's correspondence, which I discovered in Paris, permits us to compare Einstein's bold defense of the principle of relativity in the face of disconfirming data with Lorentz's own response to the same data (see Fig. 4). On 8 March 1906, Lorentz wrote to Poincaré: "Unfortunately my hypothesis of the flattening of electrons is in contradiction with Kaufmann's new results, and I must abandon it. I am, therefore, at the end of my Latin. It seems to me impossible to establish a theory that demands the complete absence of an influence of translation on the phenomena of electricity and optics." What a remarkable confession, and what a clear-cut case of falsification. After all

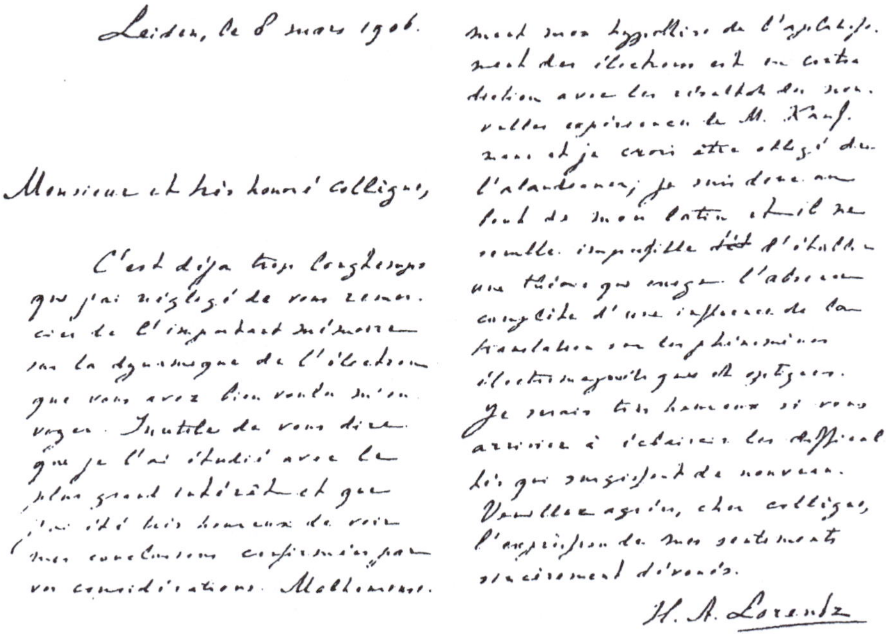

Leiden, le 8 mars 1906.

Monsieur et très honoré collègue,

C'est déjà trop longtemps que j'ai négligé de vous remercier de l'important mémoire sur la dynamique de l'électron que vous avez bien voulu m'envoyer. Inutile de vous dire que je l'ai étudié avec le plus grand intérêt et que j'ai été très heureux de voir mes conclusions confirmées par vos considérations. Malheureusement mon hypothèse de l'aplatissement des électrons est en contradiction avec les résultats des nouvelles expériences de M. Kaufmann et je crois être obligé de l'abandonner; je suis donc au bout de mon latin et il me semble impossible d'établir une théorie qui exige l'absence complète d'une influence de la translation sur les phénomènes électromagnétiques et optiques. Je serais très heureux si vous arriviez à éclaircir les difficultés qui surgissent de nouveau.

Veuillez agréer, cher collègue, l'expression de mes sentiments sincèrement dévoués.

H. A. Lorentz

FIG. 4
Letter from H. A. Lorentz to H. Poincaré, 8 March 1906.

those years of work Lorentz was willing to abandon his theory on the basis of a report of a single experiment.

Einstein's intuition served him well here, for in fact, Kaufmann's data were incorrect. In 1908 the Lorentz-Einstein prediction for the electron's transverse mass was considered to have been vindicated by the new data of Bucherer.[46] Bucherer's results were not immediately accepted by the entire scientific community, but Poincaré and Lorentz did not call for additional tests. Incidentally, three decades later it was proved that Bucherer's data were also incorrect.[47] Undoubtedly having the Kaufmann episode of 1907 in mind, Einstein in 1946 described one of two criteria for assessing a scientific theory: "The first point of view is obvious: the theory must not contradict empirical facts. However evident this demand may in the first place appear, its application turns out to be quite delicate."[48]

In conclusion, although by mid-1905 high-velocity data dominated basic physical theory, Einstein needed only familiar low-velocity data, such as electromagnetic induction, in conjunction with the data of the thought-experimenter, to bring physics into the twentieth century.

ACKNOWLEDGMENTS

Figures 1 and 2 are reproduced from Harry Woolf, ed., *Some Strangeness in the Proportion: A Centennial Symposium to Celebrate the Achievements of Albert Einstein*, Reading, Mass., Addison-Wesley, 1980, by permission of the publisher. This essay is based on research supported by a grant from the National Science Foundation's History and Philosophy of Science Program. I am grateful to the Poincaré Estate, Paris, for permission to quote from their correspondence. I thank Professor Gerald Holton for his comments on this essay.

NOTES

1. A. Einstein, "Zur Elektrodynamik bewegter Körper," *Annalen der Physik* 17 (1905), pp. 891-921. As far as we know, the editorial policy of the *Annalen* was that an author's initial contributions were scrutinized by either the editor (in 1905, Paul Drude) or a member of the Curatorium; subsequent papers were published with no refereeing. Einstein had appeared in print in the *Annalen* five times by 1905; so his relativity paper was probably accepted on receipt. I thank Mr. Allan Needell for this information, at which he arrived as a result of studying the correspondence between Planck and Wilhelm Wien.

For the purpose of serious historical analysis I had to retranslate Einstein's relativity paper from the *Annalen* version for my book, *Albert Einstein's Special Theory of Relativity: Emergence (1905) and Early Interpretation (1905-1911)* (Reading, Mass., Addison-Wesley, 1981), where the retranslation appears in the Appendix (pp. 391-415). The hitherto most frequently quoted English translation of the relativity paper is in the Dover reprint volume, *The Principle of Relativity*, trans. W. Perrett and G. B. Jeffery (New York: Dover, n.d.; reprinted from London: Methuen, 1923). But that contains some substantive mistranslations, infelicities, and outdated Britishisms. For example, Einstein's second principle of the relativity theory is mistranslated as: "Any ray of light moves in the 'stationary' system of co-ordinates with the determined velocity *c*, whether the ray be emitted by a stationary or by a moving body." The correct translation is: "Any ray of light moves in the 'resting' coordinate system with the definite velocity *c*, *which is independent* of whether the ray was emitted by a resting or by a moving body" (italics added to indicate a key phrase that was omitted in the Dover translation).

The Dover translation was made from a retypeset version of Einstein's relativity paper that had appeared in a Teubner reprint volume, thereby adding to the misprints in the original *Annalen* version. In addition, the Dover translation does not distinguish between Einstein's footnotes and those added to the Teubner edition by Arnold Sommerfeld. This state of affairs is an example of the pitfalls inherent in using a translation that was not made from the original paper and of the importance of going back to the original papers.

Hereinafter all citations to Einstein's relativity paper are to the translation in my book.

2. I discuss these materials in *Einstein's Special Theory of Relativity.* See also G. Holton, *Thematic Origins of Scientific Thought: Kepler to Einstein*, Cambridge, Mass., Harvard University Press, 1973.

3. A. Einstein, "Autobiographical Notes," trans. P. A. Schilpp, in Schilpp, ed., *Albert Einstein: Philosopher-Scientist*, Evanston, Ill., Library of Living Philosophers, 1949, p. 21.

4. See, for example, H. Poincaré, *Science and Hypothesis* (1902), trans. unknown, New York, Dover, 1952.

5. H. A. Lorentz, "La théorie électromagnétique de Maxwell et son application aux corps mouvants," *Archives Néerlandaises des Sciences Exactes et Naturelles* 25 (1892), p. 363; reprinted in Lorentz, *Collected Papers*, 9 vols., The Hague, Nijhoff, 1935-1939, vol. 2, pp. 164-343.

6. Instead of using Lorentz's original units, I take the liberty to write (1)–(5) in absolute Gaussian (cgs) units, whose usefulness was emphasized first by Max Abraham in 1902 and 1903; see the papers cited in note 12. Abraham was also the first to use the nomenclature "Maxwell-Lorentz equations," which appears in Einstein's relativity paper of 1905.

7. H. A. Lorentz, *Versuch einer Theorie der elektrischen und optischen Erscheinungen in bewegten Körpern*, Leiden, Brill, 1895; reprinted in Lorentz, *Collected Papers*, vol. 5, pp. 1-137.

8. Lorentz originally proposed the contraction hypothesis in his "The Relative Motion of the Earth and the Ether," *Koninklijke Akademie van Wetenschappen te Amsterdam* 1 (1892), p. 74; reprinted in Lorentz, *Collected Papers*, vol. 4, pp. 219-223. For further discussion of Lorentz's contraction hypothesis, see my article, "On Lorentz's Methodology," *British Journal for the Philosophy of Science* 25 (1974), pp. 29-45, and Miller, *Einstein's Special Theory of Relativity*.

9. See my "A Study of Henri Poincaré's 'Sur la Dynamique de l'Electron,'" *Archive for History of Exact Sciences* 10, nos. 3-5 (1973), pp. 207-328, and "Poincaré and Einstein: A Comparative Study," *Boston Studies in the Philosophy of Science*, vol. 31 (forthcoming).

10. W. Wien, "Über die Möglichkeit einer elektromagnetischen Begründung der Mechanik," in *Recueil de travaux offerts par les auteurs à H. A. Lorentz*, The Hague, Nijhoff, 1900, pp. 96-107; reprinted in *Annalen der Physik* 5 (1901), pp. 501-513.

11. W. Kaufmann, "Die magnetische und elektrische Ablenkbarkeit

der Becquerelstrahlen und die scheinbare Masse der Elektronen," *Nachrichten von der Gesellschaft der Wissenschaften zu Göttingen* (1901), pp. 143-155, and "Die elektromagnetische Masse des Elektrons," *Physikalische Zeitschrift* 4 (1902), pp. 54-57. For discussion, see my *Einstein's Special Theory of Relativity*, esp. chaps. 1, 7, and 12.

12. M. Abraham, "Dynamik des Elektrons," *Nachrichten von der Gesellschaft der Wissenschaften zu Göttingen* (1902), pp. 20-41, "Prinzipien der Dynamik des Elektrons," *Physikalische Zeitschrift* 4 (1902), pp. 57-63, and "Prinzipien der Dynamik des Elektrons," *Annalen der Physik* 10 (1903), pp. 105-179. For discussion, see my *Einstein's Special Theory of Relativity*, esp. chap. 1.

13. Lord Rayleigh, "Does Motion through the Aether Cause Double Refraction?" *Philosophical Magazine* 4 (1902), pp. 678-683; D. B. Brace, "On Double Refraction in Matter Moving through the Aether," *Philosophical Magazine* 7 (1904), pp. 317-329.

14. H. A. Lorentz, "Electromagnetic Phenomena in a System Moving with Any Velocity Less than That of Light," *Proceedings of the Royal Academy, Amsterdam* 6 (1904), p. 809; reprinted in Lorentz, *Collected Papers*, vol. 5, pp. 172-197, and in part in *The Principle of Relativity: A Collection of Original Memoirs on the Special and General Theory of Relativity by H. A. Lorentz, A. Einstein, H. Minkowski and H. Weyl*, trans. W. Perrett and G. B. Jeffery, New York, Dover, n.d., pp. 11-34.

15. M. Abraham, "Die Grundhypothesen der Elektronentheorie," *Physikalische Zeitschrift* 5 (1904), pp. 576-579.

16. H. Poincaré, "Sur la dynamique de l'électron," *Comptes rendus de l'Académie des Sciences* 140 (1905), pp. 1504-1508, and "Sur la dynamique de l'électron," *Rendiconti del circolo matematico di Palermo* 21 (1906), pp. 129-175. Both are reprinted in *Oeuvres de Henri Poincaré*, 11 vols., Paris, Gauthier-Villars, 1934-1954, vol. 9, pp. 489-493 and 494-550, respectively. I have discussed these papers in "Analysis of Henri Poincaré's 'Sur la Dynamique de l'Electron' "; "Poincaré and Einstein"; "On Some Other Approaches to Electrodynamics in 1905," in H. Woolf, ed., *Some Strangeness in the Proportion: A Centennial Symposium to Celebrate the Achievements of Albert Einstein*, Reading, Mass., Addison-Wesley, 1980, pp. 66-91; and *Einstein's Special Theory of Relativity*, esp. chap. 1.

17. H. Poincaré, "L'état actuel et l'avenir de la physique mathématique," paper delivered at the International Congress of Arts and Science, St. Louis, Mo., 24 September 1904, and published in *Bull. Sci. Mat.* 28 (1904), pp. 302-324; reprinted in H. Poincaré, *The Value of Science* (1905), trans. George Bruce Halsted, New York, Dover, 1958.

18. J. T. Merz, *A History of European Scientific Thought in the Nineteenth Century* (1904-1912), 2 vols., rpt. New York, Dover, 1954, vol. 1, p. 215.

19. For a discussion of Faraday's experiments with rotating magnets

and of the problem of unipolar induction, see my "Unipolar Induction: A Case Study of the Interaction Between Science and Technology," *Annals of Science* 38 (1981), pp. 155-189.

20. For discussions of the importance of *Anschauung* in the German-speaking physics community, see ibid., and my "Visualization Lost and Regained: The Genesis of the Quantum Theory in the Period 1913-27," in J. Wechsler, ed., *On Aesthetics in Science*, Cambridge, Mass., MIT Press, 1978, pp. 72-102.

21. C. L. Weber, "Über unipolare Induktion," *Elektrotechnische Zeitschrift* 16 (1895), pp. 513-514.

22. A. Föppl, *Einführung in die Maxwellsche Theorie der Elektrizität*, Leipzig, Teubner, 1894; M. Abraham and A. Föppl, *Theorie der Elektrizität: Einführung in die Maxwellsche Theorie der Elektrizität*, Leipzig, Teubner, 1904. For discussion of fundamental problems concerning electromagnetic induction, see my *Einstein's Special Theory of Relativity*, esp. chap. 3.

23. H. A. Lorentz, "Alte und neue Fragen der Physik," *Physikalische Zeitschrift* 11 (1910), pp. 1234-1257.

24. For discussions of Planck's work, see M. J. Klein, "Max Planck and the Beginnings of the Quantum Theory," *Archive for History of Exact Sciences* 1 (1962), pp. 459-479, and T. S. Kuhn, *Black-Body Theory and the Quantum Discontinuity, 1894-1912*, Oxford, Oxford University Press, 1978.

25. Einstein, "Autobiographical Notes," p. 37.

26. See my *Einstein's Special Theory of Relativity*, esp. chaps. 2 and 11.

27. A. Einstein, "Fundamental Ideas and Problems of the Theory of Relativity," address to acknowledge the Nobel Prize, delivered to the Nordic Assembly of Naturalists at Gothenburg, 11 July 1923, in *Nobel Lectures: 1901-1921*, New York, Elsevier, 1967, pp. 482-490, p. 484.

28. A. Einstein, "Über einen die Erzeugung und Verwandlung des Lichtes betreffenden heuristischen Gesichtspunkt," *Annalen der Physik* 17 (1905), pp. 132-148; translated by A. B. Aarons and M. B. Shepard in *American Journal of Physics* 33 (1965), pp. 367-374.

29. A. Einstein, "Die von der Molekularkinetischen Theorie der Wärme geforderte Bewegung von in ruhenden Flüssigkeiten suspendierten Teilchen," *Annalen der Physik* 17 (1905), pp. 549-560; reprinted in A. Einstein, *Investigations on the Theory of Brownian Movement*, trans. A. D. Cowper, with notes by R. Furth, New York, Dover, 1956.

30. Einstein, "Autobiographical Notes," p. 53.

31. Ibid.

32. For detailed development, see my *Einstein's Special Theory of Relativity*.

33. Einstein, "On the Electrodynamics of Moving Bodies," p. 392.

34. Holton, *Thematic Origins*, p. 364.

35. Einstein, "Autobiographical Notes," p. 53.

36. Einstein, "Über das Relativitätsprinzip und die aus demselben gezogenen Folgerungen," *Jahrbuch der Radioaktivität* 4 (1907), pp. 411-462.

37. Abraham and Föppl, *Theorie der Elektrizität*.

38. Einstein, "On the Electrodynamics of Moving Bodies," p. 392.

39. Ibid., p. 406.

40. Einstein, "Über das Relativitätsprinzip und die aus demselben gezogenen Folgerungen," p. 413.

41. W. Kaufmann, "Über die Konstitution des Elektrons," *Annalen der Physik* 19 (1906), pp. 487-553.

42. Miller, *Einstein's Special Theory of Relativity*, pp. 333-334.

43. Einstein, "Über das Relativitätsprinzip und die aus demselben gezogenen Folgerungen."

44. Kaufmann, "Über die Konstitution des Elektrons," p. 495.

45. Einstein, "Über das Relativitätsprinzip und die aus demselben gezogenen Folgerungen," p. 439.

46. A. H. Bucherer, "Messungen an Becquerelstrahlen. Die experimentelle Bestätigung der Lorentz-Einsteinschen Theorie," *Physikalische Zeitschrift* 9 (1908), pp. 755-762.

47. C. T. Zahn and A. A. Spees, in "A Critical Analysis of the Classical Experiments on the Variation of Electron Mass," *Physical Review* 53 (1938), pp. 511-521, demonstrated that the resolution of Bucherer's velocity filters was inadequate to distinguish between the various competing electron theories. In fact, this shortcoming pervaded every other experimental determination of the electron's mass in the period 1908-1915.

48. Einstein, "Autobiographical Notes," p. 21.

Einstein's Clocks: The Place of Time

Peter Galison

Einstein, 1933: "There are certain occupations, even in modern society, which entail living in isolation and do not require great physical or intellectual effort. Such occupations as the service of lighthouses and lightships come to mind."[1] Solitude, Einstein argued, would be perfect for the young scientist engaged with philosophical and mathematical problems. His own youth, we are tempted to speculate, might be thought of this way, the Bern patent office where he had earned a living seeming no more than a distant oceanic lightship. Consistent with this picture of otherworldliness, we have enshrined Einstein as the philosopher-scientist who, unmindful of the noise from his office work, rethought the foundations of his discipline and toppled the Newtonian absolutes of space and time.

Einstein's removal of these philosophical absolutes was more than a contribution to relativity; it has become a symbol of the overthrow of one philosophical epoch for another. To physicists such as Henri Poincaré, Hendrik Lorentz, and Max Abraham, Einstein's special relativity was startling, almost incomprehensible, because it began with basic assumptions about the behavior of clocks, rulers, and bodies in force-free motion—it began, in short, by assuming what these senior physicists had hoped to prove with starting assumptions about the structure of the electron, the nature of forces, and the dynamics of the ether. Soon a generation of

Unless otherwise indicated, all translations are my own.

1. Albert Einstein, speech delivered at the Royal Albert Hall, London, 3 Oct. 1933, in *Einstein on Peace*, ed. Otto Nathan and Heinz Norden (New York, 1960), p. 238.

Critical Inquiry 26 (Winter 2000)

physicists, including Werner Heisenberg and Niels Bohr, patterned its quantum epistemology around Einstein's quasi-operational definitions of space and time in terms of rulers and coordinated clocks. For the philosophers of the Vienna Circle, including Moritz Schlick, Rudolf Carnap, and Philipp Frank, Einstein's special relativity paper was also a turning point, an ever present banner to be flown for scientific philosophy.

For all these reasons, Einstein's 1905 "On the Electrodynamics of Moving Bodies" became the best-known physics paper of the twentieth century. Einstein's argument, as it is usually understood, departs so radically from the older, "practical" world of classical mechanics that the work has become a model of the revolutionary divide. Part philosophy and part physics, this rethinking of distant simultaneity has come to symbolize the irresolvable break of twentieth-century physics from that of the nineteenth. Recall the order of the argument. Einstein began with the claim that there was an asymmetry in the interpretation of Maxwell's equations, an asymmetry not present in the phenomena of nature (fig. 1). A magnet approaching a coil produces a current indistinguishable from the current generated when a coil approaches a magnet. In Einstein's view this was a single phenomenon (coil and magnet approach and produce a current in the coil). But in their usual interpretation Maxwell's equations gave two different explanations of what was happening, depending on whether the coil or the magnet was in motion with respect to an all-pervasive ether. When the coil moved, charge within it experienced a force due to the static magnetic field; when the magnet moved, the changing magnetic field produced an electric field that drove the charge around the stationary coil. Einstein's goal was to produce a symmetric account, one that did not distinguish between the explanation given in the frame of reference of the coil and that given in the frame of reference of the magnet. The problem, as Einstein diagnosed it, was that "insufficient consideration" had been paid to the circumstance that electrodynamics always depended on a view about *kinematics*, that is, about how clocks and rulers behaved in the absence of force.[2]

2. Einstein, "Zur Elektrodynamik bewegter Körper," *Annalen der Physik* 17 (1905): 892, hereafter abbreviated "ZE"; trans. Arthur I. Miller, under the title "On the Electrodynamics of Moving Bodies," appendix in *Albert Einstein's Special Theory of Relativity: Emergence (1905)*

Peter Galison is the Mallinckrodt Professor of the History of Science and of Physics at Harvard University. He is the author of *How Experiments End* (1987) and *Image and Logic: A Material Culture of Microphysics* (1997) and has coedited *Picturing Science, Producing Art* (1998) with Caroline Jones and *The Architecture of Science* (1999) with Emily Thompson. His current projects center on the changing understanding of scientific objectivity and on the history of theoretical physics.

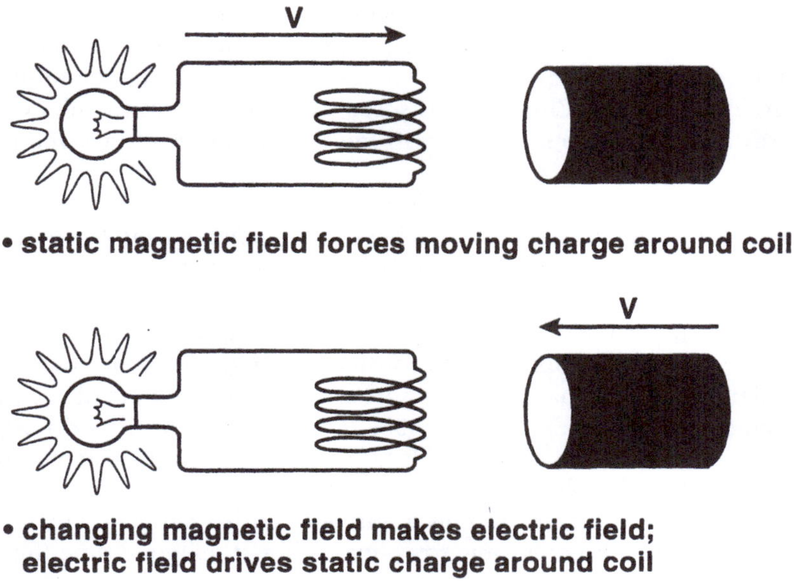

FIG. 1.—Coil and magnet.

A coordinate system was, by Einstein's lights, a system of rigid measuring rods embodying Euclidean geometry and describable with ordinary Cartesian coordinates. So far, so good. Then comes the surprising part, the reanalysis of *time* that contemporaries like Hermann Minkowski saw as the crux of Einstein's argument.[3] As Einstein put it: "We have to take into account that all our judgments in which time plays a role are always judgments of *simultaneous events*. If, for instance, I say, 'That train arrives here at 7 o'clock,' I mean something like this: 'The pointing of the small hand of my watch to 7 and the arrival of the train are simultaneous events'" ("ZE," p. 893; "OE," p. 393).[4] For simultaneity *at a point*, there is

---

and Early Interpretation (1905–1911) (Reading, Mass., 1981), p. 393, hereafter abbreviated "OE."

3. See Peter L. Galison, "Minkowski's Space-Time: From Visual Thinking to the Absolute World," *Historical Studies in the Physical Sciences* 10 (1979): 85–121.

4. We have all learned to read Einstein's papers in no small measure through the extensive work by Gerald Holton, *Thematic Origins of Scientific Thought: Kepler to Einstein* (Cambridge, Mass., 1973). I also find very helpful the more recent work by Abraham Pais, *Subtle Is the Lord* (Oxford, 1982), and Andrew Warwick, "On the Role of the FitzGerald-Lorentz Contraction Hypothesis in the Development of Joseph Larmor's Electronic Theory of Matter," *Archive for History of Exact Sciences* 43, no. 1 (1991): 29–91, "Cambridge Mathematics and Cavendish Physics: Cunningham, Campbell, and Einstein's Relativity, 1905–1911, Part I: The Uses of Theory," *Studies in the History and Philosphy of Science* 23 (Dec. 1992): 625–56, and "Cambridge Mathematics and Cavendish Physics: Cunningham, Campbell, and Einstein's Relativity, 1905–1911, Part II: Comparing Traditions in Cambridge Physics," *Studies in the History and Philosophy of Science* 24 (Mar. 1993): 1–25; Richard Staley, "On the

no problem: if an event located immediately next to my watch (say, the train engine arriving next to me) happens just when the small hand of the watch reaches the seven, then those two events are said to be simultaneous. The difficulty, Einstein insists, comes when we have to link events at a distance: what would it mean to say two *distant* events are simultaneous?

In order to address this question, Einstein advances, in a seemingly philosophical vein, a thought experiment infinitely far from the exigencies of instruments, much less the daily considerations of patent office life. How, Einstein asks, ought we to *coordinate* our clocks? "We could in principle content ourselves to time events by using a clock-bearing observer located at the origin of the coordinate system, who coordinates the arrival of the light signal originating from the event to be timed . . . with the hands of his clock" ("ZE," p. 893; "OE, p. 393; trans. mod.). Alas, Einstein notes, because light travels at a finite speed, this system is not independent of the observer with the central clock. Two events judged simultaneous with respect to one origin will not be simultaneous if the origin is moved. This epistemic straw man will not tell good time (figs. 2 and 3).

Young Einstein had a better system: let one observer at *A* send a light signal at noon to another at *B* a distance *d* away. *B* sets his clock to noon plus the time it takes a light signal to get to *B*, noon + *d*/*c*, where *c* is the speed of light. Continuing in this way, all other observers and their clocks are put in synchrony. With this system of coordination there is no special origin; there is no master clock. Here, so the account we have told ourselves goes, is the philosophical triumph of neo-Machian epistemic criticism over the fossilized absolutes of untouchable space and time. Einstein the philosopher-scientist has used thought experiments to vanquish unquestioned school dogma and a scientific-technical cadre too sophisticated to ask basic questions. But wait.

Let's go back to Einstein's train. You will recall that he wants to know what we mean by the arrival of a train in a station at seven o'clock. I have long followed Einstein himself in reading these remarks about trains and simultaneity as an instance of Einstein posing a question normally posed only "in early childhood," a matter that he, peculiarly, was still asking when he "was already grown up."[5] Such riddles about time and space appear, on this reading, to be so elementary, so basic, that they lay below the conscious awareness of the physics community. But was it, in fact, below the threshold of thought? Was no one else in 1904–5 in fact asking

Histories of Relativity: The Propagation and Elaboration of Relativity Theory in Participant Histories in Germany, 1905–11," *Isis* 89 (June 1998): 263–99; and Albrecht Fölsing, *Albert Einstein: A Biography,* trans. Ewald Osers (New York, 1997), p. 155.

5. Quoted in *Helle-Zeit-dunkle Zeit: In Memoriam Albert Einstein,* ed. Carl Seelig (Zurich, 1956), p. 71; trans. in *The Quotable Einstein,* ed. Alice Calaprice (Princeton, N.J., 1996), p. 182.

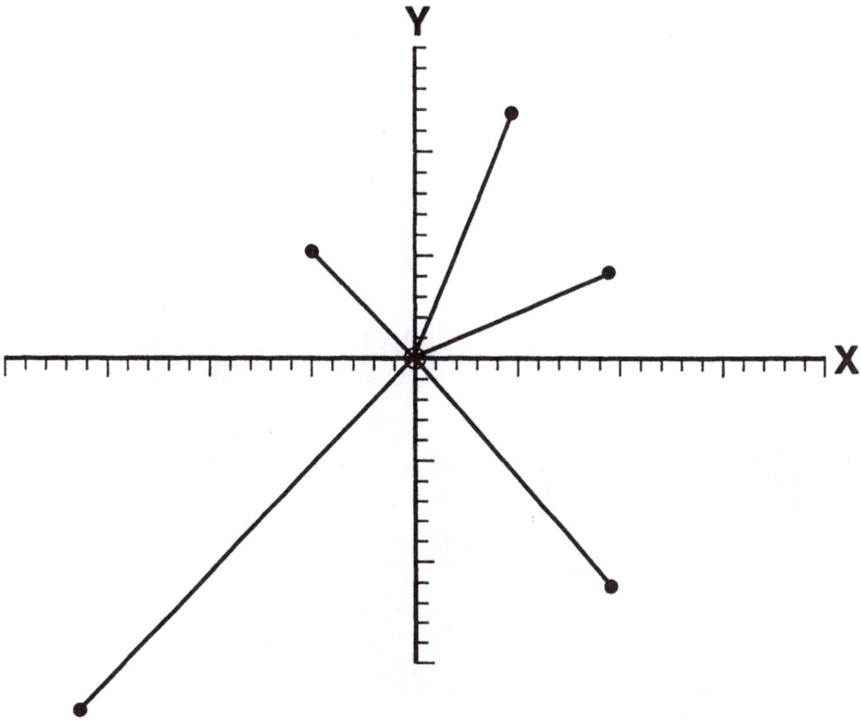

FIG. 2.—Central clock coordination.

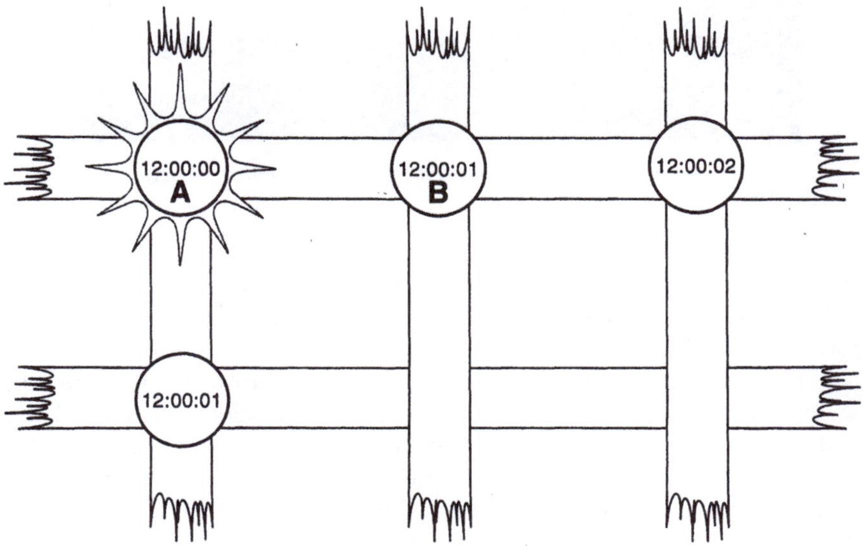

FIG. 3.—Einstein's clock coordination.

FIG. 4.—Bern train station circa 1860–65. Burgerbibliothek Bern, negative 12572.

what it meant for a distant observer to know that a train was pulling into a station at seven o'clock? Was the idea of defining distant simultaneity such a philosophical reach?

This summer I was standing in a northern European train station, absentmindedly staring at the turn-of-the-century clocks that lined the platform. They all read the same to the minute. Curious. Good clocks. But then I noticed that, as far as I could see, even the staccato motion of their second hands was in synchrony. These clocks were not simply running well, I thought, these clocks are coordinated. Einstein must have seen such coordinated clocks while he was grappling with his 1905 paper, trying to understand the meaning of distant simultaneity (fig. 4).

Already in the 1830s and 1840s Charles Wheatstone and Alexander Bain, both in England, and soon thereafter Mathias Hipp in Württemberg and a myriad of other inventors began constructing electrical distribution systems to bind distant clocks to a single central clock, called, variously, the *horloge-mère,* the *Primäre Normaluhr,* and the master clock (figs. 5 and 6).[6] In Germany Leipzig was the site of one of the first such electrically distributed time systems, followed by Frankfurt in 1859; Hipp (then director of a telegraph workshop) launched the Swiss effort at the Federal Palace in Bern, where a hundred clock faces began marching together in 1890 (fig. 7). Geneva, Basel, Neuchâtel, and Zurich followed in quick succession, each with its own clock coordination system, and railroad lines—where clock coordination was vital—were soon rigged in Switzerland for coordinated time.[7]

Without coordinated times, cities, towns, and villages functioned on their own times, marking an individuality that remained unimportant before the railroad. In England during the 1830s London time led Reading time by four minutes, marched seven minutes and thirty seconds in advance of Cirencester, and chimed fourteen minutes before Bridgewater. If you wanted to display time on the front of a major building, you needed more than one clock. The Isle Tower in Geneva boasted three: the big clockface in the center showed Geneva time (about 10:13), the face on the left showed the Paris-based time used all along the track of the Paris-Lyon-Méditerranée railroad company (9:58), and the right-hand clock boasted Bern time—a handsome five minutes in advance (10:18) (fig. 8). A few years later standardization entered the picture (fig. 9) but proceeded line by line—every railroad company defined its own proper time and did so ceremoniously. In London

> each morning an Admiralty messenger carried a watch bearing the correct time to the guard on the down Irish Mail leaving Euston for Holyhead. On arrival at Holyhead the time was passed on to officials on the Kingstown boat who carried it over to Dublin. On the return mail to Euston the watch was carried back to the Admiralty messenger at Euston once more.[8]

6. Remotely set clocks were discussed by, among others, Charles Wheatstone and William Cooke, the Scottish clockmaker Alexander Bain, and the American inventor Samuel F. B. Morse. For Wheatstone, Cooke, and Morse, clock coordination came out of their work on telegraphy. See Kenneth F. Welch, *Time Measurement: An Introductory History* (Newton Abbot, 1972), pp. 71–72.

7. For discussions of the extensive work on clock coordination before 1900, see, for example, the series of articles by A. Favarger, "L'Électricité et ses applications à la chronométrie," *Journal suisse d'horlogerie* 9 (Sept. 1884–June 1885), esp. pp. 153–58; Favarger, "Les Horloges électriques," in *Histoire de la pendulerie neuchâteloise,* ed. Alfred Chapuis (Paris, 1917), pp. 399–420; Friedrich Anton Leopold Ambronn, *Handbuch der astronomischen Instrumentenkunde,* 2 vols. (Berlin, 1899), esp. 1:183–87. On the expansion of the Bern network, see Gesellschaft für elektrische Uhren in Bern, *Jahresberichte* (1890–1910), Stadtarchiv Bern.

8. Philip S. Bagwell, *The Transport Revolution from 1770* (London, 1974), p. 125; quoted in Wolfgang Schivelbusch, *The Railway Journey: Trains and Travel in the Nineteenth Century,* trans. Anselm Hollo (New York, 1980), pp. 48–50.

FIG. 5.—L'Horloge-mère, Neuchâtel. From A. Favarger, *L'Électricité et ses applications à la chronométrie* (Neuchâtel, 1924), p. 414.

FIG. 6.—Normaluhr, Silesischer Bahnhof. From A. Favarger, *L'Électricité et ses applications à la chronométrie,* p. 470.

FIG. 7.—Coordinated clock, Federal Palace, Bern. Photo: author.

Germany, by most accounts the most advanced in its efforts to coordinate time, was still struggling with a hodgepodge of mechanical and electrical systems in 1891, when the aging Count Helmuth von Moltke came to speak, for the last time, to the Imperial German Parliament on 16 March; he died just over a month later.[9] As chief of the Prussian (and later German) general staff, von Moltke had dramatically reconceived the deployment of troops. Where earlier generations had sent armies over undependable roads, von Moltke exploited the railroads to supply, muster, and move troops over vast battlefronts. His successes with this strategy—in the Franco-Prussian War—no doubt brought the audience to attention when von Moltke pronounced on railroads, empire, and the military. In his scratchy voice he intoned:

> *Meine Herren,* . . . I will not long detain you, as I am very hoarse, on which account I have to ask your indulgence.
>
> That unity of time *[Einheitszeit]* is indispensable for the satisfactory operating of railways is universally recognized, and is not disputed. But, *meine Herren,* we have in Germany five different units of time. In north Germany, including Saxony, we reckon by Berlin time; in Bavaria, by that of Munich; in Württemberg, by that of Stuttgart;

9. The establishment of uniform time is discussed in Stephen Kern, *The Culture of Time and Space, 1880–1918* (Cambridge, Mass., 1983), pp. 11–14, and in Derek Howse, *Greenwich Time and the Discovery of Longitude* (Oxford, 1980), pp. 119–20. Simon Schaffer uses H. G. Wells's time machine as a guide through the turn-of-the-century intersection of the mechanized workplace and the literary and scientific engagement with time in "Time Machines" (unpublished manuscript, University of Cambridge).

FIG. 8.—Isle Tower, Geneva, 1880, three clocks. Centre d'iconographie genevoise. RVG N13x18 14934.

FIG. 9.—Isle Tower, Geneva, 1880, one clock. Centre d'iconographie genevoise. VG N13x18 1769.

> in Baden, by that of Carlsruhe; and on the Rhine Palatinate by that of Ludwigshafen. We have thus in Germany five zones, with all the drawbacks and disadvantages which result. These we have in our own fatherland, besides those we dread to meet at the French and Russian boundaries. This is, I may say, a ruin which has remained standing out of the once splintered condition of Germany, but which, since we have become an empire, it is proper should be done away with.

From the audience came the call "very true." Von Moltke went on to say that while this piecemeal ruin of time may only be an inconvenience for the traveler, it was an "actual difficulty of vital importance" for the railway business and, even worse, for the military. What, he asked, would happen in case of troop mobilization? There had to be a standard, one that would fall along the fifteenth meridian (about fifty miles east of the Brandenburg Gate) that would be the reference point. Local times within Germany would differ, but by a mere half hour or so on either extreme of the empire. "*Meine Herren,* unity of time merely for the railway does not set aside all the disadvantages which I have briefly mentioned; that will only be possible when we reach a unity of time reckoning for the whole of Germany, that is to say, when all local time is swept away."[10]

Von Moltke conceded that the public might dissent—wrongfully. But after some "careful consideration" the scientific men of the observatories would see things aright, and lend "their authority against this spirit of opposition." "*Meine Herren,* science desires much more than we do. She is not content with a German unity of time, or with that of middle Europe, but she is desirous of obtaining a world time, based upon the meridian of Greenwich, and certainly with full right from her standpoint, and with the end she has in view."[11] Farms and factory workers can offset their clock starting times as they wish. If a manufacturer wants his workers to start at the crack of dawn, then let him open the gates at 6:29 in March. Let the farmers follow the sun, let the schools and courts make do with their always loose schedules. Germany took action to extend its dominion of time, and much of Europe followed.

In Switzerland, the clock-world-famous Hipp (despite being arrested for consorting with anarchists) went on from his development of electrically maintained pendula to the practical deployment of a distribution system for time using low-tension circuits. Founded as a small telegraph

10. Helmuth von Moltke, "Dritte Berathung des Reichshaushaltsetats: Reichseisenbahnamt Einheitszeit," *Gesammelte Schriften und Denkwürdigkeiten des General-Feldmarschalls Grafen Helmuth von Moltke,* 8 vols. (Berlin, 1891–93), 7:38–39, 39, 40; trans. Sandford Fleming, under the title "General von Moltke on Time Reform," in *Documents Relating to the Fixing of a Standard of Time and the Legalization Thereof,* Canada Parliament Session 1891, no. 8, pp. 25, 25, 26; trans. mod.

11. Von Moltke, "Dritte Berathung des Reichshaushaltsetats: Reichseisenbahnamt Einheitszeit," p. 40; Fleming, "General von Moltke on Time Reform," p. 26; trans. mod.

and electrical apparatus factory in Neuchâtel, Hipp's company went from the establishment of the first network of public electric clocks in Geneva in 1861 to ever greater prominence; in 1889, it became A. de Peyer, A. Favarger & Cie. From 1889 to 1908 this concern extended the range of the mother clock beyond the dominion of railways to steeple clocks and even to the wake-up clocks inside hotels.[12] With the march of time into every street, methods were needed to extend indefinitely the number of units that could be branched together—a flood of patents followed perfecting relays and signal amplifiers. The Bern urban time network was inaugurated in 1890; improvements, expansions, and new networks sprouted throughout Switzerland. For not only was accurate, coordinated time important for European passenger railroads and the Prussian military, it was equally crucial for the dispersed Swiss clockmaking industry that desperately needed means of consistent calibration.[13] But it was always practical and more than practical, at once material-economic necessity *and* cultural imaginary. Professor Wilhelm Förster of the Berlin Observatory, the observatory that set the Berlin master clock to the heavens, sniffed that any urban clock that did not guarantee time to the nearest minute was a machine "downright contemptuous of people."[14]

The burgeoning technology spun off patents in each sector of the network: patents on low-voltage generators, patents on electromagnetic receivers with all their escapements and armatures, patents on contact interrupters. Fairly typical of the kind of electrochronometric work blossoming in the years after 1900 was David Perret's novel receiver that would detect and use a direct-current chronometric signal to drive an oscillating armature, which was issued Swiss patent number 30351 at 5 P.M. on 12 March 1904. Or take Favarger's own receiver that did the opposite: it took an alternating current from the mother clock and turned it into the unidirectional motion of a toothed wheel. This patent—used widely—was submitted on 25 November 1902 and issued on 2 May 1905. There were patents on remote alarms, remote regulation of pendula, telephonic—even wireless—transmission of time; other patents arrived proposing clocks for railroad departures and arrivals, in addition to patents for clocks indicating time in other time zones (fig. 10).

12. For biographical details on Hipp, see Aymon de Mestral, *Mathias Hipp, 1813–1893; Jean-Jacques Kohler, 1860–1930; Eugène Faillettaz, 1873–1943; Jean Landry, 1875–1940* (Zurich, 1960), pp. 9–34. David S. Landes's work, *Revolution in Time: Clocks and the Making of the Modern World* (Cambridge, Mass., 1983), pp. 237–337, is excellent on the Swiss watch industry, though he is not focussed here on networks but rather on clock production.

13. See Favarger, *L'Électricité et ses applications à la chronométrie,* 3d ed. (Neuchâtel, 1924), pp. 408–9.

14. Gerhard Dohrn-van Rossum, *History of the Hour: Clocks and Modern Temporal Orders,* trans. Thomas Dunlap (Chicago, 1996), p. 350. See also Ulla Merle, "Tempo! Tempo! Die Industrialisierung der Zeit im 19. Jahrhundert," in *Uhrzeiten: Die Geschichte der Uhr und ihres Gebrauches,* ed. Igor A. Jenzen (Frankfurt am Main, 1989), pp. 166–78.

All these chronometric patents—along with a great many others related to them—had to pass through the Swiss patent office in Bern, and no doubt many of them crossed Einstein's desk.[15] Einstein began work in the Bern patent office on 16 June 1902 as technical expert, third class, where he was chiefly charged with the evaluation of electromagnetic patents.[16] Standing at his wooden podium he, like the other twelve or so technical experts in the patent office, methodically went through each submission, searching for the principles that lay at its core (fig. 11).[17] Einstein's expertise on electromechanical devices came in part from the family business. Indeed, Einstein's father, Hermann, and his uncle, Jakob Einstein, had built their enterprise out of his uncle's patents on sensitive, electrical clocklike devices for measuring electrical usage. One of J. Einstein & Cie.'s electrical meters, created by Jakob and Sebastian Kornprobst, was written up prominently in the report of the 1891 Frankfurt Electrotechnical Exhibit; just a few pages before is a mechanism (typical of the time) for mounting a backup mother clock to ensure the continued operation of a system of electrical clocks. So close were the electrical measuring systems and clockwork technologies that at least one of the Jakob Einstein-Sebastian Kornprobst patents explicitly registered its applicability to clockwork mechanisms more generally (fig. 12).[18]

Einstein's own later technical work (he held a great many patents) on what he called his *Machinchen* (a device for multiplying and measuring very small electrical charges) and his studies of the Einstein-de Haas effect (leading to his atomic theory of ferromagnetic atoms) were but two further examples of his particular interest in sensitive electromechanical

15. Hundreds of relevant patents are listed in the *Journal suisse d'horlogerie* during the relevant years (1902–5). Sadly, the Swiss patent office dutifully destroyed all papers processed by Einstein eighteen years after their creation; this was standard procedure on patent opinions, and even Einstein's fame led to no exception. See Fölsing, *Albert Einstein,* p. 104.

16. The most detailed linkage between Einstein's patent work and his scientific work is on gyromagnetic compasses and Einstein's production of the Einstein-de Haas effect. See Galison, *How Experiments End* (Chicago, 1987), chap. 2; in addition, see Thomas Hughes, "Einstein, Inventors, and Invention," *Science in Context* 6 (Spring 1993): 25–42, and Lewis Pyenson, *The Young Einstein: The Advent of Relativity* (Bristol, 1985). On Einstein's assignment to evaluate electrical patents, see Max Flückiger, *Albert Einstein in Bern: Das Ringen um ein neues Weltbild: Eine dokumentarische Darstellung über den Aufstieg eines Genies* (Bern, 1974), p. 62.

17. See Flückiger, *Albert Einstein in Bern,* p. 66.

18. See J. Einstein & Cie. and Sebastian Kornprobst, "Vorrichtung zur Umwandlung der ungleichmässigen Zeigerausschläge von Elektrizitäts-Messern in eine gleichmässige, gradlinige Bewegung," Kaiserliches Patentamt no. 53546, 26 Feb. 1890; "Neuerung an elektrischen Mess- und Anzeigervorrichtungen," Kaiserliches Patentamt no. 53846, 21 Nov. 1889; "Federndes Reibrad," Kaiserliches Patentamt no. 60361, 23 Feb. 1890; and "Elektrizitätszähler der Firma J. Einstein & Cie., München (System Kornprobst)," *Offizielle Zeitung der Internationalen Elektrotechnischen Ausstellung,* no. 28 (Oct. 1891): 949. See also Viktor Yakovlevitch Frenkel and Boris Efimovitch Yavelov, *Einshtein: Izobreteniia i eksperiment* (Einstein: invention and experiment) (Moscow, 1990), pp. 75–79, and Pyenson, *The Young Einstein,* pp. 39–53.

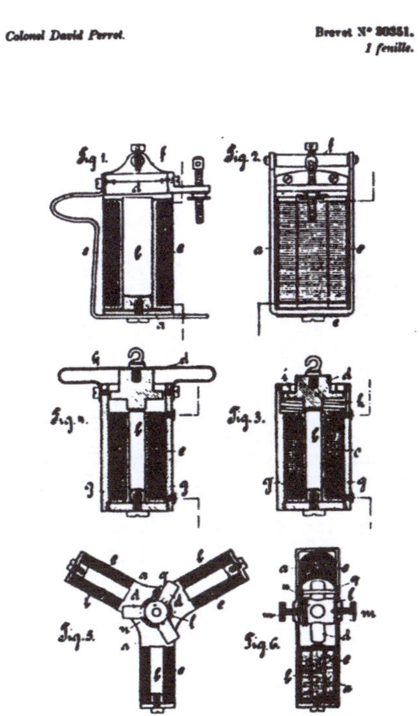

FIG. 10.—Swiss patent 30351 issued to David Perret, "Electro-aimant cuirassé," 12 March 1904.

FIG. 11.—Einstein's podium, Bern patent office. From: Max Flückiger, *Albert Einstein in Bern: Das Ringen um ein neues Weltbild: Eine dokumentarische Darstellung über den Aufstieg eines Genies* (Bern, 1974), p. 57.

FIG. 12.—Jakob Einstein & Cie. Elektrotechnische Fabrik. Elektrizität. *Offizielle Zeitung der Internationalen Elektrotechnischen Ausstellung, Frankfurt am Main, 1891*, ed. H. Massenbach and Max Quarck, 20 Oct. 1891, p. 949.

devices. Electromagnetic clock coordination patents would have been right up his alley, as they centered around means of transforming small electrical currents into high-precision rotatory movements.[19]

Heading the patent office during Einstein's tenure was Friedrich Haller, a stern taskmaster to his underlings. Early on he reproached Einstein: "As a physicist you understand nothing about drawings. You have got to learn to grasp technical drawings and specifications before I can make you permanent."[20] In September 1903 Einstein received notice that his provisional appointment was permanent, though Haller was not ready to promote him, commenting that Einstein "should wait until he has fully mastered machine technology; he studied physics." That mastery came as Einstein plunged himself into the critical evaluation of the parade of patents that came before him. By April 1906 Einstein seems to have persuaded the authorities that, physics notwithstanding, he had indeed mastered the technology, and he was promoted to technical expert, second class. Haller now judged that Einstein "belongs among the most esteemed experts at the office."[21]

Einstein's window on the electrochronometric world came at a crucial time. For despite von Moltke's resounding support and the undamped enthusiasm of the advocates of one world time, Albert Favarger, one of Hipp's chief engineers and the man who effectively succeeded him at the helm of his company, was not at all content. At the 1900 Exposition Universelle, the International Congress on Chronometry met to discuss the status, inter alia, of clock coordination efforts (fig. 13).[22] At the outset of his speech to the congress Favarger rose to ask how it could be that the distribution of electrical time was running so distressingly far behind the related technologies of telegraphy or telephony. First, he suggested, there were technical difficulties; remotely coordinated clocks could rely on no obliging friend ("ami complaisant") to oversee and correct the least difficulty, whereas the steam engine, dynamo, or telegraph all seemed to run with constant human companionship ("SD," p. 198). Second, there was a technician gap: the best technical people were staffing power and communication devices, not time machines. And finally, he lamented, the public was not funding time distribution as it should. Such lagging boosterism baffled Favarger: "Could it be possible that we have not experi-

19. On the "little machine" *(Machinchen)*, see John Stachel et al., "Einstein's 'Machinchen' for the Measurement of Small Quantities of Electricity," editorial note in *The Swiss Years: Correspondence, 1902–1914*, vol. 5 of *The Collected Papers of Albert Einstein*, trans. Anna Beck, ed. Stachel et al. (Princeton, N.J., 1995), pp. 51–55; on the Einstein-de Haas effect, see Galison, *How Experiments End*, chap. 2, and Frenkel and Yavelov, *Einshtein*, chap. 4.

20. Quoted in Flückiger, *Albert Einstein in Bern*, p. 58.

21. Quoted in Pais, *Subtle Is the Lord*, pp. 47–48.

22. See Favarger, "Sur la distribution de l'heure civile," in Congrès International de Chronométrie, *Comptes rendues des travaux, procès-verbaux, rapports, et mémoires*, ed. E. Fichot and P. de Vanssay (Paris, 1902), pp. 198–203; hereafter abbreviated "SD."

FIG. 13.—Exposition Universelle. From *Encyclopédie du siècle: L'Exposition de Paris de 1900* (Paris, 1900).

enced the imperious, absolute, I would say collective need of time exactly, uniformly, and regularly distributed? . . . Here's a question that borders on impertinence when addressed to a late nineteenth-century public, laden with business and always rushed, a public that has made its own the famous adage: Time is money" ("SD," p. 199).

As far as Favarger was concerned, the sorry state of time distribution was all out of proportion with the exigencies of modern life. He insisted that humans needed a system of exactitude and universality correct to the nearest second. No old-fashioned mechanical, hydraulic, or pneumatic system would do—electricity was the key to the future, a future that would only come about properly if humankind broke with its mechanical clock past riven by anarchy, incoherence, and routinization. In its place, a world of electrocoordinated clocks must be based on a rational and methodical approach. As he put it,

> You don't have to run long errands through Paris to notice numerous clocks, both public and private, that disagree—which one is the biggest liar? In fact if even just one is lying one suspects the sincerity of them all. The public will only gain security when every single clock indicates unanimity at the same time at the same instant. ["SD," p. 200]

How could it be otherwise? Trains often barreled through the countryside in opposite directions on single tracks, and a shunting error of timekeeping could and did lead to calamity. Remote time regulation merging observatories, railroads, and telegraphy was all that stood between a smooth ride and smoking debris. Time went up for sale, and astronomers, telegraphers, and clockmakers all profited as they sent coordinated time down the railway lines. The first time zones were these long, thin territories carved by steel tracks.[23]

Favarger reminded the assembled exposition attendees that the speed of trains roaring through Europe was mounting—100, 150, even 200 kilometers per hour. Those running the trains and directing their movements—not to speak of the passengers trusting their lives to speeding carriages—had to have correct times. At fifty-five meters per second every tick counted, and the prevalent but obsolete mechanical systems of coordination were bound to be inferior. Only the electric, automatic system was truly appropriate: "The nonautomatic system, the most primitive yet the most widespread, is the direct cause of the time anarchy that we must escape" ("SD," p. 201).

Time anarchy. No doubt Favarger's reference was in part to the anarchism that had taken a powerful hold among the Jura watchmakers, as Pyotr Kropotkin testified in his *Memoirs of a Revolutionist*. To be sure, Kropotkin recorded, the theoretical issues raised by Bakunin and others against economic despotism were important,

> but the equalitarian relations which I found in the Jura Mountains, the independence of thought and expression which I saw developing in the workers, and their unlimited devotion to the cause appealed even more strongly to my feelings; and when I came away from the mountains, after a week's stay with the watchmakers, my views upon socialism were settled. I was an anarchist.[24]

Hipp himself had been arrested, but Favarger was clearly worried about more, about the broader disintegration of personal and societal regularity. Only electrical distribution of simultaneity could provide the "indefinite expansion of the time unification zone" ("SD," p. 202). Favarger's support for distant simultaneity was, all at once, political, profitable, and pragmatic.

Should we be able to escape from this dreaded "anarcho-clockism," we would have a chance to fill a great lacuna in our knowledge of the world. For, Favarger insisted, even as the International Bureau of Weights and Measures had begun to conquer the first two fundamental quanti-

23. See Carlene Stephens, "'The Most Reliable Time': William Bond, the New England Railroads, and Time Awareness in Nineteenth-Century America," *Technology and Culture* 30 (Jan. 1989): 1–24 and "Before Standard Time: Distributing Time in Nineteenth-Century America," *Vistas in Astronomy* 28, pts. 1–2 (1985): 114–15.

24. Peter Kropotkin, *Memoirs of a Revolutionist*, trans. pub. (Montreal, 1989), p. 267.

ties—space and mass—time, the final frontier, remained unexplored (see "SD," p. 203). And the way to conquer time was to create an ever widening electrical network, enslaved to an observatory-linked mother clock that would drive relays multiplying its signals and send automatic clock resets into hotels, street corners, and steeples across continents. Tied in part to Favarger was a company that aimed to synchronize Bern's network. When, on 1 August 1890, Bern set the hands of its coordinated clocks in motion, the press hailed it as a "revolution in clocks."[25] It must not be forgotten that from many places in Bern you could clearly see several grand public clocks; when, that August, they all began running in step, time order became visible.

Swiss newspapers were not alone in seeing clock coordination as a matter of wide cultural import. To North American time booster Sandford Fleming and his allies in the 1890s the establishment of "universal" or "cosmic" time was both practical and more than practical—a boon to communication and transportation but also a "noiseless revolution" that would bring progress in all spheres of cultural and personal life.[26]

During the 1890s Einstein was not yet concerned about clocks at all; but as a young man of sixteen in 1895 he was already very much concerned with the nature of electromagnetic radiation. Even to his untutored imagination something was wrong with the customary conception of radiation as a wave in static, substantial ether. Suppose, he thought to himself, that one could catch up to a light wave—as classical physics might infer. Then, like a surfer riding an ocean wave train, he would see the electromagnetic field unfold before him oscillating in space but utterly unchanging in time. However, this corresponded to nothing ever observed.[27] Four years later Einstein was still agonizing over the nature of moving bodies and electrodynamics. To his beloved Mileva Marić he reiterated his sense that naive ether theories would simply have to go.

> D[ear] D[ollie],
> I returned the Helmholtz volume and am now rereading Hertz's propagation of electric force with great care because I didn't understand Helmholtz's treatise on the principle of least action in electrodynamics. I'm convinced more and more that the electrodynamics of moving bodies as it is presented today doesn't correspond to reality, and that it will be possible to present it in a simpler way. The introduction of the term "ether" into theories of electricity has led to the

25. Quoted in Jakob Messerli, *Gleichmässig pünktlich schnell: Zeiteinteilung und Zeitgebrauch in der Schweiz im 19. Jahrhundert* (Zurich, 1995), p. 126.

26. Fleming, *Time-Reckoning for the Twentieth Century* (Washington, D.C., 1889), p. 357. See Ian R. Bartky, "The Adoption of Standard Time," *Technology and Culture* 30 (Jan. 1989): 41 for links of Fleming to Cleveland Abbe and other meteorologists.

27. See Einstein, "Autobiographical Notes," in *Albert Einstein: Philosopher-Scientist*, ed. Paul Arthur Schilpp, 3d ed. (La Salle, Ill., 1970), p. 53.

> conception of a medium whose motion can be described, without, I believe, being able to ascribe physical meaning to it.[28]

Electricity and magnetism, Einstein concluded, would be definable as the motion of "true" electrical masses with physical reality through empty space. Ether, the centerpiece of nineteenth-century physical theories, was gone. So before Einstein set foot into the patent office crucial pieces of the relativity puzzle were in place: he knew Maxwell's equations, he was committed to a realistic picture of moving electrical charges, and he had dismissed the ether. But none of these considerations directly bore on the problem of how to treat time.

Meanwhile, some of the greatest physicists of the nineteenth century were beginning, out of desperation, to experiment with mathematical variations in the way the time variable *t* transformed in different reference frames. But all of them—Poincaré, Lorentz, Abraham—kept firmly to the notion of a true ether rest frame, and none of them accorded equal weight to these local times and the true (absolute) physical time of the ether rest system. Poincaré, Lorentz, and Abraham wanted to *begin* with special assumptions about the basic forces of nature, the forces that held together the atomic building blocks of interferometer arms, the forces that kept electrons from exploding because of their electrostatic self-repulsion. Out of such constructive, built-up theories of matter they sought to *deduce* kinematics—the behavior of matter in the absence of force. Einstein wanted none of that; he aimed for a theory that would start with simple physical principles, the way thermodynamics began with the conservation of energy and the increase of entropy. Poincaré, Lorentz, and Abraham were willing to make special assumptions about how an artificial, calculation-useful notion of time varied from frame to frame. None of them launched their detailed physics with a physical, principled set of assumptions about measured space and coordinated time.

So things stood in the years after Einstein arrived at the patent office in 1902.[29] In his patent work he had, in the instructions given by Haller,

28. Einstein to Mileva Marić, 10? Aug. 1899, in Einstein and Marić, *The Love Letters*, trans. Shawn Smith, ed. Jürgen Renn and Robert Schulmann (Princeton, N.J., 1992), p. 10. On Einstein's specific knowledge of aspects of electrodynamics, see Holton, "Influences on Einstein's Early Work," *Thematic Origins of Scientific Thought*, and Miller, *Albert Einstein's Special Theory of Relativity*.

29. Here is not the place to offer a reconstruction of all aspects of Einstein's path to special relativity. The reader is referred to an excellent short synthesis in Stachel et al., "Einstein on the Special Theory of Relativity," editorial note in *The Swiss Years: Writings, 1900–1909*, vol. 2 of *The Collected Papers of Albert Einstein*, ed. Stachel et al. (Princeton, N.J., 1989), pp. 253–74, esp. pp. 264–65, which argues that the rough sequence of Einstein's work was (1) conviction that only relative motion of ponderable bodies was significant; (2) abandonment of Lorentz's assignment of physical significance to absolute motion; (3) exploration of alternative electrodynamics justifying emission hypothesis of light relative to source; (4) abandonment of this alternative electrodynamics as Einstein assumes velocity of

the directive to be critical at every stage: "When you pick up an application, think that anything the inventor says is wrong." To follow blindly would be to court disaster by following "the inventor's way of thinking, and that will prejudice you. You have to remain critically vigilant."[30] It was advice for patent work, but it applied as well to the ethereal realms of physics. For in the electrodynamics of moving bodies Einstein had a problem that had troubled him on and off for some seven years, a problem that was with increasing force agonizing the leading physicists of the day. Meanwhile, all around him, literally, was the burgeoning fascination with electrocoordinated time. Every day Einstein took the short stroll from his house, left down the Kramgasse, to the patent office; every day he must have seen the great clock towers that presided over Bern with their coordinated clocks, and the myriad of street clocks branched proudly to the central telegraph office. After all, he had to walk under one of the most famous of them, the Zeitglockenturm, and by many others (figs. 14, 15, and 16). Sometime in the middle of May 1905—and we note that Einstein moved to the edge of Bern's unified time zone on 15 May—he and his closest friend, Michel Besso, cornered the electromagnetism problem from every side. "Then," Einstein recalled, "suddenly I understood where the key to this problem lay." He skipped his greetings the next day when he met Besso: "'Thank you; I've completely solved the problem.' An analysis of the concept of time was my solution. Time cannot be absolutely defined, and there is an inseparable relation between time and signal velocity."[31] Pointing up at a Bern clock tower—one of the famous synchronized clocks in Bern—and then to a clock tower in nearby Muri (not yet linked to the Bern mother clock), Einstein laid out his synchronization of clocks (figs. 17 and 18).[32]

Within a few days Einstein sent off a letter to his friend Conrad Habicht imploring him to send a copy of his dissertation and promising four new papers in return. "The fourth paper is only a rough draft at this point, and is an electrodynamics of moving bodies which employs a modification of the theory of space and time; the purely kinematic part of this paper [beginning with the new definitions of time synchronization] will surely interest you."[33] Ten years of thought had gone into this problem,

---

light independent of the velocity of the source; (5) critique of the usual conception of temporal and spatial intervals, and especially of distant simutaneity; and (6) physical definition of simultaneity and the construction of a new kinematic theory. Here my focus is on (5), the introduction of a conventional notion of distant simultaneity.

30. Quoted in Flückiger, *Albert Einstein in Bern,* p. 58.

31. Einstein, "How I Created the Theory of Relativity," lecture, Kyoto, 14 Dec. 1922, trans. Yoshimasa A. Ono, *Physics Today* 35 (Aug. 1982): 46.

32. See Josef Sauter, "Comment j'ai appris à connaître Einstein," in Flückiger, *Albert Einstein in Bern,* p. 156, and Fölsing, *Albert Einstein,* p. 155.

33. Einstein to Conrad Habicht, Bern, [18 or 25 May 1905], *The Swiss Years: Correspondence, 1902–1914,* p. 20.

FIG. 14.—Kramgasse Zeitglockenturm, Bern, 1890. Burgerbibliothek Bern, negative 10379.

FIG. 15.—Electrical street clock, northeast corner of Kesslergasse (today Münstergasse), Bern, circa 1893–1900. Large format postcard. Burgerbibliothek Bern, negative 10382.

but time synchronization was the final, crowning step in the development of special relativity.

In this light, Einstein's paper, completed by the end of June 1905, can now be read in a very different fashion. Instead of a pure "Einstein philosopher-scientist" merely earning his keep in the patent office, we can see him also as "Einstein patent-officer-scientist" refracting the underlying metaphysics of his relativity theory through some of the most symbolized mechanisms of modernity. The train arrives in the station at seven o'clock, as before, but now it is not just Einstein who is worried about what this means in terms of distant simultaneity. No, determining train arrival times using electromagnetically coordinated clocks was *precisely* the technological issue that had been racking Europe. Patents now raced through the system, improving the electrical pendula, altering the receivers, introducing new relays, and expanding system capacity. Time coordination in the central Europe of 1902–5 was no arcane subject; it was front and center for the clock industry, the military, and the railroad *as well as* a symbol of the interconnected, sped-up world of modernity.

By addressing the problem of distant simultaneity, Einstein was engaging a powerful and highly visible new technology that conventionalized simultaneity, first to synchronize train lines and to set longitude, and then to fix time zones. It was in this world that Einstein brought a conventional basis into his vision of a principled physics. A trace of the existing

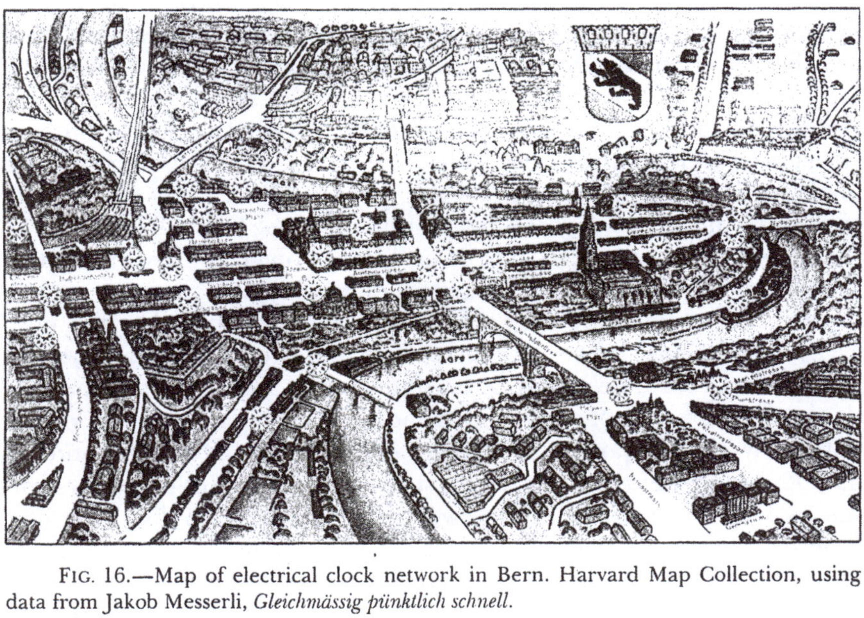

FIG. 16.—Map of electrical clock network in Bern. Harvard Map Collection, using data from Jakob Messerli, *Gleichmässig pünktlich schnell.*

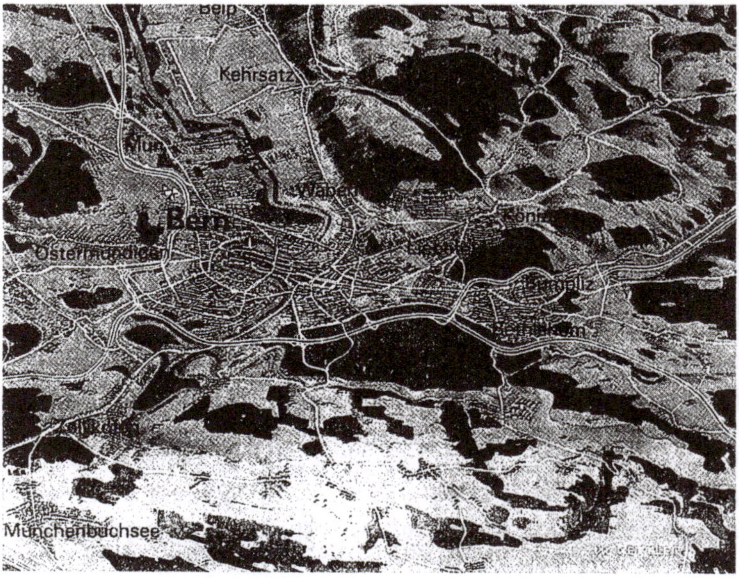

FIG. 17.—Bern-Muri map. Skorpion-Verlag.

time coordination system is there to see in the 1905 paper itself. Reconsider the scheme of coordination that Einstein *explicitly refused* to accept: an observer equipped with a clock at the center of the coordinate system. That master clock bolted to space position (0,0,0) determines simultaneity when electromagnetic signals from distant points arrive there at the same local time. But now this standard centered system no longer appears as an abstract straw man. This branching, radial clock coordination structure, visible in wires, generators, and clocks, displayed in book after book on timekeeping, *was precisely that of the European system of the mother clock along with its secondary and tertiary dependents* (see figs. 19, 20, and 21). When center-issued signals arrived at distant points, be they in the next room or a hundred kilometers away, they were *defined* as simultaneous; on that basis trains were run, troops were rousted, and telegraph messages were sent. It is even in this period that preparations were being made to send the time coordination signal out by radio waves. There was an intense burst of activity on such radio coordination systems in 1904 both in Switzerland and in France as workers tested, developed, and began deploying the new radio time system. The director of *La Nature* himself took up his pen to record new developments in the distribution of time by wireless methods. Reporting on experiments conducted at the Paris Observatory, he noted that with the aid of a chronograph distant synchronization now appeared to be possible to within two or three hundredths of a second—and the wireless technologies promised to distrib-

FIG. 18.—Muri Clocktower, circa 1900. Gemeindeschreiberei Muri bei Bern.

ute time everywhere throughout Paris and its surrounding suburbs. Not only would scientific goals be advanced, such as the determination of longitude, but freed from the constraints of physical wire, time could be broadcast out to boats at sea and even into the ordinary household.[34] By 1905 the American navy was using radio-controlled clocks, and by 1910 the Eiffel Tower station was fixing the hands of clocks across Europe (fig. 22). According to one leading radio time worker in 1911, planning for radio simultaneity had begun with radio itself, presumably sometime around 1901.[35] But whether by telegraph line or by wireless, centralized time distribution was the temporal-physical glory of the unified German empire that von Moltke wanted, made corporeal through the grand *Primäre Normaluhr* at the Silesischer Bahnhof in Berlin or the baroque and elegant *horloge-mère* of Neuchâtel.

For the telegraphers, geodesists, and astronomers, Einstein's novel time coordination scheme could clearly be understood in terms of the

34. See Henri de Parville, "Distribution de l'heure par télégraphie sans fil," *La Nature*, 30 July 1904, pp. 129–30. Experiments were conducted by G. Bigourdan, astronomer at the Paris Observatory, and presented to the Académie des Sciences on 27 June 1904; these results were printed in the *Comptes rendues de l'Académie* and quoted *in extenso* (along with the work of others including the director of the observatory at Neuchâtel) in "La Télégraphie sans fil et la distribution de l'heure," *Journal suisse d'horlogerie* 29 (Sept. 1904): 81–83.

35. On wireless time setting, see, for example, Joseph Roussel, *Le Premier Livre de l'amateur de T.S.F.* (Paris, 1922), esp. pp. 150–52. Julien Auguste Boulanger and Gustave Auguste Ferrié, *La Télégraphie sans fil et les ondes électriques*, 7th ed. (Paris, 1909) dates the Eiffel Tower radio station to 1903. Ferrié, "Sur quelques nouvelles applications de la télégraphie sans fil," *Journal de Physique*, 5th ser., 1 (1911): 178–89, esp. p. 178, indicates that planning for wireless time coordination began at the start of work on wireless; Edmond Rothé, *Les Applications de la télégraphie sans fil: Traité pratique pour la réception des signaux horaires* (Paris, 1913) discusses the details of radio-communicated time coordination procedure.

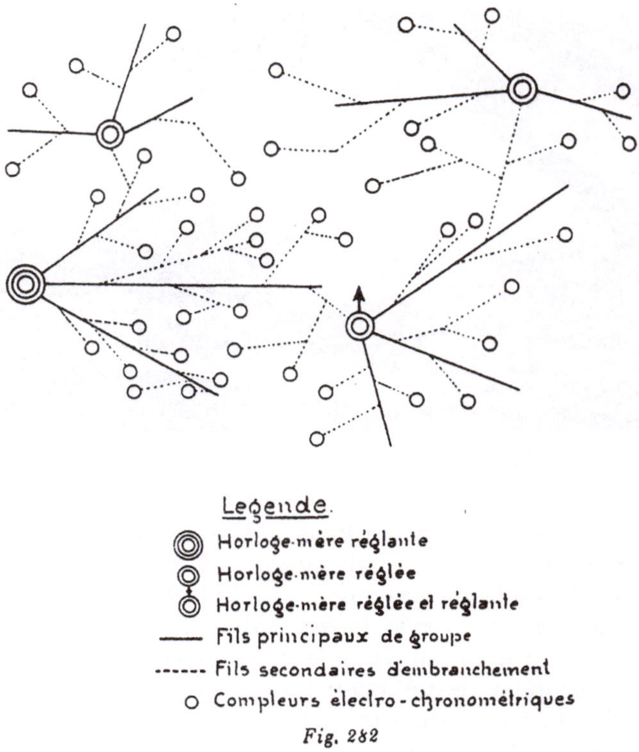

FIG. 19.—Time synchronization network I. From A. Favarger, "L'Électricité et ses applications à la chronométrie," *Journal suisse d'horlogerie* (1884–85): 320; rpt. in Favarger, *L'Électricité et ses applications à la chronométrie,* p. 394.

already extant clock coordination methods. At the École Professionelle Supérieure des Postes et Télégraphes, on 19 November 1921, when Léon Bloch sought to explain the meaning of time, he turned his audience's attention to the actual and widespread technology that they would have known like the backs of their hands:

> What do we call time on the surface of the earth? Take a clock that gives astronomical time—the mother pendulum of the Observatory of Paris,—and transmit that time by wireless to distant sites. In what does this transmission consist? It consists of noting at the two stations that need synchronization, the passage of a common luminous or hertzian signal.[36]

36. Léon Bloch, *Le Principe de la relativité et la théorie d'Einstein* (Paris, 1922), pp. 15–16. Dominique Pestre characterizes Bloch (and his brother) as physicists who were unusual for their time in France by virtue of writing textbooks that looked positively on the new physics of the early twentieth century, and who characteristically wrote using a series of progressive

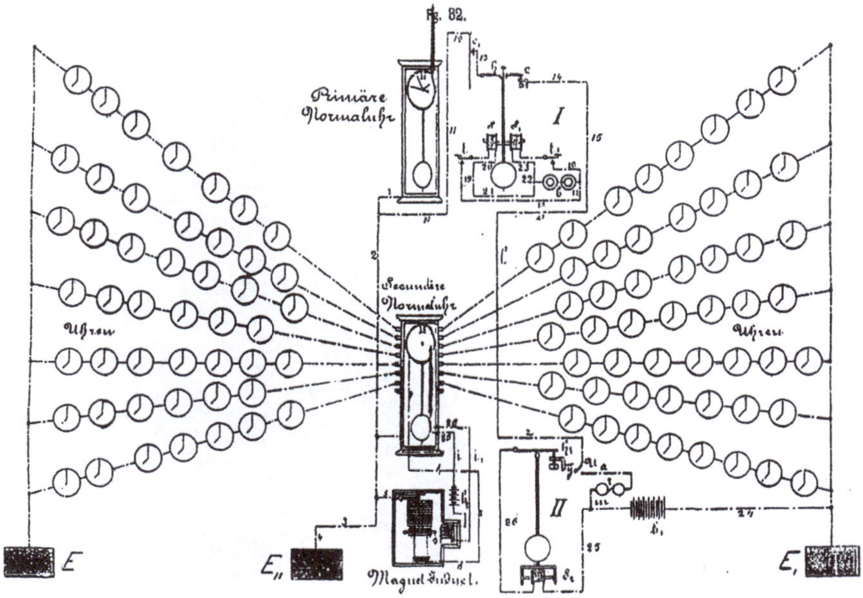

FIG. 20.—Time synchronization network II. From Ladislaus Fiedler, *Die Zeittelegraphen und die elektrischen Uhren vom praktischen Standpunkte* (Vienna, 1890), pp. 88–89.

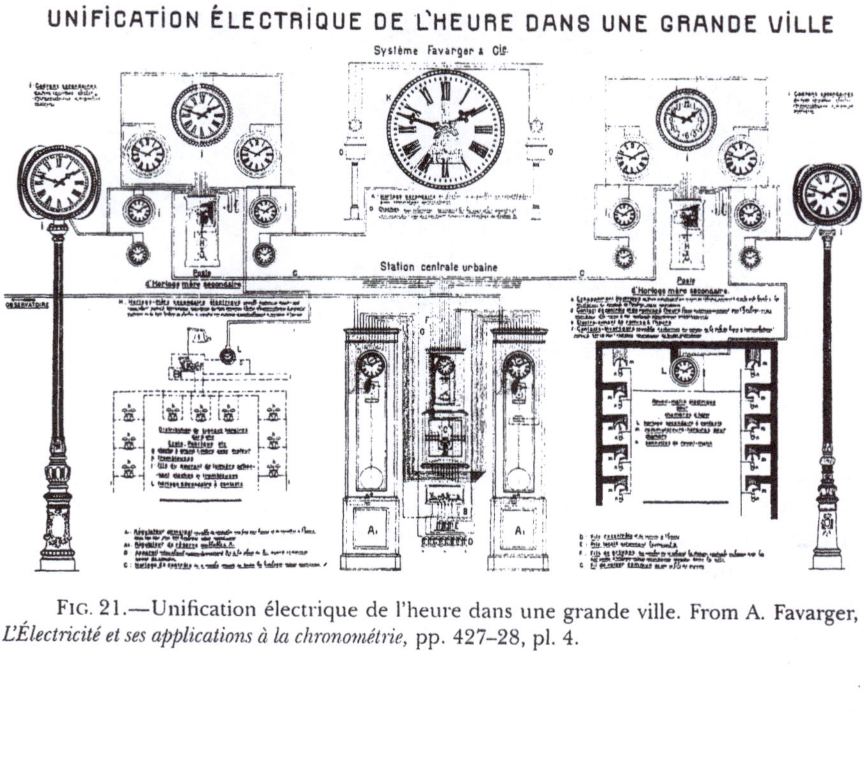

FIG. 21.—Unification électrique de l'heure dans une grande ville. From A. Favarger, *L'Électricité et ses applications à la chronométrie*, pp. 427–28, pl. 4.

FIG. 22.—Eiffel Tower radio station. From Julien Auguste Boulanger and Gustave Auguste Ferrié, *La Télégraphie sans fil et les ondes électriques*, 7th ed. (Paris, 1909), p. 429.

At least at the Postes et Télégraphes, relativity was understood through their real infrastructure of coordinated clocks. Yet more than the mere invocation of clock synchronization is involved. By 1924, and very probably some time before then, clock coordinators had begun (following Einstein) to take account of the finite velocity of radio waves. Einstein's universal time machine had rapidly drawn from the technological world,

generalizations from the concrete to the abstract (no doubt to appeal to their more experimentally oriented colleagues). See Dominique Pestre, *Physique et physiciens en France, 1918–1940* (Paris, 1984), pp. 18, 56, 117.

reshaped the physical-metaphysical one, and now begun to reconfigure machinery.[37]

Indeed, given the physical—and cultural—impact of clock coordination, and its place in such a variety of settings, one might well ask if before 1905 anyone else besides Einstein was worried about defining time through a rigorous synchronization that took on board the finite velocity of light. Astonishingly enough there was one other person, someone who perhaps quite reasonably had been a member of the French Bureau of Longitudes since 4 January 1893, ascending to its presidency in 1899 (and in 1909), longitude determination having been for centuries the domain where clock coordination had been critical. To boot he was professor at the École Professionelle Supérieure des Postes et Télégraphes from 4 July 1902; when electrocoordinated clocks came on the scene it was Postes et Télégraphes that took charge. That other was, of course, Poincaré.[38] He—like Einstein—introduced clocks coordinated by the exchange of a luminous signal.

Poincaré's first exploration of simultaneity came in 1898, when he argued that simultaneity was not an absolute concept, insisting that we have no direct intuition to any such notion. What we do have are certain rules, rules that we must invoke in order to do the quite concrete technical work of, for example, longitude determination. "When sailors or geographers determine a longitude they have precisely the problem to solve that we have been treating here; they must, without being at Paris, determine the time at Paris. How do they do it?" They could move a clock (with all the problems that attend such an effort), they could both refer to an astronomical event, or, finally, "they could make use of the telegraph. It is clear first of all that the reception of a signal in Berlin, for example, is posterior to the sending of that same signal from Paris."[39] This is not a purely hypothetical example. Commentators have often treated Poincaré's telegraphy reference as if it were an imaginary problem, an instance of abstract philosophical rumination. It was not. By 1898 an actual system of clock coordination existed in the service of longitude determination. Indeed, the geodetic and meteorological demand for coordinated time (to determine longitude) had joined with the exigencies of railroad economy and safety to launch the project of time zones.[40]

37. See Bureau des Longitudes, *Réception des signaux horaires: Renseignements météorologiques, sismologiques, etc., transmis par les postes de télégraphie sans fil de la tour Eiffel, Lyon, Bordeaux, etc.* (Paris, 1924), pp. 83–84.

38. See Ernest Lebon, *Henri Poincaré: Biographie, bibliographie analytique des écrits* (Paris, 1909), pp. 16–17.

39. Henri Poincaré, "La Mesure du temps," *Revue de métaphysique et de morale* 6 (1898): 11–12, 12, hereafter abbreviated "M"; rpt. in Poincaré, *La Valeur de la science* (1905; Paris, 1970), p. 53; trans. George Bruce Halsted, under the title "The Measure of Time," *The Value of Science* (New York, 1907), p. 35, hereafter abbreviated "MT"; trans. mod.

40. See Bartky, "The Adoption of Standard Time," pp. 25–56.

Poincaré, as noted, was a senior member of the Bureau of Longitudes in 1898, and his reference was not to telegraphy as an abstract instance of finite signal propagation but rather, explicitly, as a means of clock-coordinated longitude work:

> In general, the duration of the transmission [from Paris to Berlin] is neglected and the two events are regarded as simultaneous. But, to be rigorous, a little correction would still have to be made by a complicated calculation. In practice such a correction is not made because it would be smaller than observational errors, but its theoretical necessity is not diminished from our point of view, which is that of providing a rigorous definition [of simultaneity]. ["M," p. 12; "MT," p. 35; trans. mod.]

"Theoretically necessary" was the recognition that simultaneity was *always* conventional. And in this instance, at least, Poincaré's conventionalism was tied more than homonymically to the grubby world of real conventions that were, at precisely this conjuncture, setting time zone simultaneity, railway simultaneity, and national time unification.

Sometime after the spring of 1902 Einstein might have read Poincaré's "The Measure of Time." We know from Einstein's friend Maurice Solovine that their little discussion group (grandly titled the Olympia Academy) definitely read a later Poincaré work, *Science and Hypothesis,* that cited the 1898 work.[41] Critical as Poincaré was of any attempt to pretend that time-coordinating conventions were intuitive absolutes, on *practical* grounds, as he suggested in the remarks above, he did not militate for the abandonment of Newtonian kinematics. Ordinary rules of simultaneity, "the fruit of an unconscious opportunism," were "not imposed upon us and we might amuse ourselves in inventing others; but they could not be cast aside without greatly complicating the enunciation of the laws of physics, mechanics and astronomy" ("M," p. 13; "MT," p. 36). Corrections to the Newtonian world, Poincaré believed, would be small and complicated, so "theoretical necessity" would be trumped by the demands of simplicity. Einstein looked at distant simultaneity almost exactly as had Poincaré. But where Poincaré saw the new light-signal synchronization as leading to inevitable complexity, Einstein saw it as the harbinger of a vastly *simpler* physics.[42]

41. "Poincaré's *Science and Hypothesis* . . . engrossed us and held us spellbound for weeks" (Maurice Solovine, introduction to Einstein, *Letters to Solovine,* trans. Wade Baskin [New York, 1987], p. 9). See, for example, "La Mécanique classique," chap. 6 of Poincaré, *La Science et l'hypothèse* (Paris, 1902), esp. p. 111: "Not only don't we have a direct intuition of the equality of two durations, but we don't even have one of the simultaneity of two different events that occur in two different sites; this is what I explained in an article titled 'The Measure of Time.'"

42. Between 1900 and 1904 Poincare kept his programmatic statements about simultaneity largely separate from his explorations into the details of electrodynamics. But even

The turn-of-the-century Euro-American world was crisscrossed with overlapping networks of coordination: webs of train tracks, telegraph lines, meteorological networks, longitude surveys under the watchful, increasingly universal, clock system. In this context, the clock coordination system introduced by Einstein was, in a nontrivial sense, a world-machine, a vast, at first only imagined network of clocks. At the risk of seeming self-contradictory, there is a sense in which Einstein's special theory of relativity was a machine, an imaginative one, to be sure, but one built nonetheless on a real skein of wires that synchronized time machines by the exchange of electromagnetic signals.

Such a *technological* reading of this most *theoretical* paper suggests one final observation. It has long struck scholars that the style of "On the Electrodynamics of Moving Bodies" does not even look like that of an ordinary physics paper. There are essentially no footnotes, very few equations, no mention of new experimental results, and a lot of banter about simple physical processes that seem far removed from the frontiers of science.[43] Pick up a typical paper from the *Annalen der Physik* and a very different form appears in nearly every article: they typically begin with an experimental problem, a calculational correction; they are filled with references to other papers. But read Einstein's paper through the eyes of the patent world and suddenly it looks not so strange—at least in style. As one recent author has commented, patents are *precisely* characterized by their refusal to lodge themselves among other patents through footnotes—that would compromise the entrepreneurial advantage the author seeks. The simplistic banter is not unusual; patents are in fact written for a "person skilled in the art" (as patent language had it), not specialist readers.[44] And

---

when Poincaré did introduce his notion of local time into his electrodynamics to insist on the conventionality of judgments of simultaneity, he did not, as Einstein did, use light-signal coordination to reorganize mechanics and electrodynamics in such a way that force-free analysis of space and time clearly begin before any considerations of electron deformations and molecular forces come into play. For Einstein, it was precisely the point that kinematics, the play of temporal and spatial measures, would enter before dynamics. But here is not the place to sort out the relative contributions of these two physicists. Compare Henri Poincaré, "Relations entre la physique expérimentale et la physique mathématique," in *Rapports présentés au Congrès International de Physique,* ed. Ch.-Éd. Guillaume and L. Poincaré, 4 vols. (Paris, 1900), 1:1–29, and Henri Poincaré, "L'État actuel et l'avenir de la physique mathématique," *Bulletin des sciences mathématiques* 28 (1904): 302–24. For a comparison of Einstein and Poincaré's understanding of the electrodynamics of moving bodies, see Miller, *Albert Einstein's Special Theory of Relativity,* and Pais, *Subtle Is the Lord.*

43. This has been pointed out many times, for instance by Leopold Infeld in *Albert Einstein: His Work and Its Influence on Our Times* (New York, 1950), p. 23; Holton, "Influences on Einstein's Early Work," in *Thematic Origins of Scientific Thought;* and Miller, *Albert Einstein's Special Theory of Relativity* and "The Special Relativity Theory: Einstein's Response to the Physics of 1905," in *Albert Einstein: Historical and Cultural Perspectives,* ed. Holton and Yehuda Elkana (Princeton, N.J., 1982), pp. 3–26.

44. Greg Myers, "From Discovery to Invention: The Writing and Rewriting of Two Patents," *Social Studies of Science* 25 (Feb. 1995): 77.

the story told by the author aims to describe a procedure rather than point the way in a long train of work.

Given the embedding of this scientific-technological intervention in the wider world of time coordination another puzzle arises. In a sense Einstein—the same Einstein who at sixteen had abandoned his German citizenship and who over a lifetime lambasted the "herd life" of "the military system"—had, ironically, at age twenty-six, completed von Moltke's project.[45] Time was identified with timekeeping, and *Einheitszeit* became the technopolitical endpoint of establishing, procedurally, distant simultaneity in an ever expanding domain. Einstein's clock synchronization system, like its predecessors, reduced time to procedural synchronicity, tying clocks together by electromagnetic signals. And in Einstein's scheme clock unity extended beyond city, country, and empire, beyond continent, indeed, beyond the world, to the infinite, now pseudo-Cartesian, universe as a whole.

But the irony inverts. For while Einstein's clock coordination procedure built on at least fifteen years of intense efforts towards electromagnetic time unification, he had, devastatingly, removed one crucial element of von Moltke's vision. There was, in Einstein's infinite clock machine, no *Primäre Normaluhr,* no *horloge-mère,* no master clock. His was a coordinated system of infinite spatiotemporal extent, but its infinity was without center—no Silesischer Bahnhof linked upwards through the Berlin observatory to the heavens and downwards to the edges of empire. By infinitely extending a time unity that had originally been grounded in the imperatives of German national unity, Einstein had both completed and subverted the project. He had opened the zone of unification, but in the process he not only removed Berlin as the *Zeitzentrum* but also designed a machine that upended the very category of metaphysical centrality. With time coordination now defined by the exchange of electromagnetic signals, Einstein could finish his description of the electromagnetic theory of moving bodies with no spatial or temporal reference to any specially picked out ether rest frame. He had a theory of relativity in which the asymmetry between reference frames was gone.

* * *

Times change. Einstein left the Bern patent office in 1909 for the University of Zurich, then went to Prague, and eventually in 1914 took up his post at the University of Berlin. After World War I Favarger, that avatar of Swiss chronometric unity, published his technical 550-page third edition of his treatise on electrical timekeeping, framing it, once again, in broadly cultural terms. The Great War, he argued, had contributed powerful technical developments, but it also destroyed a great part

45. Einstein, "The World as I See It," *Ideas and Opinions,* trans. Sonja Bargmann, ed. Seelig (New York, 1954), p. 10.

of the human wealth that sustained peace had created. The world remaining was by contrast "a heap of ruins, of miseries, and of suffering."[46] To exit this disaster, at least materially, work would save humanity, and work, mechanics teaches us, is a product of time and force. Time, he goes on to say, "cannot be defined in substance; it is, metaphysically speaking, as mysterious as matter and space." (Even stolid Swiss clockmakers were driven to metaphysics by time.) All the activities of man, whether conscious or unconscious—sleeping, eating, meditating, or playing—take place in time; without order, without specified plans, we risk falling into anarchy—into "physical, intellectual and moral misery." The remedy: the precise measurement and determination of time with the rigor of an astronomical observatory. But measured time cannot remain in the astronomers' redoubt; time rigor must be distributed electrically to anyone who wants or needs it: "we must, in a word, popularize it, we must *democratize* time" for people to live and prosper. We must make every man "*maître du temps,* master not only of the hour but also of the minute, the second, and even in special cases the tenth, the hundredth, the thousandth, the millionth of a second."[47] Distributed, coordinated time was more than money for Favarger; it was each person's access to orderliness, interior and exterior. Throughout the late nineteenth and early twentieth centuries, coordinated clocks were never just gears and magnets.

My hope in exploring the material culture of clock coordination is to set Einstein's place in a universe of meaning that crossed mechanisms and metaphysics. More generally, my hope here—and elsewhere[48]—for studies in the material culture of science is to avoid two equally problematic positions on the relation of things to thoughts. On one side we have a long tradition of unregenerate materialism or empiricism, a view that ideas emerge causally and univalently from the disposition of objects and the impressions they make upon us. In the history of physics, empiricism of a specifically logical positivist sort directly and unambiguously shaped an inductive, observation-centered account of scientific development, codified in the *Harvard Case Histories in Experimental Science* but present throughout the 1950s and into the 1960s. In that frame, theory, and the philosophy with which it was associated, was an always provisional addition, not the bulwark of science. Einstein here appears as having taken the inexorable next step in an inductive process that gradually drove out the ether: the ether couldn't be measured to first order in the ratio of velocity to the speed of light *(v/c),* it wasn't there to second order in *v/c,* and therefore (so the argument went), Einstein concluded that the ether was superfluous.[49] No doubt there is much to be said for this experiment-

46. Favarger, *L'Électricité et ses applications à la chronométrie,* p. 10.

47. Ibid., p. 11.

48. See Galison, *Image and Logic: A Material Culture of Microphysics* (Chicago, 1997).

49. Representations of Einstein's relativity as a culmination of increasingly accurate "no ether" measurements are rife; perhaps the most scholarly attempt to locate Einstein's

grounded Einstein—his fascination with the detailed conduct of the fast electron experiments and his gyrocompass work at the Physikalisch-Technische Reichsanstalt reveal a theorist with a clear sense of laboratory procedure and the operation of machines. Things structured thoughts.

On the other side, and characteristic of the inverse project, was the antipositivist movement of the 1960s and 1970s that largely aimed at reversing the previous generation's epistemic order: programmes, paradigms, and conceptual schemes came first, and these reshaped experiments and instruments all the way down. Now thoughts fully structured things. Einstein on the antipostivist screen appears as the philosophical innovator who dispensed with the material world altogether in a sustained drive for symmetry, principles, and operational definitions. Much truth here, too; from the antipositivist reaction we learned to be sensitive to those moments when Einstein was chary of experimental results, dubious, for example, about supposed laboratory refutations of special relativity and about nominal astronomical contradictions with the general theory.

Granting both historiographical traditions their due, I am not proposing to split the difference, and I am certainly not advocating a technological reductivism. Instead, it seems to me that we have, in the form of a philosophically informed and historicized material culture, a way out of this binary oscillation between a historiography of implicit idealism or implicit materialism. Work on telegraphs, steam engines, scientific instruments, and astronomical observation over the last years has set questions that refuse the untenable either/or of things and thoughts.[50] In each instance we can explore the philosophical issues engaged with historically specific values and symbols.

When Einstein came to the Bern patent office in 1902 he entered into a world in which the triumph of the electrical over the mechanical was already symbolically wired to dreams of modernity. He found a world in which clock coordination was a practical problem (trains, troops, and telegraphs) demanding workable, patentable solutions in exactly his area of greatest concern and professional occupation: precision electromechanical instrumentation. The patent office was anything but a deep-sea

---

formulation as a mere variant of the early ether-electron theories is to be found in Edmund Whittaker, *A History of the Theories of Aether and Electricity* (London, 1953), where the chapter "The Relativity Theory of Poincaré and Lorentz" includes the remark: "Einstein published a paper [in 1905] which set forth the relativity theory of Poincaré and Lorentz with some amplifications, and which attracted much attention. He asserted as a fundamental principle the *constancy of the velocity of light* . . . which at the time was widely accepted, but has been severely criticised by later writers" (p. 40). See Holton, *Thematic Origins of Scientific Thought*, esp. chap. 5, and Miller, *Albert Einstein's Special Theory of Relativity*.

50. See Schaffer, "Late Victorian Metrology and Its Instrumentation: A Manufactory of Ohms," in *Invisible Connections: Instruments, Institutions, and Science*, ed. Robert Bud and Susan E. Cozzens (Bellingham, Wash., 1992), pp. 23–59; M. Norton Wise, "Mediating Machines," *Science in Context* 2 (Spring 1988): 77–114; and Galison, *Image and Logic*.

lightship. No, the office was a grandstand seat for the great parade of modern technologies. And as coordinated clocks went by, they weren't traveling alone; the network of electrical chronocoordination signified political, cultural, and technical unity all at once. Einstein seized on this new, conventional simultaneity machine and installed it at the principled beginning of his new physics. In a certain sense he had completed the grand time coordination project of the nineteenth century, but by eliminating the master clock and raising the conventionally set time to a *physical principle,* he had launched a distinctively modern twentieth-century physics of relativity.

Hypersymbolized—by which I mean many competing interpretations were in play—the regulated coordination of *Einheitszeit* meant, alternately, imperial empire, democracy, world citizenship, and antianarchism. What they held in common was a sense that each clock signified the individual and that clock coordination came to stand in for a logic of linkage among people and peoples. As such, the project for country or worldwide regulation elicited specific conditions of possible technocultural moves—that is, moves that would at once bear both scientific-technological and cultural significance.

It has become a commonplace over the last thirty years to pit bottom-up against top-down explanations. Neither will do. Borrowing a medieval saying aimed at capturing the links between alchemy and astronomy, we might put it this way: In looking down—to the electromagnetically regulated clock networks—we see up—to images of empire, metaphysics, and civil society. In looking up—to the metaphysics of Einstein's operationalized distant simultaneity, to the shifting culture of space, time, and motion—we see down—to the wires, gears, and pulses passing through the Bern patent office. We find metaphysics in machines, and machines in metaphysics.

Understanding the history of special relativity

STANLEY GOLDBERG. *Understanding relativity. Origin and impact of a scientific revolution.* Boston, Basel, Stuttgart: Birkhäuser, 1984. xvii, 494 pp., appendices, bibl. essay, index.

ARTHUR I. MILLER. *Albert Einstein's special theory of relativity. Emergence (1905) and early interpretation (1905–1911).* Reading, Mass.: Addison-Wesley Publishing Company, 1981. xxviii, 466 pp., appendix, bibl., index.

LEWIS PYENSON.. *The young Einstein. The advent of relativity.* Bristol and Boston: Adam Hilger Ltd., 1985. xiv, 255 pp., index.

IN THE SPRING of his *annus mirabilis*, 1905, Albert Einstein informed his friend and colleague Conrad Habicht of four papers he promised to send in exchange for Habicht's dissertation. One concerned the properties of atoms, another the motions of particles in liquids or Brownian motion. The third paper proposed the existence of light-quanta, "and is very revolutionary." "The fourth work," he continued, "lies at hand in draft and is an electrodynamics of moving bodies, making use of a modification of the theory of space and time; the purely kinematic part of this work will surely interest you."[1]

Einstein's paper on special relativity, alias the electrodynamics of moving bodies, invites superlatives. It is perhaps the most influential physics paper of this century. It is also among the best written. Yet it was one of the most difficult to comprehend for its physics and remains so today for its history. The paper contained only four footnotes, few references to experimental data, no citations of literature,

The following abbreviations are used in the notes and the appended bibliography: *AJP, American journal of physics; AS,* P.C. Aichelberg and R.U. Sexl, eds., *Albert Einstein: His influence on physics, philosophy and politics* (Braunschweig, 1979); *BJPS, British journal for the philosophy of science;* French, A.P. French, ed., *Einstein. A centenary volume* (Cambridge, Mass., 1979); *HE,* Gerald Holton and Yehuda Elkana, eds., *Albert Einstein. Historical and cultural perspectives* (Princeton, 1982); *JSHS, Japanese studies in the history of science.* Woolf, Harry Woolf, ed., *Some strangeness in the proportion* (Reading, Mass., 1980);

1. Quoted by Seelig (Bib. 79), 124–125. The fourth work: Einstein, "Zur Elektrodynamik bewegter Körper," *Annalen der Physik, 17* (1905), 891–921.

notes, few references to experimental data, no citations of literature, and it utilized only basic mathematics. It demanded a conceptual transformation in the thinking of its readers that was recognized and assimilated only slowly and with great difficulty.

Unlike many theoretical works, Einstein's paper started from aesthetic concerns, an examination of the practical meaning of such seemingly elementary concepts as space and time, and the enunciation of two fundamental postulates: the principle of relativity and the constancy of the speed of light. It resulted in new formulations of electrodynamics and mechanics and a strange new world in which clocks and rules behaved differently for different moving observers. By the time that Einstein's colleagues fully understood what he had done, the author was already extending his theory to include gravitation and accelerated motion.

1. RELATIVISTIC HISTORIOGRAPHY

The historiography of Einstein's paper also invites superlatives. It has probably attracted more scholarly attention than any other piece of 20th century physics, and, as the three books under review here may suggest, historical treatments and interpretations have differed very widely. The books' authors disagree about almost everything: the origin, significance and reception of Einstein's theory; its implications about the ways science progresses and innovations occur; the role of the individual scientist; the impact of discipline formation; and the purpose of history. The disagreement is abetted by a near total absence of primary documentation, by the authors' philosophical and historiographical commitments, and by their reluctance to build upon the work of others.

Einstein himself is also a factor. First and foremost a physicist, not a historian or a philosopher, he was skeptical of the purpose and ultimate success of efforts to reconstruct the thoughts of lone individuals like himself. Moreover, Einstein's recollections, often decades later, of how he arrived at the theory of special relativity are sometimes tempered by contemporary concerns and occasionally contradictory on essentials. Perhaps he soon forgot precisely how he arrived at the theory. Yet his recollections do agree on one point: that the theory did not suddenly emerge in the spring of 1905, as some evidence might imply, but that elements of it probably evolved over years of independent study and reflection. He began assembling them during his studies in Zurich, or perhaps earlier; they fell almost into place shortly after the publication of Planck's quantum theory in 1901; and they came together along with the theories of Brownian motion and lightquanta, over the next several years.[2] It is certainly a very

2. This is gleaned from Einstein, "Autobiographical notes," in P.A. Schilpp, ed., *Albert Einstein: Philosopher scientist* (LaSalle, Ill., 1949), 1–94, and "Autobiographische

complicated story, and not fully accessible to historians. We are thus left more than usual to our own inclinations.

For Arthur Miller, Einstein is a "philosopher-scientist" and the relativity paper a product of epistemology. Miller's treatise is first and foremost technical.[3] A long and informative section retracing the main events in electrodynamics from 1890 to 1905 is followed by a chapter summarizing "Einstein's philosophical viewpoint in 1905," a viewpoint largely reconstructed from Einstein's recollections. This serves as an orientation and transition to the bulk of the work to follow: a paragraph-by-paragraph exegesis of the relativity paper treated as a "window" on the physics of the period. A translation of the relativity gospel with numbered lines is appended for easy reference. The theological metaphor is not ill conceived. If one of the aims of history is to discern the "truth" through the study of human behavior and events of the past, whereas one of the aims of theology is to interpret human behavior and events in light of the "truth," then Miller's exposition leans toward theology. The topics and assertions in Einstein's paper direct and inform the content and chronology of Miller's treatment of physics in Einstein's era, rather than the reverse.

Miller's exegesis is the most thorough explication to date of Einstein's text, and, in grand scholastic manner, it also furnishes valuable summaries of published research relating directly or indirectly to Einstein's work. The accounts of unipolar induction and of experimental studies stimulated by Einstein's work are singularly exhaustive. Miller's approach yields some oddities, however: a discussion of the reception of the paper before the paper is fully received, and some overly speculative thought experiments on Einstein's presumed thoughts.

Besides exegesis, one of Miller's aims seems to be to demonstrate the uniqueness and singular importance of the relativity paper relative both to Einstein's own work and to the physics of his day. Historians who have tried to place it within broader contexts receive little notice.[4] For example, Miller ignores suggestions that Einstein's Brownian motion paper and his "very revolutionary" lightquantum paper should be considered of a piece with their coeval partner on relativity; he discusses them not so much as positive contributions within Einstein's overall research program than as the last proofs of the inadequacy of pre-relativistic physics.[5]

Skizze," in C. Seelig, ed., *Helle Zeit—dunkle Zeit*, (Zurich, 1956), 9–17.

3. A less ambitious account is provided by Pais (Bib. 67).

4. E.g., McCormmach (Bib. 46, 47); Klein (Bib.39); Holton (Bib. 31). Cf. Miller, 132–135.

5. Miller writes that Einstein "sought to demonstrate in the first two papers [on Brownian motion and light] the insufficiency of mechanics, electromagnetism, and ther-

According to Miller, the fundamental breaks with the past and present required for the creation of relativity theory occurred wholly within the realm of pure reflection: Einstein's foundational analysis of physical theory went far beyond science as it is normally conceived; from an analysis of electromagnetic induction into an analysis of sensations, and then into an analysis of thinking itself. Guidance from the neo-Kantian view that was predicated upon the usefulness of organizing principles such as the second law of thermodynamics, allowed Einstein to rise above the current state of physical theory (p. 191). From these organizing principles—the two postulates—"everything flows like the pristine form of a geometrical treatise."[6] Thus Miller's theological Einstein.

2. CULTURAL RELATIVISM

Stanley Goldberg aims to reduce reliance on pure thought and genial thinkers in accounting for scientific change by demonstrating the impact of cultural and social factors on the creation and acceptance of scientific ideas. Declaring individual creative thought an impenetrable "mystery," Goldberg seeks instead to demystify science itself for the cultured masses.

Goldberg's history of special relativity serves "as a case study to illustrate and support the general argument" (p. xiii), which he derives from the methodological gap between theory and experiment, from the fact that "no logical way leads from the phenomena to the basic principles of the theory," as Einstein put it in 1918.[7] The problem was exhibited by special relativity itself. Lorentz' electrodynamics and Einstein's relativity differed greatly in their theoretical contents and aims but, within the experimental capabilities of the time, yielded identical empirical predictions. Therefore, experimental tests could not be invoked during the early years to decide between the two theories. According to Goldberg, alogical and extrascientific factors must at first have guided preference for either of the two theories.[8]

---

modynamics in volumes where fluctuation phenomena could not be overlooked; and then to propose, in the third paper of vol. 17, a theory of principle as a new beginning for problems concerning the constitution of matter" (pp. 135–136).

6. Miller (Bib. 58), 102. Miller's portrayal of Einstein as thinker agrees with, and is largely derived from, Einstein's later pronouncements on the matter, especially in Einstein, "Notes" (ref. 2).

7. Einstein, "Motive des Forschens," in E. Warbur et al., *Zu Max Plancks sechzigsten Geburtstag: Ansprachen, gehalten am 26. April 1918 in der Deutschen Physikalischen Gesellschaft* (Karlsruhe, 1918), 29–32, on 31; inadequately exerpted in C. Seelig, ed., Einstein, *Mein Weltbild* (Frankfurt a.M., 1981), 107–110.

8. This circumstance is examined by Schaffner (Bib. 72).

The problem is still with us. Since positivistic philosophies of science have yet to yield a theory-free observational language or a universally valid logic of discovery, some scholars have invoked alogical factors to explain a scientist's choice of one set of "freely created" hypotheses over other sets that account equally well for a given body of data. Among the alogical factors proposed are the overall judgement of the scientific community, in-bred individual and group aesthetic commitments, and purely cultural, economic and institutional influences. This has opened contemporary historiography of science to charges of irrationalism, cultural relativism, and anarchy by some defenders of the rationalist faith.

Goldberg's commitment is to the overwhelming influence of culture.[9] He argues that, since no scientific method exists to insure objectivity, scientific truth should be regarded as culturally conditioned, that is, derived from individual beliefs that are "culturally shaped and subject to cultural currents" (p. 37). Science, thus dethroned, is accessible to the average cultured citizen and, equally important, subject to the policy decisions of an informed public rather than the whims of the technically informed few.

Miller and Goldberg diverge especially in their versions of Einstein's route to special relativity.[10] Simultaneity and the constancy of the speed of light yield the most obvious difference. Relying upon Einstein's late-in-life recollections of a thought experiment in 1895 or 1896 on how a light wave might appear to someone running along side it,[11] Miller traces the origins of most of Einstein's later assertions about the nature of time, electromagnetic waves, and "thinking itself" to this youthful inquiry (p. 189). New concepts of time and simultaneity, evolved in Einstein's inner thoughts, are thus for Miller prior and fundamental to the second postulate (the constancy of the speed of light) and to Einstein's formulation of special relativity. Miller takes the order of presentation of the paper on relativity to be the order of Einstein's creative thought.

Goldberg eschews reliance on Einstein's recollections,[12] and leans instead on the difference between the relativity paper and Lorentz's paper on electrodynamics. According to Goldberg, Lorentz's work provided a dynamical theory of matter and radiation, and Einstein's a "heurisitic" theory of measurement. The crucial element in the for-

9. Another recent example of this viewpoint is Andrew Pickering, *Constructing quarks: A sociological history of particle physics* (Chicago, 1984).

10. It is also evident in their differing interpretations of Poincaré. Compare Goldberg (Bib. 15, 19) and Miller (Bib. 53, 64).

11. Einstein, "Notes" (ref. 2), 52, and "Skizze" (ref. 2), 10.

12. Goldberg (Bib. 21).

mulation of special relativity was, for Goldberg, the substitution of a finite, constant speed of light for the infinite signal speed of classical theory. The constancy of the speed of light—an essentially untestable assertion, raised of necessity to a postulate—occasioned, together with the principle of relativity, redefinitions of space, time and simultaneity.[13] The temporal sequence is the opposite of Miller's and leads to an opposite interpretation of Einstein's paper; it "did not reflect the manner in which he originally thought through the theory."[14] Differing a priori aims have led our authors to differing readings.

3. ECO-BIOGRAPHY

Technical minutiae are not a feature of Lewis Pyenson's contribution. "These points are, and have been for many years, well-treated in the literature" (p. xii). Pyenson seeks instead to comprehend some of Einstein's more general intellectual predilections—as displayed in his relativity theory and in his subsequent career. Without invoking historical causality, he traces the sources of Einstein's early predilections to the "environment" created by the circumstances of his youth, family, and education. Pyenson demonstrates a remarkable resourcefulness and a facility in the use of diverse perspectives: from psychobiography, the sociology of outsiders, Wilhelmine business history, and pedagogy. In so far as it can be successful, Pyenson's approach is the modern biographer's answer to Einstein's skepticism and an essential element of any comprehensive biographical study with or without abundant primary sources. It is both complementary and contradictory to Miller's recent appeal to Piaget and cognitive psychology.[15]

Pyenson's book is an assembly of one new and eight slightly revised studies of the early Einstein in relation to his special theory of relativity. The first three articles attempt to resolve apparent contradictions in Einstein's orientations by appeal to three early environments: his mathematics education, his family's electrotechnical business, and his southern German Jewish roots.[16] The last of these is particularly provocative. Utilizing Simmel's sociological notion of the

13. Goldberg argues that the second postulate is empirically untestable because one-way measurements of the speed over a given path would require simultaneous readings of clocks at each end of the path. For a different analysis, based upon Poincaré's views, see Miller, 186–189.

14. Goldberg (Bib. 21), 244. Goldberg finds support in Einstein's later account in Einstein, *Uber die spezielle und die allgemeine Relativitätstheorie, gemeinverständlich* (Braunschweig, 1917).

15. Miller (Bib. 64), part 3.

16. Pyenson dismisses the supposed influence of Zurich's "bohemian" environment on Einstein argued by Feuer (Bib. 13).

"Fremder," Pyenson attributes much of Einstein's sense of being an "outsider" (*Einspänner*) to his family's Jewish heritage, and perhaps to his own early religiosity. At the same time his family's failing business ventures and his fin-de-siècle secondary education predisposed the future theorist to esteem experimental data and apparatus and to view mathematics as a useful, but limited, tool in attaining insights into nature.[17] This placed him at odds with the prevailing view among German mathematical physicists, who subsequently provided the initial reception—and the initial misunderstanding—of relativity theory.

4. RECEPTIVITY TO RELATIVITY

Pyenson and Goldberg agree that alogical factors "are central for understanding the reception of Einstein's work because, on the level of experimental results of predictions, there is very little difference between Einstein's etherless special theory of relativity and Lorentz's electron theory..." (Pyenson, p. 69). And Pyenson and Miller agree that Einstein inclined, not to abstract mathematics, but to what Miller calls "the perceptual-linguistic Kantian notion of mental imagery."[18] But they disagree on nearly every other aspect of the reaction to relativity.

Goldberg draws from his dissertation in comparing the receptions that the theory received in four major western countries.[19] Of the cultural factors influencing different receptions, Goldberg, a concerned teacher, focuses upon educational systems. For specialists, the discussion is unsatisfying. Goldberg oversimplifies or stereotypes national characterisitics: French education is reduced to Poincaré's preferences, the Germans are made overly rigid, and the Americans are characterized by the usual "contempt for abstraction" (p. 260). He refers specialists (for whom the book is not ostensibly written) to his other works for fuller and more nuanced accounts.[20] Miller is more helpful here in giving expositions of Kaufmann's experiments and experiments on the relativistic motion of rigid bodies, and discussing de-

17. Research has revealed at least two unfortunate errors in Pyenson's account: Einstein never had Adolf Sickenberger as a mathematics teacher; Einstein's father was not the only Hermann Einstein in business in Munich around 1890 (there was also an Antiquitätenhändler). Pyenson unduly restricts his attention to only two of Einstein's teachers at the Zurich Polytechnical School (ETH), the mathematician Minkowski and H.F. Weber, a leading technical physicist. Cf. McCormmach (Bib. 49).

18. Miller (Bib. 64), 120.

19. Goldberg (Bib. 17).

20. Goldberg (Bib. 17, 18, 20). Shinn (Bib. 74) offers a broader account of the French research system. Reactions of some Germans to relativity theory are recounted by Hermann and Benz (Bib. 23) and by Kleinert and Schönbeck (Bib. 44).

bates to which they gave rise. His conclusion that the Germans were more receptive to relativity because of their preference for Kantian "mental imagery"[21] clashes with Pyenson's claim that they did not readily accept the theory precisely because of it!

Most of Pyenson's book concerns the interplay of mathematics and physics in the reaction to relativity in Göttingen, the most influential center of German mathematics. In Pyenson's account, Göttingen's reaction was influenced by over-commitment to pure mathematics as the royal road to physical reality and by belief in a "pre-established harmony" between mathematics and physical reality that insured that physical theories could be created by manipulations of mathematics. This notion, enunciated in theological terms by Leibniz, was introduced into German education during the *Schulkrieg* of the 1890s. In that period, neo-humanism, and with it the supposed virtues of pure mathematics, lost favor to practical *Realstudien*. The powerful Göttingen mathematician Felix Klein sought to adapt pure mathematics to the new situation by establishing research programs and institutes in various areas of applied mathematics and mathematical physics. Appeal to the pre-established harmony was intended to protect the primacy of pure mathematics.[22]

The resulting environment in Göttingen accounted both for the initial misunderstanding of Einstein's special relativity there and for its mathematical reformulations and extensions by Minkowski, Hilbert, and others. Unlike the Göttingen mathematicians, Einstein was a "theoretical physicist," a new kind of creature for whom the book of nature was no longer written in mathematical symbols but in human language: "Nature's language is man's language" (p. 73). Here was a different kind of harmony, between human abstraction and physical data, unmediated by mathematics. Perhaps one of its clearest expressions may be found in Einstein's paper on special relativity.[23]

21. Miller (Bib. 64), 120. Strict Kantianism would, of course, have prejudiced adherents against relativity.

22. See also Pyenson (Bib. 70). Klein's efforts are discussed by Karl-Heinz Manegold, *Universität, Technische Hochschule und Industrie: Ein Beitrag zur Emanzipation der Technik im 19. Jahrhundert unter besondere Berücksichtigung der Bestrebungen Felix Kleins* (Berlin, 1970).

23. Pyenson (p. 152) cites Einstein's own reference to "pre-established harmony" in 1918 as evidence for the notion's pervasiveness. Einstein (ref. 7). But Einstein's use of the term differs from that used earlier in Göttingen. Also, as Howard (Bib. 35) recently argued, Einstein's version apparently grew out of his intervening work on general relativity and his interaction with the philosopher Moritz Schlick.

5. THEORY DESPITE MATHEMATICS

A new conceptual language, a new physics, not primarily a new mathematics, was, as Pyenson argues, an essential element of special relativity. This was the point the Göttingen scientists failed at first to grasp. As Russell McCormmach emphasized over a decade ago, and as Pyenson reiterates, Einstein and special relativity must be placed within the context of the rise of a new subdiscipline, theoretical physics, of which they were early representatives.[24]

Theoretical physics, which began to emerge from mathematics during the last decades of the nineteenth century, only gradually made room for itself in the sphere of established physics.[25] In Wilhelmian Germany physics meant experimental physics, presided over by powerful heads of well-staffed and well-equipped academic laboratories.[26] By the turn of the century, the influence of academic experimentalists was under increasing challenge from inside and outside the profession. The demands of rapidly expanding technical industries and an increasing cultural and scientific malaise surrounding the decline of the prevailing metaphysical world view—peculiarly essential for Germans—brought new tasks to German science. The situation exerted at least two influences on young physicists. The fully occupied, hierarchical structure of German academe encouraged talented youngsters to enter less populated boundary disciplines such as theoretical physics;[27] and theoretical physics attracted thinkers like Planck and Einstein, who were more intrigued by the search for universals and the fundamentals of a new unifying world view than the solutions of mathematical formulae or the behavior of industrial materials.[28]

Although a Swiss citizen and resident when he invented relativity, Einstein's personal inclinations and academic situation made him in both regards typical of the new German theoretical physicists. But theoretical physicists were uncommon during the first decade of this century. According to data presented by Forman, Heilbron and Weart, of a total of 3170 physicists throughout the world in 1909, only 17 defined themselves as theoretical physicists, and 10 of these were Germans.[29] The history of relativity and of Einstein, even when

24. McCormmach (Bib. 46, 49).

25. Ibid., and Olesko (Bib. 66). See also the forthcoming treatise by Christa Jungnickel and Russell McCormmach.

26. See, for example, Cahan (Bib. 6).

27. This phenomenon was pointed out by Joseph Ben-David, *The scientist's role in society: A comparative study* (Englewood Cliffs, N.J., 1971). It occurred in other disciplines as well.

28. See Einstein,"Notes"(ref. 2), and Max Planck, *Wissenschaftliche Selbstbiographie* (Leipzig, 1955), 7.

29. Paul Forman, J.L. Heilbron, and Spencer Weart, "Physics circa 1900: Personnel,

viewed against the wider backgrounds of Einstein's life and work, is thus the study of an important but minuscule minority of physicists and physics problems at the turn of the century.

A comprehension of the nature of all of physics at that time is essential for a full understanding of minority contributions such as relativity theory. Cross fertilizations of ideas and influences no doubt occurred through the training of future theorists in traditional disciplines, through the wide publicity of work in other areas, such as radioactivity, and through the awareness among professional theorists that many of their problems and solutions were not shared or easily understood by their colleagues.[30] While the reactions of experimentalists to relativity were guided, at least in part, by their attitudes toward the theoretical physics profession, theorists were not unmindful that their profession was part of a much larger discipline dominated in terms of commitments, authority, and jobs by persons outside their immediate specialty. The response of professional theorists to this situation may be reflected in their work, the problems they chose, and the way they chose to present both.

Another essential feature of German theoretical physics, one that lent it enormous disciplinary definition and status, was its function as the locus of the search for a new scientific world view. This task helps to account for the relatively large number (10) of theorists residing in Germany in 1910. It was an important task, for both scientists and the public required a replacement for the declining and fragmented mechanical view, as defined by nineteenth-century physics, and Einstein fully participated in the effort.[31] His researches through 1905 made clear to him the inadequacy of available choices and the need for alternatives.[32] Special relativity, he apparently thought at first, might even lead to a new "relativity world view."

World views were ultimately at stake in the German debate over relativity. Einstein's equal treatment of mechanics and electrodynam-

funding, and productivity of the academic establishments," *HSPS*, 5 (1975), 1–185, Table A.7, pp. 34–35. In the countries surveyed for 1900, the number of posts in theoretical and mathematical physics as a percentage of all physics posts varied from 0%(Japan) to 25%(Netherlands), with 15% for Germany (ibid., table A.5, p. 31), but the absolute numbers were very small. Germany offered the largest number of such posts, 16.

30. Even within theoretical physics differentiation occurred according to specialties, such as quantum theory, relativity, and atomic physics.

31. McCormmach (Bib. 46, 47); Hirosige (Bib. 27); Holton (Bib. 32). Cf. Miller, 131–132. Mechanical explanation and criticism of it are discussed by Klein (Bib. 40), and J.L. Heilbron, "Physics at the fin-de-siècle," in C.G. Bernhard et al., eds., *Science, technology, and society in the time of Alfred Nobel* (Oxford, 1982), 51–73 .

32. See the works in ref. 4.

ics after several years of struggling with both seemed to some adherents of electromagnetism a throw-back to mechanism, to others a reformulation of Lorentz's electrodynamics.[33] Even more attention to the commitment of Göttingen's "electron theory" seminar to the electromagnetic world view would enrich Pyenson's account of Göttingen's reception of relativity. Only after scientists understood that each of the world views used by Einstein was inadequate but essential were the full implications of special relativity realized and accepted.[34] From this point of view, Lorentz's precipitous renunciation of his theory in 1906 in a letter to Poincaré seems less a case of instant falsification as Miller claims (p. 334) than an expression of exasperation at trying to reconcile his deformable electron with the electromagnetic orthodoxy espoused by Kaufmann and Abraham.[35]

Some elements of the interplay between pure mathematics, experimental physics and the metaphysical demands of the broader culture in attitudes toward relativity appeared in varying degrees outside Germany too. But any international comparison must take account, not only of the content of relativity theory as perceived in each nation, but also of its German-speaking origins and supporters. Pervasive cultural nationalism during the last decades before the First World War is evidenced, for example, by the public competition for Nobel Prizes, a circumstance that certainly played a role in the reception of relativity within and beyond German-speaking lands.[36]

6. TEXTBOOK REVIEWS

Goldberg's study in the last third of his book of the American reception of relativity as revealed by textbooks is uniquely useful for the reasons just reviewed and fascinating in its own right. Americans competed less directly with Germans in most matters of national prestige than did Europeans, and Americans perhaps therefore accepted German science more readily than others did. But they translated it into their own terms, as Goldberg's analysis of textbooks shows. He

33. McCormmach (Bib. 47,48); Hirosige (Bib. 27).

34. Cf. Hirosige (Bib. 27). The world views of that era require much further definition and study. For instance, the place and use of atomism in these views, especially as atomic parts and particles began to proliferate, are still uncertain.

35. Abraham's electron deformed in uniform motion in the transverse direction, while Lorentz's electron deformed longitudinally. Both were considered dynamical effects of the electromagnetic field. The technical details are discussed by Miller, the world-view implications by McCormmach (Bib. 48).

36. Elisabeth Crawford, *The beginnings of the Nobel Institution: The science prizes, 1901–1915* (New York and Paris, 1984). Lack of facility in reading Einstein's German may also have posed a problem for foreigners.

concludes that as late as 1980, when his survey ends, that American physics students were shielded from the full import of Einstein's theory.

Since American culture and science emphasized empiricism, practicality and restrained abstraction, relativity theory was presented, not as a kinematic theory of measurement, but as a modified form of Lorentz's electrodynamics, a semi-empirical, dynamical theory about nature itself. Goldberg finds that the two postulates were usually presented as empirically tested, and that the Michelson-Morley experiment was universally cited as a motivation for the theory. The wide popularity of Bridgman's operationism, originating from his interpretation of relativity, is indicative of the empiricist propensity in American theoretical physics.[37] Goldberg concludes that a Kuhnian revolution required by Einstein's relativity has yet to occur in American physics. But American practicality allowed a less pretentious assimilation.

As valuable and revealing as Goldberg's textbook analyses may be, textbooks are, of course, only one indication of the reception and use of a theory, and Einstein's theory in particular requires further study in the American context. Einstein's visits to America before settling in Princeton, his activities after arriving there, and the impact of other German theories and theorists after World War I, greatly influenced and altered American perceptions of relativity and of theoretical physics in general.[38] Much remains to do on the latter problem, as on the entire history of relativity theory and its author.

Goldberg, Miller, and Pyenson have, together with numerous other scholars, provided invaluable indications and interpretations. Nevertheless, as indicated throughout this essay, understanding the history of special relativity still requires much fuller accounts of many facets of that history than are presently available. It also requires *better utilization of existing studies of the same material.* Only by diversifying, coordinating, and learning from one another can we ever hope to comprehend this immense subject, the history of Einstein's theory of relativity and its place in modern physics.

37. Miller and Goldberg diverge again over Bridgman. Compare Miller (Bib. 63) and Goldberg, 293–307.

38. This is especially emphasized by Sayen (Bib. 71). See also Holton (Bib. 33); Woolf (Bib. 91); and Katherine Russell Sopka, *Quantum physics in America, 1920–1935* (New York, 1980). Current American historiography of theoretical physics in the United States is nicely summarized by Albert E. Moyer, "History of physics," *Osiris*, series 2, *1*, (1985), 163–182.

Selected Bibliography

Items are selected for their relevance to topics discussed in the essay. For abbreviations used see footnote on page 177.

1. Balàzs, Nandor L. "Einstein: Theory of relativity." *Dictionary of scientific biography, 4*, 319–332.

2. Bergia, Silvio. *Einstein e la relatività.* Bari, 1978.

3. ______. "Einstein and the birth of special relativity." In French, 65–89.

4. Bergmann, Peter G. "The development of the theory of relativity." *AS*, 1–16.

5. Byrne, Patrick H. "Statistical and casual concepts in Einstein's early thought." *Annals of science, 37* (1980), 215–228.

6. Cahan, David. "The institutional revolution in German physics, 1865–1914." *HSPS, 15* (1985), 1–65.

7. Costa de Beauregard, Olivier. "Einstein créateur en relativité et en méchanique quantique." *Einstein, 1879–1955. Colloque du centenaire, Collège de France, 6–9 juin 1979.* Paris, 1980. Pp. 49–55.

8. Cullwick, E.G. "Einstein and special relativity. Some inconsistencies in his electrodynamics." *BJPS, 32* (1981), 167–176.

9. Cushing, James J. "Electromagnetic mass, relativity, and the Kaufmann experiments." *AJP, 49* (1981), 1133–1149.

10. Cuvaj, Camillo. "Henri Poincaré's mathematical contributions to relativity and the Poincaré stresses." *AJP, 36* (1968), 1102–1113.

11. ______. *A history of relativity. The role of Henri Poincaré and Paul Langevin.* Ph.D. thesis, Yeshiva University, 1971.

12. ______. "Paul Langevin and the theory of relativity." *JSHS, 10* (1971), 113–142.

13. Feuer, Lewis. *Einstein and the generations of science.* New York, 1974.

14. Galison, Peter Louis. "Minkowski's space-time. From visual thinking to the absolute world." *HSPS, 10* (1979), 85–121.

15. Goldberg, Stanley. "Henri Poincaré and Einstein's theory of relativity." *AJP, 35* (1967), 934–944.

16. ______. "The Lorentz theory of electrons and Einstein's theory of relativity." *AJP, 37* (1969), 982–994.

17. ______. *The early response to Einstein's special theory of relativity, 1905–1911. A case study in national differences.* Ph.D. thesis, Harvard University, 1969.

18. ______. "In defense of ether. The British response to Einstein's special theory of relativity, 1905–1911." *HSPS, 2* (1970), 89–125.

19. ______. "Poincaré's silence and Einstein's relativity. The role of theory and experiment in Poincaré's physics." *British journal for the history of science, 5* (1970), 73–84.

20. ______. "Max Planck's philosophy of nature and his elaboration of the special theory of relativity." *HSPS, 7* (1976), 125–160.

21. ______. "Albert Einstein and the creative act: The case of special relativity." In Rutherford Aris et al., eds., *Springs of scientific creativity. Essays on founders of modern science.* Minneapolis, 1983. Pp. 232–253.

22. Gutting, Gary. "Einstein's discovery of special relativity." *Philosophy of science, 39* (1972), 51–68.

23. Hermann, Armin, and Ulrich Benz. "Quanten- und Relativitätstheorie im Spiegel der Naturforscherversammlungen 1906–1920." In H. Querner and H. Schipperges, eds., *Wege der Naturforschung 1822–1972 im Spiegel der Versammlungen Deutscher Naturforscher und Ärzte.* Berlin, 1972. Pp. 125–137.

24. Hirosige, Tetu. "Electrodynamics before the theory of relativity, 1890–1905." *JSHS, 5* (1966), 1–49.

25. ______. "Theory of relativity and the ether." *JSHS, 7* (1968), 37–53.

26. ______. "Origins of Lorentz' theory of electrons and the concept of the electromagnetic field." *HSPS, 1* (1969), 151–209.

27. ______. "The ether problem, the mechanistic worldview, and the origins of the theory of relativity." *HSPS, 7* (1976), 3–82.

28. Holton, Gerald. *Thematic origins of scientific thought. Kepler to Einstein.* Cambridge, Mass., 1973.

29. ______. "What, precisely, is 'thinking'? Einstein's answer." In French, 153–164.

30. ______. "Einstein's model for constructing a scientific theory." *AS*, 109–136.

31. ______. "Einstein's scientific program. The formative years." In Woolf, 49–64.

32. ______. "Einstein's search for the *Weltbild.*" American Philosophical Society. *Proceedings, 125* (1981), 1–15.

33. ______. "The formation of the American physics community in the 1920s and the coming of Albert Einstein." *Minerva, 19* (1981), 569–581.

34. ______. "Introduction. Einstein and the shaping of our imagination." In *HE,* vii–xxxii.

35. Howard, Don. "Realism and conventionalism in Einstein's philosophy of science. The Einstein-Schlick correspondence." *Philosophia naturalis, 21* (1984), 616–629.

36. Janich, Peter. "Die erkenntnistheoretischen Quellen Einsteins." In H. Nelkowski et al., eds., *Einstein Symposion Berlin. Aus Anlass der 100. Wiederkehr seines Geburtstages, 25. bis 30. März 1979.* Berlin, 1979. Pp. 412–427.

37. Kanitscheider, Bernulf. "Einstein's treatment of theoretical concepts." *AS*, 137–158.

38. Keswani, G.H. "Origin and concept of relativity." *BJPS, 15* (1965), 286–306; *16* (1966), 19–32.

39. Klein, Martin J. "Thermodynamics in Einstein's thought." *Science, 157* (1967), 509–516.

40. ______. "Mechanical explanation at the end of the nineteenth century." *Centaurus, 17* (1972), 58–82.

41. ______. "Einstein and the academic establishment." In French, 209–214.

42. ______. "Einstein, Albert." *Dictionary of scientific biography, 4,* 312–319.

43. ______. "Some turns of phrase in Einstein's early papers." In Abner Shimony and Herman Feshbach, eds., *Physics as natural philosophy. Essays in honor of Laszlo Tisza on his 75th birthday.* Cambridge, Mass., 1982. Pp. 364–375.

44. Kleinert, Andreas, and Charlotte Schönbeck. "Lenard und Einstein. Ihr Briefwechsel und ihr Verhältnis vor der Nauheimer Diskussion von 1920." *Gesnerus, 35* (1978), 318–338.

45. Lanczos, Cornelius. *The Einstein decade, 1905–1915.* New York, 1974.

46. McCormmach, Russell. "Editor's foreword." *HSPS, 2* (1970), ix–xx.

47. ______. "Einstein, Lorentz, and the electron theory." Ibid., 41–87.

48. ______. "H.A. Lorentz and the electromagnetic view of nature." *Isis, 61* (1970), 459–497.

49. ______. "Editor's foreword." *HSPS, 3* (1971), ix–xxiv.

50. ______. "On academic scientists in Wilhelmian Germany." *Daedalus, 103* (1974), 157–171.

51. ______. "Editor's foreword." *HSPS, 7* (1976), xi–xxxv.

52. ______. "Lorentz, Hendrik Antoon." *Dictionary of scientific biography, 8,* 487–500.

53. Miller, Arthur I. "A study of Henri Poincaré's 'Sur la dynamique de l'électron'" *Archive for history of exact sciences, 10* (1973), 207–328.

54. ______. "On Lorentz's methodology." *BJPS, 25* (1974), 29–45.

55. ______. "Albert Einstein and Max Wertheimer. A Gestalt psychologist's view of the genesis of special relativity theory." *History of science, 13* (1975), 75–103.

56. ______. "On Einstein, light quanta, radiation and relativity in 1905." *AJP, 44* (1976), 912–923.

57. ______. "The physics of Einstein's relativity paper of 1905 and the electromagnetic world picture of 1905." *AJP, 45* (1977), 1040–1048.

58. ______. "On the history of special relativity theory." *AS*, 89–108.

59. ______. "On some other approaches to electrodynamics in 1905." In Woolf, 66–91.

60. ______. "Unipolar induction. A case study of the interaction between science and technology." *Annals of science, 38* (1981), 155–189.

61. ______. "The special relativity theory. Einstein's response to the physics of 1905." In *HE*, 3–26.

62. ______. "On Einstein's invention of special relativity." Philosophy of Science Association. *Proceedings, 2* (1982), 377–402.

63. ______. "Introduction. P.W. Bridgman and the special theory of relativity." In P.W. Bridgman. *A sophisticate's primer of relativity.* 2nd ed. Middletown, Conn., 1983. Pp. vii–xlviii.

64. ______. *Imagery in scientific thought. Creating 20th-century physics.* Boston, 1984.

65. Nagel, Ernest. "Relativity and twentieth-century intellectual life." In Woolf, 38–45.

66. Olesko, Kathryn M. *The emergence of theoretical physics in Germany. Franz Neumann and the Königsberg school of physics, 1830–1890.* Ph.D. thesis, Cornell University, 1980.

67. Pais, Abraham. *"Subtle is the Lord...." The science and the life of Albert Einstein.* Oxford, 1982.

68. Pyenson, Lewis, and Douglas Skopp. "Educating physicists in

Germany circa 1900." *Social studies of science, 7* (1977), 329–366.

69. Pyenson, Lewis. "La place des sciences exactes en Allemagne à l'époque de Guillaume II." *Europa, a journal of interdisciplinary studies, 4* (1981), 187–217.

70. ______. *Neohumanism and the persistence of pure mathematics in Wilhelmine Germany.* Philadelphia, 1983.

71. Sayen, Jamie. *Einstein in America. The scientist's conscience in the age of Hitler and Hiroshima.* New York, 1985.

72. Schaffner, Kenneth. "Einstein versus Lorentz. Research programmes and the logic of comparative theory evaluation." *BJPS, 25* (1974), 45–78.

73. ______. "Space and time in Lorentz, Poincaré, and Einstein. Divergent approaches to the discovery and development of the special theory of relativity." In P.K. Machamer and R.G. Turnbull, eds., *Motion and time, space and matter. Interactions in the history of philosophy of science.* Columbus, Ohio, 1976. Pp. 465–507.

74. Shinn, Terry. "The French science faculty system, 1808–1914. Institutional change and research potential in mathematics and physical sciences." *HSPS, 10* (1979), 271–332.

75. Schöpf, Hans-Georg. "Albert Einsteins annus mirabilis, 1905." *NTM. Zeitschrift für Geschichte der Naturwissenschaft, Technik und Medizin, 15:2* (1978), 1–17.

76. Schwartz, H.M. "Poincaré's Rendiconti paper on relativity." *AJP, 39* (1971), 1287–1294; *40* (1972), 862–872, 1282–1287.

77. ______. "Einstein's comprehensive 1907 essay on relativity." *AJP, 45* (1977), 512–517, 811–817, 899–902.

78. Scribner, Charles, Jr. "Henri Poincaré and the principle of relativity." *AJP, 32* (1964), 672–678.

79. Seelig, Carl. *Albert Einstein. Leben und Werk eines Genies unserer Zeit.* Zurich, 1960.

80. Stachel, John, and Roberto Torretti. "Einstein's first derivation of mass-energy equivalence." *AJP, 50* (1982), 760–763.

81. Stachel, John. "Einstein and Michelson. The context of discovery and the context of justification." *Astronomische Nachrichten, 303:1* (1982), 47–53.

82. Suvorov, S.G. "Einstein. The creation of the theory of relativity and some gnosiological lessons." *Soviet physics uspekhi, 22* (1979), 528–554.

83. Swenson, Loyd S. *The ethereal aether. A history of the Michelson-Morley-Miller aether-drift experiments, 1880–1930.* Austin, 1962.

84. ______. *Genesis of relativity. Einstein in context.* New York, 1979.

85. Tonnelat, M.A. "La relativité avant Einstein." *Organon, 2* (1965), 79–103.

86. ______. *Histoire du principe de relativité* (Paris, 1971).

87. Treder, Hans-Jürgen. *Über Prinzipien der Dynamik von Einstein, Hertz, Mach und Poincaré. Zur Geometrisierung der Relativität der Beschleunigung und der Trägheit.* Berlin, 1974. (Akademie der Wissenschaften, Berlin. Forschungsbereich kosmische Physik. *Veröffentlichungen, 4*).

88. ______. "Die Einheit der Physik bei Helmholtz, Planck und Einstein." In Hans-Jürgen Treder, ed., *Einstein-Centenarium 1979. Ansprachen und Vorträge auf den Festveranstaltungen des Einstein-Komitees der DDR. Berlin, 1979.* Pp. 240–255.

89. Vizgin, V.P. *Relativistskaia teoriia tiagoteniia. Isotoki i formirovanie 1900–1915.* Moscow, 1981.

90. Whittaker, E.T. *A history of the theories of aether and electricity.* 2 vols., revised ed. London, 1951–1953.

91. Woolf, Harry. "Albert Einstein. Encounter with America." In Woolf, 28–37.

LEWIS PYENSON

THE RELATIVITY REVOLUTION IN GERMANY

In 1905, at around the time that the first of Albert Einstein's publications of the year was to appear, the young theoretical physicist wrote to admonish a close friend of his for not having remained in touch. The friend, Conrad Habicht, was along with Einstein a member of a small study group, the Olympia Academy, which had been meeting in Berne for several years. The tone of the letter is light-hearted; its prose is littered with clumsy attempts at irony and self-mockery. Einstein asked to see Habicht's doctoral dissertation, as he would be one of the few people around who could understand it; Einstein assessed Habicht a stiff increase in his study-group fees as a penalty for non-attendance. Then there is mention of Einstein's own work. Among the four works of his that would be appearing in 1905, Einstein indicated that one – the paper on the light quantum – set forth "very revolutionary" consequences.[1] He had the opportunity of identifying the theory of relativity as a revolutionary doctrine, and he declined to do so.

Einstein's private remark suggests that, in 1905, he saw his first paper on special relativity as a preliminary and tentative approach toward resolving some of the fundamental problems associated with the electron theory. He certainly had firm reasons for believing that his approach yielded results superior to those of other theories being discussed, but he thought that more work was required to settle the point. He sought to contribute to a revolution that was already underway. Events moved quickly. By 1912 Einstein was the master of the relativity revolution.

In the following pages, I reconsider the relativity revolution in Germany from the standpoint of social history of science. The technical content of Einstein's work and that of his colleagues (for which good analyses have long been available) recedes in favor of seeking broader explanations for the reception accorded to relativity.[2] The focus is placed on the *revolutionary* dimensions of what happened. I begin by affirming how, in the decades before 1914, physicists and other commentators spoke in unambiguous terms about the revolution that was then in progress. The thread of rhetoric regarding revolution, commencing before what we now identify as Einstein's revolutionary thought, suggests that physicists in Central Europe were particularly disposed to seeking revolutionary solutions for the scientific problems that they faced. After having identified the revolutionary urge, I indicate its incarnation, as well as its rejection, in the various positions regarding relativity which were taken by physicists over the first two decades of the twentieth century. The inspiration for this recounting is found in the French Revolution of the late eighteenth century. Once the course of relativity's reception has been delineated, the issue of a social matrix for the events is considered, with special attention devoted to patterns of publishing and apprenticeship.

The literature surrounding scientific revolutions, especially in the generation following the publication of Thomas Kuhn's book on the topic, is enormous. Kuhn is largely silent, however, on the matter of revolutionary structure – of what happens during a revolutionary crisis and whether there are stages of development in the period falling between two paradigms. Some scientific revolutions have been studied with the object of delineating their structures, and broad remarks have been directed to comparing scientific revolutions in general with political revolutions in general.[3] Examining an unambiguous scientific revolution from the perspective of one of the best known political revolutions, however, seems not to have been undertaken recently. It should be clear at the outset that the author of any such attempt is well advised to take account of Max Weber's strictures, offered in 1909, about uncritically transferring forms of thinking from one discipline to another.[4]

The task of a writer seeking to fathom the twentieth-century revolution in exact sciences has recently been outlined by the Japanese historian of science Chikara Sasaki. He follows Kuhn in believing that science proceeds by nonlinear and drastic changes, or revolutions. During such scientific revolutions, there is a complete transformation in both the conceptual and institutional environments.[5] Sasaki believes that the core of the matter lies in social symbols, which condition the way people view nature. Because science is a product of everyday life, scientific revolutions seem to parallel social and political ones.[6] The two concurrent revolutions are, however, only synchronic; their forms are not necessarily homologous. Although Sasaki believes that changes in social reality precede changes in the image of nature, he affirms that scientific discourse cannot be a mere byproduct of social intercourse. Neither can institutional support be a starting point for scientific ideas.[7]

Factors external to scientific discourse do not determine disciplinary content. If internal history cannot be deduced from external history, external history is nevertheless more than a mere framework for internal history. The two structures are interactive and indissociable.[8] There is an organic relationship between ideas and environment.

In his massive investigation of scientific revolutions, Sasaki suggests, as much by what he writes as by what he leaves out, the difficulty of revealing such organic unity. The parameters of the ideational content of a scientific revolution in progress, especially, seem difficult to capture. Sasaki disallows the notion, often attributed to Kuhn, that a scientific revolution simply replaces one dogma by another. He argues for the primacy of historical specificity: To speak of dogma in general, he reminds us, is to engage in scholasticism. He allows on this point that something like the notion of a permanent revolution might provide a materialist response to idealist formulations.[9] For Sasaki, Marx and the French *annalistes* provide a useful corrective to Kuhn.[10] Because his is a materialist treatment, Sasaki avoids considering whether events in politics can provide an analogical or abstract structure for changes in ideas and institutions.[11]

I have introduced Sasaki's work at some length because it is one of the most suggestive of recent publications on scientific revolutions. Other treatments, notably I. Bernard Cohen's *Revolution in Science*, might have similarly been plumbed. The exercise is not without interest, but it would not bear directly on the question at hand: how to understand what was perceived then and now as a revolution in physics. For this reason, the present text, finding profit and inspiration in the work of Sasaki and others, does not refer to their work in a systematic way.

The revolution here is that of relativity in Germany, but in what follows it will be apparent that many of the figures were not German. Some, like Ludwig Boltzmann, Philipp Lenard, Menyhért Palágyi, Gustav Herglotz, and Jakob Laub, were born and raised in various parts of the Austrian Empire. Others, like Walter Ritz and Jakob Kunz, spent many of their early years as physicists in their native Switzerland. Einstein himself, from even before his years at the Patent Office in Berne until the late 1930s in Princeton, travelled as a Swiss citizen. All these people and many others published extensively in German journals or with German presses, and most held positions of one kind or another in Germany. Borders in German-speaking Europe were easily crossed before 1914. The relativity revolution in Germany owes much to foreign infiltrators and agitators, and for this reason although the present text focuses on Germany, it does not do so exclusively.

If the cast of characters is seen as large and somewhat heterogeneous, the period of time under consideration will be found indistinct at both beginning and end. Revolutionary *discourse* in physics extends back before the discovery of the electron, and uncertainty among senior and otherwise distinguished physicists about the basic tenets of special relativity pushes forward into the 1920s. Such a continuum, I would suggest, is intrinsic to any attempt at providing a nuanced social history of ideas which deals with relatively sudden innovations as well as with patterns of activity reflected in the *longue durée*. Introducing his major synthesis of renascences in Western art, Erwin Panofsky emphasized:

> The very definition of a period as a phase marked by a 'change of direction' implies continuity as well as dissociation. We should, moreover, not forget that such a change of direction may come about, not only through the impact of one revolutionary achievement which may transform certain aspects of cultural activity as suddenly and thoroughly as did, for example, the Copernican system in astronomy or the theory of relativity in physics but also through the cumulative and, therefore, gradual effect of such numerous and comparatively minor, yet influential modifications as determined, for example, the evolution of the Gothic cathedral from Saint-Denis and Sens to Amiens.[12]

Panofsky, who was wont to think in terms of epochs spanning half a millenium, would naturally have seen Copernicus and Einstein as triumphing in a trice. He continued, in the passage just cited, to underscore that the stock of human ideas suffered attrition as well as replenishment:

> A change of direction may even result from negative rather than positive innovations: just as more and more people may accept and develop an idea or device previously unknown, so may more and more people cease to develop and ultimately abandon an idea or device previously familiar.

For the historian of physics, this passage calls to mind the electromagnetic ether, which, into the twentieth century, had a steadily decreasing circle of admirers as it proved to be more trouble than it was worth. The fate of the ether is only one story that may be informed by Panofsky's vision. The difficulties, but also the rewards, of his approach are reflected in the following pages.

1. THE REVOLUTIONARY URGE AND THE REVOLUTIONARY IDENTIFICATION

Germans may choose from a wide range of possibilities in order to bring nuance to the English notion of revolution. There is, of course, *Revolution*, coming directly from the French term. One also has *Grundstürzung*, a revolution with destructive or radical connotation. Then come the transfigurational senses: *Umsturz*, an overthrow, downfall, or ruin; *Umwälzung*, an upheaval or overthrow; *Umgestaltung*, *Umwandlung*, *Umänderung*, a transformation or metamorphosis; *Umstellung*, a transformation or conversion; *Umschwung*, a sudden change, like a giant swing on a gymnastic high bar. In the period under study, all these words were used to describe what was happening in physics. It was a choice of vocabulary special to Central Europe. French writers did not often refer to a *révolution* in physics. British commentators, too, did not speak in revolutionary terms.[13]

Revolution is unmistakably a part of the vocabulary of physicists, mathematicians, and scientifically-minded writers who were active in Germany in the years around 1900. Sought was a new vision that would be able to synthesize the specialized knowledge that had, for generations, been issuing from the pens of the nation's learned men. The mathematician and poet Felix Hausdorff, in reviewing one of Wilhelm Ostwald's books on energeticism, pointed to the issue with signal clarity in 1902:

> Whoever understands the signs of the times cannot fail to recognize that the enterprise of natural science has recently undergone an *Umgestaltung*. The period of unquestioned, unqualified specialization, the signature of the last decade, appears close to its end, while the synthetic, comprehensive, universalist spirit dares emerge from its long winter's sleep. The present yearns for a unified worldview in which the great profusion of isolated knowledge will be built into a unified, systematic whole.[14]

The researcher had to prepare himself for the possibility of completely changing his understanding of things, Paul Drude concluded in his 1894 inaugural address at the University of Leipzig. Truth was subjective, in his opinion, and one's approach to it could not be rigid. "The researcher will not despair, in striving toward the truth," Drude went on, "if a new tool requires *Umwälzungen* in familiar views."[15] In reviewing the development of pure mathematics over the preceding fifteen years, Siegfried Aronhold noted, in 1899, the *Umwälzung* in physical theories in molecular physics, acoustics, optics, thermodynamics, and electricity. These theories were "in part completely revised" due to the applications of mathematics.[16]

The preceding extracts, having been advanced in the passive mode, obscure the extent of opinion on the structure of scientific revolutions. Revolutions were, for a number of writers, the result of one man's revolutionary insight. According to the young neoidealist philosopher Ernst Cassirer in his first book-length publication of 1902, Leibniz sought "not a *Revolution* in the outcome of thought but rather in the way of thinking."[17] The young astrophysicist Karl Schwarzschild lectured the Cassel Naturforscherversammlung in 1903 on Henri Poincaré's "*revolutionierenden* law" of 1890 in celestial mechanics.[18] Schwarzschild's Göttingen colleague, mathematician Felix Klein, spoke in 1904 about Heinrich Hertz's thoroughgoing *Umänderung* of the intuitive view of mechanics, which, although keeping to the basic ideas of Helmholtz and Maxwell, replaced the notion of potential energy by that of kinetic energy of hidden motion.[19] When, in the same year, Klein pointed to the successful *Umschwung* in mathematics instruction at the secondary schools and chided his colleagues in higher education to bring about a similar *Umschwung*, he sought to claim the mantle of leadership in a pedagogical revolution.[20] Klein no doubt knew that just two years previously one of his supporters, the mathematician Paul Staeckel, had addressed the *Umschwung* in mathematics instruction which had taken place during the middle of the preceding century.[21]

Revolutions, commentators noted, produced counterrevolutionaries. In considering the relationship between mechanical engineering and science and life, Otto Kammerer emphasized in 1899 that the new technological man was the product of revolutionary Europe at the end of the preceding century: The "storm of the *Revolution* against the worn-out edifice of the state" resulted in the freeing of a profession, although the "new ideas" were stiffled by *Reaktion*.[22] The theoretical physicist Ludwig Boltzmann, in fact, identified himself, in 1899, somewhat ironically on the side of the reaction. He noted that "when I look back on all these developments and *Umwälzungen* [that had taken place since his student days], I feel like a monument to ancient scientific memories." He felt that he was a "*Reaktionär*, one who has stayed behind and remains enthusiastic for the old classical doctrines as against the men of today."[23] Returning to Austria in 1902, Boltzmann then changed the terms of revolutionary reference. In his inaugural lecture at the University of Vienna, he noted the *Reaktion* against the mechanical explanation of nature. He portrayed himself as the liberating revolutionary leader arrayed against unrepentant recidivism.[24]

Giving Boltzmann cause for alarm were the partisans of the electromagnetic view of nature – those hoping that some improvement in the laws of electrodynamics would be found to lie at the base of all physical theory.[25] The industrial and commercial applications of electricity contributed to its popular image as a Promethean liberator – the image portrayed, for example, in the official promotional illustrations for the Frankfurt electrotechnical exhibition of 1891.[26] This would reasonably have been the reference intended by writer Paul von Zech, who in 1892 referred in the *Deutsche Revue* to the "stupendous *Revolution*" of nineteenth-century electrical theory.[27] Zech's view extended to university lecture halls. Entirely familiar with laboratory experiments of the day, Hermann von Helmholtz emphasized in his lectures of 1888/89 that a "complete *Umwälzung*" was brewing in electromagnetic theory, once the contemporary crisis in the field had been resolved.[28]

In such a climate of revolutionary optimism, it would be normal that the startling results and experiments surrounding the electron theory, beginning to appear about 1899, were seen as signaling the arrival of a new vision of physics. Walter Kaufmann, one of the first experimentalists to focus all his attention on the electron, emphasized in 1901 that the history of science experienced, from time to time, major changes in long-accepted views. One such example of this kind of change was the *Umwälzung* in views regarding electrical processes which had occurred over the past decades.[29] The various theoretical formulations of electron dynamics, however, even after having taken into account Einstein's and Hermann Minkowski's seminal work, left unanswered questions regarding the basic shape and constitution of the elemental charge of negative electricity and, which is more noteworthy from our perspective, experimental verification of the electron's variation in mass as a function of velocity. Especially in view of the latter, unresolved question, the old Göttingen experimentalist Eduard Riecke, in speaking to the Naturforscher in 1909, qualified his sense of the revolution that had just occurred. He observed that future physicists would speak of the early twentieth century as the epoch of the electron theory. The theory had not fulfilled the expectation that it would be able to solve all the problems of physics, but it "concluded a movement which lifted the whole structure of our mechanics off its hinges and which threatens to *umzustürzen* deeply rooted views."[30] Partisans of the electromagnetic view of nature, in the throes of elaborating a dynamics for the electron, also expressed reserve about the new point of view. Wilhelm Wien, writing on the electron theory in 1904, cautioned: "The great *Umschwung* that physics has undergone in recent times doubly admonishes us to look to other theoretical possibilities and thus always to leave open the possibility of a change in the chosen path."[31] A Dutch physicist, Cornelis Harm Wind, writing in the *Physikalische Zeitschrift* in 1905, emphasized that the new revolution could itself be overthrown. Reflecting on the "crisis" in mathematical physics announced by Henri Poincaré at the St. Louis World's Fair in 1904, Wind wondered whether the invariability of mass with respect to velocity, recently proposed, would *umstürzen* the electron theory."[32]

We have seen that Einstein, in 1905, wrote to his friend Conrad Habicht about how his forthcoming paper on the light quantum was revolutionary – a judgment explicitly withheld from his papers on relativity and Brownian movement. The next year, he wrote ironically to Maurice Solovine that he had succumbed to stationary and sterile old age, when *revolutionäre* feelings were given over to youth. In Einstein's letter, the word revolutionary is written over a false start that begins "reak. . . ."[33] Einstein, it seems, was taken here by the same polarity that Boltzmann expressed on several occasions. Einstein's appeal to revolutionary rhetoric does indeed erode with time; it is likely that the adjective does not appear in his later scientific publications or correspondence at all. The word drops from Einstein's vocabulary, however, just before his identification in the popular literature as a scientific revolutionary for his work on special relativity.[34]

Foremost in arranging Einstein's apotheosis was Max Planck, Germany's most distinguished theoretical physicist and the senior editor of the world's most prestigious physics journal, the *Annalen der Physik*. Planck, a conservative physicist both politically and scientifically, described to a popular readership in 1910 how theoretical physics had experienced a major movement "of such a radical, *umwälzender* kind," which had given rise to such scientific controversy, that it could be compared only to the Copernican achievement. Planck expanded, later in his article, on the "*revolutionären* consequences of the principle of relativity," on the interpretation of the concept of time, whereby the exact hour and date always have a physical sense, depending on the motion of the observer.[35]

Planck's view was extended by those who had studied with him.

Ernst Lamla, in the dissertation that he wrote under Planck on a hydrodynamical formulation of the principle of relativity, emphasized in 1912:

The principle seems so simple and it describes such far-reaching *Umwälzungen* in all fields of physics that one is forced to give up or to entirely revise many views and concepts of the old physics to which one has become accustomed.[36]

In 1914, Planck's former student Hans Witte, an *Oberlehrer* at the Wolfenbüttel Gymnasium and *Privatdozent* for physics at the Brunswick Institute of Technology, compared the principle of relativity with Copernicus's achievement. It was "a complete *Umstellung* in our grasp of the world and especially of the essence of time." Witte indicated to his readers that he had previously proposed a general mechanical theory of electricity (he had surveyed the state of such mechanical theories in his 1906 dissertation under Planck's direction), but that he recanted, proclaiming Einstein, next to Hendrik Antoon Lorentz, as the discoverer of the new relativity principle. This principle, in his view, had *grundstürzenden* meaning for the essence of space and time. The greatest question of physics was how far the principle extended, whether it was really the long-sought "universal principle."[37]

Planck was not alone in his judgment of 1910. Others who were not especially productive researchers expressed similar views. Johannes Classen, professor of physics at the state physical laboratory in Hamburg, noted in a periodical reaching secondary-school teachers in 1910 that no one could fail to observe the *Umschwung* of late in certain basic views. A number of sentences further along Classen was clear to identify the origins of the *Umwälzung* in the principle of relativity.[38] A related message was echoed in the same year by Otto Berg, a young physicist writing in Göttingen philosopher Leonard Nelson's journal of neofriesian persuasion:

Certainly the principle of relativity places stronger demands on the capacity for abstraction than any other law of theoretical physics, and none other demands that one breaks so basically with traditional notions.[39]

Philosopher Max Frischeisen-Köhler, who did not especially share Nelson's outlook, elaborated on the revolutionary nature of relativity in 1912: "The *Umwälzung* in the field of the physical worldview brought about by this principle is comparable only to the introduction of the Copernican system."[40]

The sense in which the revolution depended on authority and belief came clearly to one of Planck's colleagues. Max Bernhard Weinstein, titular professor of physics and *Privatdozent* at the University of Berlin, emphasized in a general article of 1913 that many serious researchers saw in the principle of relativity a threat to the sound development of science, and for this reason they were demanding a return of earlier visions. In this view they were right, insofar as

> the meaning of this principle has often been extended so far and without reflection that an intolerable intolerance of other opinions has developed to the most silly assertions, which is almost comparable to a medieval constraint of belief [*Glaubenszwang*].[41]

Weinstein, unlike Planck, saw Minkowski as the largest revolutionary figure. In his exhaustive, 1913 summary of relativity physics, Weinstein observed that, in Minkowski's formalism, complex equations belonged to reality, in contrast to the way that, previously, complex equations really had no proper association with physical objects. In his view, "that certainly means one of the greatest *Umwälzungen* in our familiar views."[42] Others concurred in this judgment. Also in 1913, mathematician David Hilbert, writing to a high administrator in the Prussian Kultusministerium, underscored that the consequences of Einstein's "'relativity principle' were *revolutionierend* for mathematical and physical science," although in effect he attributed the revolution to Minkowski's four-dimensional "world mechanics."[43]

As the quotations from Weinstein and Hilbert will have suggested, all physicists of the time were not sympathetic with the new revolutionary spirit regarding relativity. One notable reactionary visited Germany from abroad to stake his claim as an architect of the new mechanics. When French mathematician Henri Poincaré spoke in Berlin around 1910, he ironically addressed the revolution in modern physics that seemed to have called into question the surest foundations of classical mechanics. He used the word *Revolution* in deference to current usage, he implied, for he thought that the revolution was really only a "menacing spectre, because it is quite entirely possible that in the short or long term some old, tested, Newtonian dynamical principles will emerge from this struggle as the winner." Poincaré thought that there was something positive to be gained from all this talk about revolutions, however. It provided the opportunity for a valuable lesson in conventionalist epistemology.[44] Less accomplished than Poincaré, but also having a vested interest in associating his own ideas with the current rage, was Greifswald physicist Gustav Mie, a firm believer in the electromagnetic ether. Mie emphasized in 1911 that over the recent past physics had seen the beginning of a process that in all likelihood would lead "to a complete *Umwandlung* of the basic concepts of science." The new development was "almost imperceptable to nonspecialists, but it will become important in the foreseeable future also for wider fields of learning." It concerned the concept of matter itself.[45] Mie who was working toward his own four-dimensional ether-based, gravitational theory, like Poincaré chose not to mention Einstein. Much farther down the scale was the Hungarian philosopher of science Menyhért (or Melchior) Palágyi, one of Ernst Gehrcke's fellow-travellers. Palágyi rejected Einstein's relativity because of the "indeterministic grasp of nature" that in his view resulted from it; relativity was to have broken the "causal bond" between the inner motion of a system and its collective motion. The relativity theories of Minkowski and Einstein, Palágyi wrote in 1914, "are now in a state of crisis" and were in danger of being treated as a theory of current fashion (*Modetheorie*); such a sensational theory "is raved about by young, overzealous researchers, only immediately to fall into oblivion." This was precisely "the fate of certain *revolutionärer* but unfruitful theories."[46] And the Leipzig experimental physicist Otto Wiener similarly worried about the ongoing revolution in physics. He underscored in his contribution to the major, encyclopedic collection, *Kultur der Gegenwart*, in 1915:

> To the uninitiated who hear of the *grundstürzenden* new theories of physics, the impression may be given that physics forever presents new structures from philosophical theories and these destroy the old theories. That is in no way the case. The old theories are completely integrated into the new ones.[47]

In Wiener's view, the mechanical theory of light had been surpassed by the electromagnetic theory of light, but in reality nothing had changed except the names. The views of Wiener, Palágyi, Mie, and Poincaré are reactionary ones. By 1915, the revolutionists had come to dominate the Convention of physics in Germany.

2. A STRUCTURE FOR THE REVOLUTIONARY ACTION

A discussion of revolution in Wilhelmian Germany would have brought to mind events in Paris during the last decade of the eighteenth century. The French Revolution and its imperial successor intervened directly in the course of German history, both in political and in intellectual realms: Kant interrupted his daily walk to ponder news from Paris; the battle of Jena finds an echo in Hegel's *Phenomenology of Mind*. Of all the features of the French upheavals, furthermore, the notion of Jacobin excesses remained as a clear warning throughout the nineteenth century; in the assemblies and conventions of the early 1790s, class struggle – not unlike that which threatened to erupt in Germany – was nakedly revealed.

When physicists and mathematicians wrote about revolution, they generally avoided analogies to events in the realm of politics. Yet given the anxieties of the aristocracy and the bourgeoisie, one naturally comes to wonder about allusions conjured up by calls for a revolution in physics during the years around 1900. The question assumes a greater intensity when it is recalled that the beginning of the twentieth century has consistently been viewed as the time of a revolution in exact sciences: Writers from Gaston Bachelard to Lewis Feuer have used events and circumstances from this period – largely extracted from the German-language scene – to inform models of scientific change.

In view of rhetoric about the relativity "revolution," then and now, it is of more than passing interest to consider the extent to which what happened in Germany, around 1900, may be said to parallel events in the French Revolution. Such an enquiry, of course, should be seen as a hazardous undertaking. It would lie beyond the bounds that Marc Bloch established for comparative history. It would conflate bloody events in politics with texts in the scientific literature. It would cross time, space, and cultural frontiers.[48] The exercise retains its interest, nevertheless, especially given recent writings in sociology of science regarding how scientific ideas change. Some commentators, in fact, have come to assert that scientific discourse reflects nothing other than the same sorts of venial and selfish personal interests that determine, no doubt, the course of political affairs.[49]

Critics have cautioned us not to expect to find many features of political revolutions in the world of science, but it is surprising that there has been no consideration of the extent to which any particular scientific revolution follows or departs from one or another political revolution. The following pages respond to this lacuna by reconsidering the relativity revolution in Germany from the point of view of the French Revolution. Developments in physics, considered on their own terms, will be cast following the institutions and structures that dominated Paris during the 1790s. It will be seen that although a number of features of the relativity revolution find no counterpart in Revolutionary France, the parallel can be extended to great lengths. Just as in Bloch's comparative history, heuristic rewards follow from seeking to document one set of events by reflecting them against what happened elsewhere.[50]

Political revolution, deriving as it does from perceived injustice, requires organization and technology to succeed, for flagrant oppression and systematic abuse in themselves can readily become structural features of a political regime. Revolutionaries, those who define injustice, if lacking in organization are condemned to live as oracles; revolutionary organizers without technological means are quickly liquidated. When Paris deprived the Bourbons of power, France was Europe's richest country; it was vital, both intellectually and commercially. The revolutionaries established a national political infrastructure and defended their gains through the actions of a technologically superior army.[51]

Physics in Germany at the end of the nineteenth century, under the old regime of classical mechanics, was in its own way no less vital than life in old-regime France. Magnificent new institutes broadcast the power and prestige that the subject commanded. Hundreds of young physicists, newly certified with doctorates, labored at highly specialized research in the hope of receiving a permanent position. They and their senior colleagues published their results, of enormous variety, in dozens of learned periodicals. Foreigners flocked to sit at the feet of the German masters who knew the compass of nature's laws. German researchers led the world assembly of physicists.[52]

Discontent with classical mechanics was building in the 1890s, and it gave expression to precipitous formulations like Wilhelm Ostwald's and Georg Helm's energetics. The act that symbolized a break with the past came unambiguously and suddenly: In 1897 Joseph James Thomson discovered the electron in cathode rays. Others had intimated the discovery before Thomson, but the director of the Cavendish Laboratory at Cambridge had the acuity to identify the electron as the object radiating spectral lines, the constituent element of atoms, and the smallest unit of electrical charge. Thomson's discovery was the storming of the Bastille; it provided a definitive break with the old regime of physics.[53]

In rapid sequence came theoretical formulations of electron dynamics and measurements of the ratio of electrical charge to electron mass.

Under the influence of the elegant syntheses of Hertzian electrodynamics produced by Henri Poincaré and especially Hendrik Antoon Lorentz, German researchers scurried to elaborate how electrons behaved both in their pure state and bound up in matter. Previously, under classical mechanics, electrodynamics was but a large part of physics, intended to be subsumed under the fundamental laws of mechanics. Within several years after the discovery of the electron, however, electrodynamics was cut free from its mechanical underpinnings. Proponents of a new vision, the electromagnetic view of nature, argued that Maxwell's equations or some modification of them were to replace Newtonian mechanics as the fundamental starting point for physics.[54]

The electrodynamicists – members of the Third Estate of physics in Germany – disputed the traditional powers of the other two estates: mechanics and thermodynamics. The role of Mirabeau – aristocrat and concilliator – in the early phase of the revolution is reflected in the activity of Dutch theoretician Lorentz. Fluent in German and by 1900 a senior statesman of physics, Lorentz circulated easily in Central Europe. His eloquence and consistent equanimity of spirit allowed him to speak with authority before physicists of all persuasions.

The fall of the Bastille and the agitation of revolutionary firebrands did not in themselves define a major change in law and administration. The latter was manifest by the convening, in 1791, of the Legislative Assembly, a single deliberative body with places for thinkers representing all factions and opinions. Previously, between 1787 and 1791, there had been a National Constituent Assembly that ironed out the form of a constitutional monarchy, but its accomplishments were not sufficient to forestall its demise. The Legislative Assembly, charged with defining a new course for civic life, was France's first revolutionary government. It is not without some justification, then, that one can associate the Assembly's first meetings in the Manège with discussions and debate following publication, in 1905, of Einstein's first papers on the special theory of relativity. A tour of the seats in the Legislative Assembly of German respondents to relativity will place this homology in clearer focus.

On the extreme right wing of the Assembly were the traditionalists, supporters of the old regime of classical physics. Following French-Revolution terminology, we may call these men royalists, or blacks. There were a number of prominent speakers. First came Ludwig Boltzmann, adamant in his defiance of the electron revolutionaries while at the same time inspiring, in his seminal work on statistical mechanics over the preceding decades, the early reflections of men like Einstein, Paul Ehrenfest, and Fritz Hasenöhrl. Boltzmann was La Fayette to the assembled physicists – a veteran of another major transformation and a man who, although a confirmed traditionalist, led revolutionary forces into battle. A second prominent royalist was Vilhelm Bjerknes, a Norwegian physicist with strong German ties who was a partisan of various classical, hydrodynamical, atomic models.[55] In this role, Bjerknes paralleled the Duc d'Orléans. Along with Philipp Lenard, the 1905 Nobel Laureate at Heidelberg, Bjerknes believed in the physical reality of the electromagnetic ether.[56] Their point of view was supported by other physicists, notably Arthur Korn and the talented Swiss-American researcher Jakob Kunz.[57]

Astronomy figures intimately in Einstein's general relativity, and for this reason it is significant to observe that another prominent royalist appeared in the form of Hugo von Seeliger, Germany's doyen of astronomy, who sought to explain deviations from Newton's law of gravitation by the hypothesis of dust rings in the solar system. He opposed everything about the electron theory and Einstein's relativity: "Seeliger *hates* this most recent development with all the sincerity and impulsiveness of his being," Arnold Sommerfeld exclaimed to Karl Schwarzschild in 1910.[58] Following Seeliger's lead were other astronomers, for whom the entire program of the electron revolutionaries was suspect. Egon R. von Oppolzer, professor of astronomy at Innsbruck, wrote to Schwarzschild in 1902 that he had just read about Max Abraham's electron theory in the *Physikalische Zeitschrift* reports of the Carlsbad Naturforscherversammlung. "I do not understand how one can accept such a thing," Oppolzer protested. "I hope that we will soon have this erroneous concept behind us." Oppolzer was astonished that Abraham – and by extension the other contributers to the electron theory – would have confused "mol ions and gas ions with, what the devil, lovely little particles." It was all a dead end, "sterile ashes." Oppolzer hoped to be alive in twenty years "to see what remains of all the skeletons without flesh and blood." In his view, "everything is a trivial consequence of hydrodynamical theory."[59]

Moving to the right of center, we find those who were prepared to change classical mechanics, but not to dethrone it and abandon the ether. They may be called the monarchists, or, to use the name given supporters of an eighteenth-century English-style bicameral, constitutional system in the French Revolution, the *Anglomanes*. The monarchist physicists were anglophile in following the nineteenth-century English ether theorists, while retaining a healthy respect for the mathematical methods of classical mechanics. Here sat Gustav Mie, a long-time advocate at Hertzian electrodynamics;[60] physicist Emil Kohl at Vienna; and Emil Wiechert, an important theoretical voice in the first years after the electron's discovery. Joining them was Johannes Stark, a firm believer in the ether whose sympathetic feelings toward the mechanist right later drew him into a political alliance with Lenard.[61] None of the monarchists worked on the electron theory with single-minded intensity; each published substantial contributions in other areas.

In dead center it is convenient to place a rather diffuse group of experimentalists. The men who measured the variation of electron mass with velocity, of course, approached their tasks with dispositions and prejudices that supplemented their degree of technical competence. Walter Kaufmann labored with overly soft vacuums at Göttingen to verify Max Abraham's rigid-sphere electron model; Alfred Bucherer's unreliable experiments verified not his own electron theory but that of Lorentz and Einstein; Adolph Bestelmeyer, Kurt Wolz, Erich Hupka, and Wilhelm Heil were all cautious about embracing Einsteinian relativity, even though they worked with the encouragement of relativity-supporter Max Planck at Berlin.[62] In general, the experimentalists were willing to adjust their theoretical allegiances as a function of laboratory results. Kaufmann, like Bucherer, eventually abandoned a rival theory in favor of Einstein's relativity. Heil, in his doctoral dissertation of 1909, expressed great caution in considering whether experiment spoke in favor of Einstein.[63] Polymath Jakob Laub, Einstein's early collaborator and faithful supporter, believed in 1910 that special relativity had been verified in the laboratory; yet in the same year he undertook experiments, as Lenard's less-than-loyal postdoctoral assistant, to place bounds on the density of the electromagnetic ether![64]

From the center to the left in the physicists' Assembly sat the relativist patriots, those who heralded relativity as the dawn of a new age. The largest block of patriots circulated around the University of Göttingen. These were the electron theorists. Mathematicians or mathematically-inclined physicists, they sought to see just how far the electromagnetic view of nature could be extended. In matters of physical credence they were generally agnostic, believing only in the overwhelming heuristic power of pure mathematics for calling into being new physical laws. In the years around 1905, they applied classical action principles and potential theory to the Lorentz electron theory, and they regularly commented on the physics of faster-than-light motion. None seemed especially eager to banish the electromagnetic ether to obscurity. Among the most important Göttingen-inspired electron-theorists we find Hermann Minkowski, Karl Schwarzschild, Gustav Herglotz, Arnold Sommerfeld, and Paul Hertz. As a group they may be said to have valued, above all, the techniques of classical mechanics, which they were not prepared to abandon without a struggle. Unlike the monarchists, they envisioned a secure place for the remnants of the old order. Their homologues in the French Revolution are the *Feuillants*, or the Constitutionals.

Beyond the *Feuillants* in the Legislative Assembly were those who sought, from the point of view of physical laws, a clean break with classical mechanics. Some of these Democrats, or *Jacobins*, were uncertain about the path to follow: Wilhelm Wien, a former supporter of the electromagnetic view of nature, became an early follower of Einstein without himself undertaking much innovative research in the field. The majority of the Democrats proposed one or another alternative to classical mechanics or seriously questioned whether conventional physics could be the whole story. Max Abraham, indeed, belongs here: Although he initially opposed Lorentz's electron theory and then seriously queried both Minkowski's electrodynamics and Einstein's general relativity, he wrote with a brilliantly clear vision of Einstein's special relativity as a physical theory. The same may be said of Walter Ritz, who proposed an emission theory of light in opposition to Einstein's first postulate of relativity. Here, too, sat Hans Witte, Planck's student who abandoned mechanical electrodynamics in favor of relativity, and Boltzmann's student Ehrenfest. The strongest Democrat of all was Max Planck, who set his Berlin entourage to explore relativity from all conceivable points of view. Around 1905, Planck's position was in the minority, but it was he who, more than any other physicist, provided crucial support for Einstein.

It is well to emphasize that around 1906 there was no general agreement regarding what special relativity consisted of and how it related to the electron theory. At the Stuttgart Naturforscherversammlung in that year, Bruno Strasser and Max Wien observed that acceptance of the principle of relativity, which argued against speaking of motion relative to the ether, was "in flux."[65] At the same meeting, Max Planck referred to the "Lorentz-Einstein" electron as the *Relativtheorie* – the 'relative theory' – a term by which special relativity would often be known over the next five years. Planck found it more congenial than Max Abraham's spherical electron theory, which was, in his view, purely electrodynamical.[66] Arnold Sommerfeld injected a generational issue into the discussion: Men under forty supported the electrodynamical postulate (the electrodynamical view of nature); men over forty were in favor of the mechanical-relativistic postulate of Lorentz and Einstein.[67] Sommerfeld, having by 1906 been forced to reverse his position on a number of points in the electron theory, still saw this theory as the leading edge of research in physics.[68] For Sommerfeld, Einstein was behind the times.

After 1906, when Planck, Germany's leading theoretician, came out completely in favor of Einstein's relativity theory, the Legislative Assembly dissolved. It was simply no longer possible for the royalists and the monarchists to participate with the Democrats in the relativity revolution. Traditionalists like Mie – those holding to the fundamental principles of classical mechanics – withdrew to consider their next move. In place of the Legislative Assembly came the National Convention of the First Republic, wherein all delegates sought a new unifying vision of physics. Here began the romantic phase of the relativity revolution, where its ideas were sent to do battle against traditionalists from abroad and at the same time diffused within Germany. Just as in the French Revolution, the relativity Convention fell into three sections.

The first section, and at the beginning the dominant one, was that of the heritors of the electron-theory *Feuillants*. Hermann Minkowski's firey oratory and new four-dimensional formalism swayed dozens of uncommited physicists and above all mathematicians to relativity's cause. Minkowski's imperializing message was not lost on his supporters and followers, many of whom came to speak of the relativity revolution with little, if any, mention of Einstein. Advancing Minkowski's formalism came to be seen as a worthy, revolutionary end in itself; repeating Minkowski's rhetoric was taken as a declaration of faith. Much of the activity of Minkowski's followers led to little in the way of concrete innovation: Max Born, Gustav Herglotz, Waldemar von Ignatowsky, Wilhelm Behrens, Erich Hecke, Ludwig Föppl, Percy Daniell, and Gunnar Nordström (a Finnish physicist strongly influenced by the Göttingen school) ran into dead ends when trying, after Minkowski's death in 1909, to formulate a relativistic notion of classical rigidity;[69] the four-dimensional formalism completely seduced others, like the ageing mathematician Felix Klein, his younger colleague Fritz Noether, and Gustav Jaumann. As might be expected, the followers of Minkowski were generally unsophisticated about the physical bases of relativity theory. In fact, many continued to believe in some form of the electromagnetic ether. They constituted, by analogy with the Convention of the French Revolution, the *Gironde*. Minkowski, leader of the *Gironde*, finds a reflection in Vergniaud, the independent-minded *Girondin* orator; and perhaps, too, there is something of the *Girondin* philosopher and mathematician Condorcet in Minkowski's make-up.

Arrayed against the *Girondins*, high at the back of the seats in the Manège, sat the Mountain – the advanced revolutionaries who were more or less firmly convinced of Einstein's revolution. The Mountain fell under the sway of Planck, his students, and his associates. It was Planck, the relativist Danton, who in 1910 elevated Einstein to the status of a new Copernicus. Among his students, Planck counted Max von Laue, author in 1911 of the first definitive textbook on special relativity; Ludwik Silberstein, author of the second such textbook, published in 1912 in English; Hans Witte, the reconverted mechanist; the experimentalists Kurd von Mosengeil, Erich Hupka, and Wilhelm Heil. Among his critical charges as coeditor with Wilhelm Wien of the *Annalen der Physik* was winnowing acceptable manuscripts on relativity.

Planck's student Max Abraham also sat as a *Montagnard*. He played the role of a firebrand Marat, functioning as a safety valve for the revolutionaries. Just as Marat belonged to Danton's club, the *Cordeliers*, so Abraham was of Planck's physicalist school. Neither accepted nor well liked, Abraham spoke for many by voicing stern disapproval of the *Girondin* penchant for mathematical abstraction, and he probed, perhaps more critically than any other physicist, for logical inconsistencies in the various formulations of relativity which appeared from month to month in the scientific literature. Like Marat, Abraham was less dangerous within the Convention than outside it, and like Marat, Abraham did not survive the revolution. In 1909 he took up a chair in Milan.[70]

The third and largest section of the National Convention of relativists was the Plain, or the Marsh. Here sat followers of the revolution who watched the debates between the Göttingen *Girondins* and the Berlin *Montagnards*. They would have been attuned, initially, to the great success of the *Girondins* in dealing with threats from counterrevolutionary critics. Poincaré, in 1910, addressed Berlin physicists on the conventionality of all scientific ideas and the regularity of progress in physics; in his sardonic introductory remarks, as we have seen, he studiously avoided mentioning Einstein's name. Mie, in 1910, published a treatise on ether mechanics, and Lenard thundered forth in the same year on how the ether related to matter. Emil Wiechert, an early architect of the electron theory, serialized a small book in the pages of the *Physikalische Zeitschrift* in 1911; his aim was to recast relativity in a kind of absolute space. Energeticist Georg Helm came to Wiechert's defense. Experimentalist Ernst Gehrcke called relativity a mathematical-physical "jest"; mathematician *Oberlehrer* Alois Höfler and physicist *Oberlehrer* Friedrich Poske argued against introducing relativity – a "passing fashion"in Höfler's view – into the schools; Leo Gilbert saw relativity as "foppery"; Italian physicist Michele La Rosa sought to persuade his German colleagues to resurrect the ether.[71]

Poincaré here plays the role of La Fayette, a founder of the *Feuillants* who in 1792 went over to the revolution's enemies; Mie corresponds to La Rochejaquelein, leader of the Vendée counterrevolutionaries; Lenard and company, along with Seeliger, represent the Spanish and German armies arrayed against France; and Wiechert is Dumouriez, a revolutionary general who turned coat in 1793. The revolution suffered a backlash from traditionalist physicists just when senior *Jacobins* like Planck, Wien, Sommerfeld, and Einstein decided to devote the first Solvay Congress to the problems of quantum physics because to their mind relativity posed no fundamental problems. The assertion of victory signaled an irrevocable break with classical physics.

In his chronological tabulation of the number of publications dealing with relativity, Maurice Lecat revealed a sudden dip around 1911. This, the single exception to a monotonically rising curve, provides support for the contention that the decline of the *Girondin* electron theorists was largely independent of the advent of general relativity. The *Gironde* fell from power, in fact, because of its initial success. Its work was greatly suggestive for mathematicians, but it was finally unable to arrive at experimentally verifiable propositions. It exuded little more than ebullient joy in proposing mathematical elaborations or refinements – epicycles on Minkowski's *oeuvre* – before Einstein and Marcel Grossmann introduced differential geometry to deal with gravitation in 1912.

With the publication of Einstein and Grossmann, the Convention ended. The Mountain, under the leadership of Planck, pressured the formalist *Gironde* to retire by flogging Einstein's publications. Around

1912, Einstein, the Robespierre of the small core who determined the progress of the coming treatment of gravitation, was personally unassailable. Over the next three years he pursued his program of general relativity with constant devotion. Planck, the eloquent Danton, invited Einstein to sit in Berlin as an extra-university oracle, and Einstein accepted. There he joined Planck's former student Laue and others, like astronomer Erwin Findlay Freundlich, whom he found congenial. From Berlin he easily disposed of the alternative theories advanced by his revolutionary colleagues Nordström and Abraham. As his attention turned inward, he dismissed out of hand the counterrevolutionary work of Gustav Mie; he omitted answering other contenders, such as Jun Ishiwara, with whom he had personal contact. Einstein left Planck behind with general relativity, not completely unlike the way that Robespierre superseded Danton: Although the senior theoretician continued to evaluate manuscripts on the subject for the *Annalen der Physik*, he declined to engage differential geometry himself; he was not entirely comfortable with light bending in general relativity. Between 1912 and 1916 nothing withstood Einstein's withering gaze. When Gunnar Nordström objected to general relativity on the grounds that the velocity of light would depend, in it, on the gravitational field – in his view this marked "at least a complete *Umwälzung* in the foundation of the previous theory of relativity – Einstein easily rejected the alternative that Nordström proposed.[72] These were the years of Einstein's Terror.

Late in 1915, mathematician David Hilbert, a close associate of the old *Girondin* electron theorists, published the same covariant field equations of general relativity as Einstein, and nearly at the same time. Hilbert drew strength from Mie's rather traditionalist theory of gravitation; he proceeded from an entirely nonrelativistic point of view by postulating the ether, absolute space, and special properties of the electron. His publication was widely hailed as a Thermidor, an end to the relativity revolution's excesses. Mie predictably greeted Hilbert's work with enthusiasm while ignoring Einstein's contribution entirely. He wrote to Hilbert in 1916:

> I have read your work on the foundations of physics with growing wonderment and enrapture. Although I must naturally take much trouble in doing so, still I am excited by what I have already understood. It is to me as if I had been feeling in the dark; the theory that you have given is certainly in the right direction. But I lack the mathematical force to replace these obscure feelings with clear concepts, and I didn't hope to venture that such a powerful step as you have taken could be made.[73]

Hilbert's triumph stimulated Felix Klein to take up relativity once more. Seeliger and his astronomer colleagues, notably Hermann Struve, took the occasion to strike a blow against Einstein by removing Einstein's coworker Freundlich from his position at the Berlin-Babelsberg observatory.[74] Planck remained loyal to Einstein, even if he did not completely support general relativity, by processing a dozen or so new authors of works on the topic through the *Annalen*, and foreigners like Lorentz elaborated on general relativity in an Einsteinian vein.[75] The mathematicians, however, came under the influence of the old Göttingen formalists, and they dominated the field between 1916 and 1919. Hermann Weyl, in his treatise on the unified field theory, set the tone for much of what followed in the theoretical response to general relativity in Germany.

The old Göttingen formalists paraded as *Muscadins* and *Incroyables*, while in the new revolutionary Directory sat Hilbert as Barras and Klein as Carnot. Poor Freundlich, beginning in 1911 the only German observational astronomer who worked to verify general relativity, was Babeuf. Carnot betrayed Babeuf to the Directory's counterrevolution; Klein – Freundlich's old doctoral dissertation adviser – did not intervene to save the young astronomer from Seeliger and Struve's wrath.

Viewed in the large, the Directory of Hilbert and Klein was ineffective in bringing about a consensus on relativity. It little more than sanctified a long-standing pattern, where pure mathematicians felt able to jump in and deal with one or another aspect of the electron theory. Things might have been different had the clouds not parted on Principe to permit Arthur Eddington's photographs, for in that case William Wallace Campbell might have published his results from the 1916 Goldendale eclipse – results that seemed to speak against Einstein's light-deflection prediction. As it turned out, however, British astronomers used the Principe results to vindicate not Hilbert, who had signed the notorious manifesto to the civilized world of 1914, but Einstein – the Jewish pacifist of Swiss nationality who had only moved to Berlin a short while before the war began. The announcement of the Principe results was the 18th Brumaire for Hilbert and Klein's Directory.

A new Einstein emerged around the time of the announcement that general relativity had been verified. The theoretician became reified – purged, as it were, of observationalist sentiment. He rejected modernist positivism in favor of critical idealism. He read Kant with admiration and Schopenhauer with empathy. He enjoyed Lichtenberg's aphorisms. He came to hold, along with Leibniz, that pure mathematics could prefigure the laws of the physical universe. Einstein indeed flirted with the spirit of the Göttingen Directory. In 1928, he was welcomed into the fold of its successors when he became an editor of the *Mathematische Annalen*.

It would be unkind to ask a generous reader who followed the revolution thus far to go ahead through the 1920s. An argument could be made for the extent to which the positions taken in debates over special and general relativity recur in the development and early reception of quantum mechanics; the story of Lenard and Stark's furious campaign against Einstein's 'Jewish' physics might have been limned; the tortuous progress of attempts to perform the three 'classical' tests of general relativity might have been reconsidered. These themes have received a certain amount of attention and no doubt could benefit from a closer look. In the 1920s, however, the drama of the relativity revolution had run its course among physicists in Germany.

The meaning of an analogy between relativity in Germany and the French Revolution may not be set aside so lightly. The dissymmetries have been, surely, as striking as the parallels. Both Boltzmann and Poincaré have been cast as a La Fayette at different stages in the revolution. A large and heterogeneous group of experimentalists have been placed at the center of the Manège – a conventional position for the traditional adjudicators of theoretical questions, to be sure, but one that tends to undervalue the extent of the experimentalists' theoretical prejudices and stylistic inclinations. Minkowski is cast as the leader of the electron-theorist *Girondins*; it must be remembered, though, that the leaders of the *Gironde*, Vergniaud and Condorcet among others, were liquidated by the Terror, while Minskowski died early in 1909 at the height of the electron-theorists' prestige. Casting Einstein as Robespierre risks a quarrel, for although his star rose like that of the revolutionary *Jacobin*, Einstein always had the friendliest of relations with his colleague and supporter Max Planck – relativity's Danton. Should Ernst Mach be perceived as a disconsolate forerunner of relativity, a revolutionary who later came to regret the theory's excesses?[76] And which revolutionary shall Karl Schwarzschild be – the electron theorist of 1903 who in 1913 sought to find the gravitational reshift in the solar spectrum and in 1916 produced a lovely result with Einstein's covariant field equations?[77] When one expects, by analogy with 1799, to see an unprincipled Bonaparte take charge in the aftermath of the British eclipse results, none other than Einstein – albeit somewhat transfigured – reemerges from the shadow of the Thermidor to become the Man of the Hour.

It is well to emphasize that, in setting down the analogy, the focus has been on men and their scientific thought. Counterparts to the Parisian *sans-culottes* or the provincial *chouans*, insofar as such comparisons would have any meaning, are reserved for the last section of the present text. Women actors have not been identified, although in an extended analogy one would do well to remark on the activity of women of critical social importance, for example, Käthe Jerosch Hilbert, Helena Joseph Weyl, and Elsa Einstein, and to compare them with the salon keepers Tallien, de Staël, and Recamier. Politics and war have been kept in the background in order for essential parts of the analogy to be constructed: In the political realm one can find few congruences between the conservative nationalist Max Planck and the democratic pacifist Einstein, two men who stood side by side on the issue of relativity.

Despite the apparent limitations of analogy, rephrasing the relativity revolution in Germany in terms of the French Revolution furnishes a model of considerable explanatory power. It clearly places Einstein in the revolutionary tradition of the electron theorists without in any way detracting from his unique and prodigious accomplishments. It illuminates the longstanding disagreements between Einstein and those who seriously grappled with the physical implications of relativity, on the one hand, and formalist, mathematically oriented researchers, on the other hand. It suggests that Max Abraham (a student of Planck's) was closer to Einstein's general way of thinking than was Max Born (an assistant of Minkowski's), despite the fact that Abraham persistently questioned Einstein's theories of relativity while Born acquiesced to them. The analogy preserves Einstein's own notion that special relativity was 'in the air' around 1905 whereas general relativity was his special achievement. We see that many contributors to special relativity who styled themselves as revolutionaries were rather anti-Einsteinian in their beliefs, just as the range of revolutionary sentiment among the *Jacobins* ran from bourgeois republicanism to popular democracy. The analogy suggests that Einstein's final formulation of the covariant field equations of general relativity was challenged by a counterrevolutionary Thermidor. Einstein's understanding of general relativity triumphed after the stunning and partisan coup brought about by the British astronomers, although the Einstein of 1919 was not the Einstein of 1905.

The analogy is of greatest interest because of the structure that it imposes on the relativity revolution. There is seen to be a progression of change – several transformations – not unlike the case in France at the end of the eighteenth century. To seek *which* development was the essential one, from this point of view, is to engage in a historicist enterprise. Who was, after all, the 'essential' actor of the French Revolution? Was Mirabeau more of a significant force than Condorcet, Carnot, Marat, Robespierre, Barras, Babeuf, or Napoléon? The answer – insofar as a meaningful answer exists – depends on one's ethical perspective.

The structure that has been suggested by the analogy allows for deep, religious-style conversions as well as other changes of mind within a fixed or slowly changing ensemble of prejudices and methodological inclinations. Many actors fundamentally changed their opinions without having a deciding effect on the course of the revolution. Hans Witte came to reject mechanistic explanations of electrodynamics; Walter Kaufmann came to lose confidence in his experiments verifying Max Abraham's electron theory; Max Born lost interest in attempting to extend Hermann Minkowski's formalism to account for rigid-body motion; Gunnar Nordström abandoned Minkowski's relativistic formalism in favor of Einstein's special relativity, and then rejected his own gravitational theory in favor of Einstein's general relativity; Max Abraham directed his immense talent toward finding logical inconsistencies with Einstein's work; Einstein himself oscillated on a number of basic propositions. All the foregoing researchers changed their minds without changing their basic approaches to their discipline. It is certainly the case that Max Planck, Wilhelm Wien, Max von Laue, and other early supporters of Einstein's work accepted his interpretation of special relativity after having been schooled in a contrary interpretation of things, and there is a story to be told in understanding how they did so. Biographers have long been interested in the making of great minds – the moment in his life when, for example, Danton became a democrat. From the present perspective, however, when the structure of a revolution is under study, such a question seems less relevant than asking about Danton's relationship with Marat, Robespierre, and the *Girondins*.

The structure revealed here, then, seems compatible with Thomas

Kuhn's understanding of scientific revolutions. It can accommodate his notion of a *Gestalt* switch in the mind of a scientist who has been converted to a new point of view. It also suggests the sense in which Kuhn is right in emphasizing that many scientists willingly attempt to contribute to their own understanding of a new point of view without having undergone such a change of mind. Neither conclusion should astonish us. After all, with regard to formalism, there is little to distinguish Lorentz's electron theory from Einstein's special theory of relativity and nothing to separate Hilbert's gravitational theory from Einstein's general relativity. In the revolutionary process, as Kuhn would allow, one finds a continual adapting of innovative principles – great and small – to create a new synthesis, rather than a struggle fought between two clearly demarcated forces on the plains of a scientific Armageddon. Finally, relativity, revolution is portrayed as a self-conscious process called into being by the historical actors themselves. The revolutionary call is taken up and repeated at the same time that new syntheses are being proposed and debated. The requirement of an explicitly *revolutionary* alternative, indeed, appears in the last days of the old regime, before an acceptable solution has been set down. Far from being fostered on a complacent scientific community by the unexpected appearance of a work of genius, the revolution is willed into being by communal exertion. From this perspective, both revolutionaries and reactionaries wander through a period of intense disagreement and confusion – the revolutionary break – before the establishment of a new scientific regime.

All structures and their resulting hagiography depend on the historian's choices. One may ask whether it is more than fortuitous that, in following the high road of personalities in the French Revolution, homologies have been found in the relativity revolution in Germany. Modern historiography surely takes a dim view of studies that begin and end with men at the top. It exhorts us to look for the big picture of life in society at large, to focus on actors with weak voices and slender means, to seek structures in patterns of daily activity rather than in the vagaries of parliamentary seating. Let us, then, for the moment set aside critiques and contentions emanating from the legislature sitting in a converted riding stable. Let us turn to consider the relativity revolution from below.

3. THE LONG VIEW OF REVOLUTIONARY CHANGE

Personalities, the short of history, are fixed to persons who act in surroundings that, taken in the large, are quite unrecognizable without further explanation. Whereas qualities of personality – courage, perseverance, love, honesty, loyalty – are read directly (if not always accurately) in historical lives, constraints on and motivation for action can require an appreciation of conditions special to time and place. To gauge innovation and originality, indeed, it is necessary to reach an understanding of contemporaneous, long-term strategies of rather ordinary people in a community or society. The most innovative historical questions of the past generation have concerned the dimensions of such phenomena as the revolutionary crowd in Paris and the counterrevolutionary crowd in the Vendée; village life in provincial, medieval France; joint-stock investors in seventeenth-century England and proletarians in the first, English, industrial revolution. Historians of science have not been immune to the claims of a 'bottom-up' perspective, where the King's honours list assumes less importance than the grocer's balance sheets. From this perspective, the significance of even spectacular, contingent events, such as the discovery of X-rays or the publication of Darwin's *Origin of Species*, tends to recede into the background.

In this section, I take a longer view of the relativity revolution by considering two of its structural features: the field of literature and the process of appprenticeship. These are not the only themes from social history which may inform the nature of the scientific discourse in question. They are of special interest here, however, because of the way that each helped to determine the course of relativity's reception in Germany.

a. The Field of Literature

In their degree of persistence with the task at hand, electron-theory and relativity revolutionaries were not unlike their counterparts in the French Revolution. Some, like Max Abraham, dedicated themselves exclusively to elaborating the meaning of the new vision of the world. Others, like Einstein, Planck, Laue, and Hilbert, ranged across physics to consider a number of burning issues, notably radiation and the light quantum. Still other men contributed to the early states of the revolution (Walter Kaufmann, Hendrik Antoon Lorentz) and then turned to consider less revolutionary matters. To judge, indeed, from the level of quality in publications of the time, the intense field of discussion was accessible to virtually any interested party; one was able to gratify the urge to appear in print by commenting briefly or at length in either popularizing glosses or highly technical treatises.

It will be evident that the publishing restrictions of the Ancien Régime have no real counterpart in Wilhelmian Germany, where authors heaped ridicule on scientific adversaries shamelessly and without fear of legal reprisals. The jury is out on whether knowledge circulated in eighteenth-century France more widely and among a wider readership than in other countries. Yet with what we know about the dimensions of the literary establishment and the literary underground, it seems not unreasonable to think that the reach of the printed word was longer in France than elsewhere. This long reach, from the top of society to the bottom, must figure, as Robert Darnton has emphasized, among the intellectual origins of the French Revolution.[78]

Taken as a whole, publishing outlets in Wilhelmian Germany proved rather accommodating to the community of physicists. Colorful debate appeared to an extent that, one thinks, would not be tolerated today in the world of specialist journals. Quite unlike the situations in Great Britain and France, where relativistic debate occurred – to the extent that it did occur – within the confines of a small circle of elite periodicals, significant contributions appeared in as many as a score of German journals. With one partial exception – Johannes Stark's *Jahrbuch der Radioaktivität und Elektronik* – the journals sought to appeal to a general readership of physicists, scientists, philosophers, or educated onlookers. Unlike the case in the United States and France, journals devoted to special parts of physics did not proliferate in Germany until after the First World War.

Though the field of periodical publication in Germany is polyvalent and complex, one may distinguish in it three broadly-defined kinds of journals: those controlled by learned corporations, those run by broad-based societies, and those issued by individual proprietors. It is well to consider each kind in turn.

There is no clearer indication of the extent to which science has remained an elite activity than the persistence, for three centuries, of high-quality academy proceedings. By the end of the nineteenth century, many of these offered fellows and their protégés a secure means of laying claim to an idea or to a discovery. A communication, made public by having been read in a sitting of the academy, could appear in print in the academy's *Verslagen, Sitzungsberichte*, or *Comptes-Rendus* within a matter of weeks. By virtue of exchange ententes with other academies around the world, the publication would circulate quickly among the high and mighty. Furthermore, academies were empowered to grant their authors substantial quantities of reprints, which could then be sent among less eminent, but perhaps more active, researchers. Academies at Berlin, Göttingen, Vienna, and Amsterdam, especially, published large numbers of works dealing with the electron theory and relativity. But because academies only published what passed through the hands of their fellows, their engines of publication depended on the presence of a full-time stoker. Relatively little on the electron theory and relativity appeared, for example, from sister academies at Munich, Königsberg, Heidelberg, and Haarlem, and from junior cousins like the Physical-Medical Society at Würzburg.

The second kind of periodical, that published by broad-based scientific societies, generally avoided the question of personal sponsorship. Much depended, though, on the society in question. A society like the Naturforscher or the German Physical Society would be receptive to including in its *Verhandlungen* the text of just about any communication delivered orally at one of its meetings.[79] Limitations of space naturally prevented most authors from producing a sustained argument. Other societies, such as the German Mathematicians' Society, the astronomical society, and the mathematical societies in Berlin and Hamburg, were catholic but principled in accepting contributions to their *Jahresberichte, Vierteljahrsschriften, Sitzungsberichte*, or *Mitteilungen*; the editor and his friends no doubt sought more to prevent an embarrassment from appearing than actively to seek long and innovative articles. Authors in a hurry would often knock on other doors first, for, with the exception of the astronomical society, a manuscript could wait for a long time before being printed in the society's outlet. Still other societies, such as the Silesian Society for the Civilization of the Fatherland or any of the various school-reform associations, were self-selecting in the kinds of material that would come their way. In the latter category should be placed small institutional publications, like the *Abhandlungen* of the Göttingen Fries-Schule, which existed to promote one or another philosophical or social point of view.

Important as the elite-academy and democratic-society publications were in the relativity revolution, most attention focused on the third kind of periodical, the proprietary journal. Such journals were controlled by an editor or coeditors backed up by a small circle of disciplinary colleagues. Some proprietary journals (so called because they existed to make money for the owner, or at least to break even as a result of subscriptions and advertisements) had affiliation with a disciplinary society – the most famous of these affiliations being between the *Annalen der Physik* and the German Physical Society – but in no sense was the journal a house organ. The proprietary journals solicited contributions from anyone – German or foreigner, academic or amateur – in the hope of thereby appealing to a broad community of interest. Indeed, the overwhelming majority of foreign contributions to the relativity literature in Germany appeared in proprietary journals. Different from their counterparts who had been appointed by learned corporations, editors of proprietary journals – Max Planck and Wilhelm Wien of the *Annalen der Physik* and Emil Bose and Felix Krüger of the *Physikalische Zeitschrift* – would not infrequently revise contributors' prose or impose other conditions on having a manuscript accepted. Editors and publisher developed a clear idea about the profile of prospective readers, and they wanted above all to retain the special character of their journal. Some proprietary journals, like the philosophically-oriented *Kantstudien* and *Annalen der Naturphilosophie*, no doubt received few submissions that required much deliberation. Others, like the *Physikalische Zeitschrift* and the *Zeitschrift für Mathematik und Physik*, to judge from their content, seemed to publish nearly anything that came their way. Johannes Stark's *Jahrbuch der Radioaktivität und Elektronik* published timely reviews of the literature which the editor usually solicited. *Himmel und Erde* sought material vaguely related to astronomy. *Die Umschau* sought to present scientific questions before intelligent laymen.

One of the valuable functions of proprietary journals was to keep researchers informed about the latest monographic treatments in their field. From the very beginning, the electron theory generated vast, book-length treatises devoted to one or another point of view. Some of these, like Max Bernhard Weinstein's summaries of relativistic physics, were tedious exercises in algebra. Others, like Emil Cohn's, Alfred H. Bucherer's, and Gustav Mie's books, were attempts to set forth a general philosophy for dealing with electromagnetic phenomena. Still others – Johannes Classen's or Joseph Petzoldt's lectures, for example – were philosophical elaborations of recent scientific debates. A number of monographic treatments, however, became classics that a serious researcher ignored at his peril. Into this class fall Lorentz's contributions in 1903, 1904, and 1909 on electromagnetism and the electron theory; the two editions, in 1904 and 1909, of Abraham's book on the electron theory; Laue's book on special relativity in 1911; Silberstein's recapitulation, in English, of 1914; and Pauli's encyclopedic treatment of 1919. No one less than Einstein looked to monographs as authoritative summaries. As might be expected, references to books sometimes appeared in journal articles.

Editors manipulated their own publishing outlets featuring works on relativity for special ends. *Oberlehrer* Friedrich Poske used his *Zeitschrift für den physikalischen und chemischen Unterricht*, and unabashedly so, as a forum for promoting science instruction in the schools in particular and a vision of a humanistic-scientific worldview in general. Mathematician Felix Klein and his colleagues used the *Nachrichten* of the mathematical-physical section of their academy, the Royal Scientific Society at Göttingen, as a rapid means of publishing hastily composed, interim reports and mathematical speculations about the electron theory; they usually excluded reflective reviews of the literature and general overviews. Of greatest interest here, however, is the editorial policy of the *Annalen der Physik*, without doubt the premier physics journal of the period.[80]

Beginning in 1877, with the appointment of Gustav Heinrich Wiedemann as its editor, the *Annalen der Physik und Chemie* enjoyed financial support from the Berlin Physical Society. The Society exacted a measure of control over the proprietary journal, published by the firm of Johann Ambrosius Barth, as the price of its support: It delegated Germany's most distinguished physicist, Hermann von Helmholtz, as its factotum in the editorial office, and it made sure that Helmholtz's name appeared right under Wiedemann's, in smaller type, on the title page. When Helmholtz died in 1893, Wiedemann named his son as a coeditor, but the Berlin Society (from 1898, the *German* Physical Society) still wanted to have an overseer; in 1895 it designated Max Planck to succeed Helmholtz in this role. Wiedemann died in 1899 and his son declined to continue alone. The Physical Society turned to Paul Drude, professor of physics at the small University of Giessen and someone at ease with both experiment and theory, as the new editor. This time the Society named four experimentalists, however, to join Planck as an editorial council, and it dropped the word 'chemistry' from the journal's title.

After five years of editing the *Annalen*, Drude was called to become professor of experimental physics at the University of Berlin. It was the most prestigious post in the discipline. Drude lasted one year. Overwhelmed by teaching, administrative, and editorial responsibilities, he shot himself in 1906.

Into Drude's editorial chair stepped Max Planck. In an attempt to distribute some of the burden, Planck asked Wilhelm Wien to become coeditor. Wien, professor of physics at Würzburg, was like Drude both an experimentalist and a theoretician. In the sharing of tasks, though, Planck remained the senior partner. Either editor could, it appears, accept a paper without setting conditions on the material, but the coeditors agreed to contact each other about submissions to be revised or rejected. Over the next decade or so, the two men seem to have accepted at least around 80% of what crossed their desks.

With regard to papers on special and general relativity, the coeditors were generous, firm, and principled: generous to first-time authors with something new or innovative to present; firm about rejecting tedious summaries of previously published work or long, formalistic digressions offering no new results; principled in seeking to avoid, at all costs, personal invective and bad feelings. In rejecting a manuscript of Austrian physicist Emil Kohl's, Planck wrote to Wien that among all Kohl's equations there was not "a single one in which a new relationship between measurable quantities is provided." Urging rejection of a submission by industrial physicist Emil Arnold Budde, Planck wrote that Budde's work contained "no original thought that is not already found in the scientific literature, and done better there." On a paper of F. Grünbaum's, eventually rejected, Planck remarked that the argument was "correct, but it includes nothing really new and its physical interest is only very indirect." And let us listen in on Planck's evaluation of a manuscript by a young Polish physicist, Felix Joachim Wiśniewski: "The author defines every last thing in a formal way and assumes that, behind it all, these definitions have physical meaning. But nothing new comes from it."[81]

Planck's generosity, however, extended to more serious and sophisticated submissions. He published contributions by such authors as Walter Ritz, Philipp Frank, Ferencz Jüttner, and Gunnar Nordström, even though he was not convinced that their formulations of relativity would turn out to be correct. He was glad to handle the submission of secondary-school teacher Ernst Reichenbächer, in 1916, on the interrelationship between electricity and gravitation. Planck reacted to Reichenbächer's theory with skepticism. "The value of such a theory," he felt, "lay in what it finally delivers." Of greatest importance in any theory were "simplicity and intuitiveness and above all whether it has such characteristic consequences that can be tested by experiment."[82] Reichenbächer's work failed to meet these criteria. Planck returned the manuscript to the author. Reichenbächer submitted a revision. Planck urged Reichenbächer to contact Einstein. A meeting took place, and Planck soon published Reichenbächer's paper as the first unified field theory following Einstein's covariant field equations of general relativity. Planck, the senior statesman of physics in Germany and the relativity revolution's Danton, sought to accommodate any serious and flexible author.

b. Patterning and the Schools

Between 1890 and 1910 a major transformation occurred in pedagogy in German-speaking Europe. Before this time, graduates of the *Gymnasien* (the classical schools offering Greek and Latin) retained many official privileges not extended to graduates of the *Realgymnasien* (the semi-classical schools offering Latin but not Greek) and the *Oberrealschulen* (the non-classical schools offering neither Greek nor Latin). In place of Greek, the *Realgymnasien* offered modern languages and a bit more mathematics and natural science, a substitution pushed farther by the *Oberrealschulen*. In nineteenth-century Germany, however, passage through the *Gymnasium* was held to be necessary for anyone with broader cultural ambitions – and certainly for anyone who sought to contribute to the world of science and scholarship. As a result, although in the early years of the German Empire, the Prussian *Realgymnasien* were awarded the right to send students to the universities to study modern languages and natural sciences and in 1882 the same privilege was extended to the *Oberrealschulen*, until 1900 most university programs and many governmental positions remained closed to students trained in these *Realanstalten*. Only in 1901 did the Prussian educational authorities grant the *Realanstalten* virtual parity with the *Gymnasien*, a policy taken up as well by the other German states. A similar liberation of spirit occurred in 1899, when the German institutes of technology were given the right to award a doctorate of engineering. German educational authorities acknowledged the modernist basis of their Empire's technological strength, then, more than a generation after the advent of the second industrial revolution.[83]

The transformation in pedagogy may be called a revolution. It involved more than simply breaking the monopoly on higher culture that had been held for nearly one hundred years by neohumanist-minded educators in the classical *Gymnasien*. It involved, as well, as restructuring of the curriculum offered in all secondary schools – both in terms of content and in terms of spirit. Beginning around 1900, elements of differential and integral calculus were increasingly taught in secondary-school mathematics, especially in the *Oberrealschulen*. Many modernist schools introduced pupils to natural history, chemistry, and electrotechnology, while at the same time devoting increasing attention to foreign languages like French and English. Pedagogical technique adapted to the new ethic of usefulness and practicality. In place of rote memorizing and systematic regurgitation of assorted facts came an emphasis on organic learning, where pupils were encouraged to discover meaning for themselves in their studies. Secondary-school instructors, the *Oberlehrer*, shed their nineteenth-century image of scholarship and idealism – an image savagely caricatured as nothing more than pedantry; they became guides and seers, essential interpretors of a complicated modern world who attempted to mold the personalities of their charges into harmonious wholes.

The revolutionary changes followed much agitation from three constituencies in the academic world. The first comprised science and mathematics *Oberlehrer*, who increasingly saw themselves as the purveyors of a new, scientific humanism that was appropriate for the technological age. A second constituency came from practising engineers and their *Dozenten* in the polytechnics and institutes of technology, who saw themselves as holding the keys to mastery of the universe. Third and last to come out in favor of educational changes were mathematics and science professors at the universities. The *Oberlehrer* and the engineers organized pressure groups, associations, and house organs to demand that *Realen* – practical arts and useful skills – be granted equality in education with *Idealen* – abstract and ennobling studies unrelated to the world of industry and commerce. The German universities, based as they were on the notion of pure learning, were not eager to abandon their claim that education be at its base *allgemein* and *wissenschaftlich* – general and scholarly or scientific – rather than *fachlich* and *angewandt* – specialized and applied. And, as may be expected in entrenched institutions with special privileges, even professors who were intimate with the needs of industry and technology, men like Helmholtz and Bunsen, were reluctant to see science students entering their classes without proper grounding in Latin and Greek.

The most visible proponent of *Realen*, among university professors, was mathematician Felix Klein. Since coming to Göttingen in 1887, he labored to implement a vision of mathematical universalism, where pure mathematicians worked together with scientists to solve the pressing questions of physical theory and industry. Klein used his considerable influence with Friedrich Althoff, the Prussian authority in charge of universities, to bring together at Göttingen a large circle of sympathetic and brilliant colleagues and to obtain funding for a handful of new physical institutes. A portion, by no means the largest portion, of the cost of this new development was supplied by Klein's industrialist friends. At the same time that he was setting his Göttingen empire in place, Klein presented himself to reform-minded *Oberlehrer* and engineers as a supporter of their cause. Perceptive engineers immediately realized that Klein sought to bring them into his orb and domesticate them, essentially reserving for the universities the privilege of carrying out most of the fundamental research and instruction related to applied problems. Major confrontations ensued between Klein and engineers like Alois Riedler. The *Oberlehrer*, though, were generally charmed by Klein's attentions, and most of them did not realize that their university patron was striving to revise the secondary-school curriculum in order the better to place, as mathematics and science teachers, young university graduates.

Klein's imperialist vision, designed to preserve the dominant position of the universities in the new materialistic age, also reserved a special place for pure mathematics, his cherished specialty. Pure mathematics, a discipline flourishing in the late nineteenth century and one identified with the *Wissenschaftlichkeit* of the neoclassicists, was along with other forms of *Idealen* threatened by the demands of practical-oriented reformers. Reformist schoolteachers and engineers viewed mathematics as a tool for understanding and transforming the world; they were not sympathetic to the classical vision of pure mathematics as a schooling of the intellect. Klein conceived of a way for pure mathematicians to respond to the demands of scientists and engineers without compromising the integrity of their 'abstract' discipline. Pure mathematicians were in Klein's view to become the general staff of an industrial army, condescending to draw up tactics and plan strategy for dealing with problems in the real world. Under no circumstances were pure mathematicians to abandon their calling, however, and flock to the new field of applied mathematics.

The organizing skill of Klein and his fellow reformers combined with institutional inertia to guarantee a secure place in the secondary schools for graduates of university programs in both pure and applied mathematics. The curriculum in both the schools and the universities shifted to respond to the new interest in mathematization. The first decade of the twentieth century saw an overture – an openmindedness – toward applying the structures of pure mathematics in physical situations. It was as if pure mathematics extended the promise of resolving longstanding scientific questions by the force of deductive reasoning alone. In the schools, especially in the *Oberrealschulen*, calculus made strong inroads; in the universities, group theory, integral equations, differential geometry, and finite difference equations were turned to problems in the electron theory, geophysics, and spectroscopy.

Given the variation in individual trajectories both on the secondary-school and university levels, the effects on physicists of this transformation in mathematics instruction are difficult to assess. We know that whereas the class structure of young physics graduates remained unchanged between 1900 and 1914, the percentage of physicists educated in *Realanstalten* increased from around 20% to around 60%, an increase greater than that for other fields of study, although only a fraction of this young physics community engaged relativity theory in print.[84] An attempt can be made, nevertheless, to correlate positions taken during the relativity revolution with the secondary instruction that physicists and mathematicians received.

The special relationship between secondary-school education and positions taken from the benches of the revolutionary Convention may be expressed best in a negative way. There is *in the large* no association between having attended a *Realanstalt* and supporting or rejecting Einstein's theories of relativity: After all, such diverse intellects as Ritz, Debye, Lenard, Wiechert, Bucherer, and Einstein himself received the *Abitur* from a non-classical secondary school. Nor can one observe, across the board, a marked proclivity for *Realanstalt* graduates, distinct from *Gymnasium* graduates, to engage in complicated mathematical techniques when grappling with relativity: Mie, Witte, Herglotz, Sommerfeld, Wien, Silberstein, Laub, and Ehrenfest all graduated, for example, from *Gymnasien*, and their attitudes toward relativity diverged in a number of ways. The principal distinction with regard to secondary education, however, seems to lie in views regarding *Minkowski's* approach to the electron theory. There seems to be no *Realanstalt* graduate who wholeheartedly followed Minkowski's lead and joined the *Feuillantiste* and then the *Girondin* faction of the revolutionary assemblies.

The process at work here is that of displacement. Minkowski's call centered around the notion of a preestablished harmony between pure mathematics and the physical world.[85] A *Gymnasium* education was predicated on the notion that abstract and ideal values would as a matter of course find application in daily affairs. In the classical *Gymnasium*, the accent was placed on values central to Roman and especially Greek authors. It required little effort, then, for a young mathematician or physicist to find in pure mathematics a replacement for Greek and Latin. Such a displacement did not become equally manifest in all young students. One must bear in mind that a number of *Realanstalt* graduates – Wiechert, Ritz, Lamla, Blaschke, and also Einstein – used highly sophisticated mathematics in extending or in criticizing the electron theory or relativity; it is reasonable to identify these researchers as having been exposed to the new pedagogy in mathematics. The new pedagogy extended to many *Gymnasium* graduates who were partisan to Einstein's way of viewing relativity, or at least informed in their analysis of it: To mind come Erich Kretschmann, Jakob Laub, Kurd von Mosengeil, Wolfgang Pauli, Walter Schottky, Karl Schwarzschild, Ludwik Silberstein, Wilhelm Wien, Paul Ehrenfest, and Erwin Finlay Freundlich. A number of these men in fact passed through Klein's Göttingen: Schwarzschild was a professor and Laub, Ehrenfest, and Freundlich were students there.

Pedagogical styles and academic trajectories exhibited much variation in late Wilhelmian Germany, and any thesis of educational determinism would be of limited significance in explaining the development of scientific ideas. One is nevertheless struck by what may be called the 'physicalist' thought apparent in the approach of a number of researchers educated outside Germany in modernist secondary schools. The centrality of an intuitive grasp of physical laws is seen in the work of both Einstein and Philipp Lenard, two graduates of foreign *Realanstalten* who squared off against each other as early as 1910.

Just as Einstein was marked by the lessons of his mathematics and science instructors in Munich and Aarau, so Lenard found much to admire in his secondary-school instructor of physics, Virgil Klatt, at the Pressburg *Realschule*. Having taught himself calculus and spectroscopy, Lenard came to spend time in Klatt's physical cabinet, "a rather small room with 3 windows," two stories up in the *Realschule* building. There teacher and pupil would spend their Sunday mornings together. Later, on vacation from his university studies at Heidelberg, Lenard returned to Klatt's laboratory where he tried out what he had learned from Robert Wilhelm Bunsen. In his autobiography, Lenard devoted many pages to Klatt; he never mentioned his doctoral dissertation or even the name of his dissertation adviser. Lenard took pains to describe, however, how his progress in German universities was impeded by not having had formal instruction in Latin. He had to pass a special examination on translating Livius in order to receive a doctorate from Heidelberg. When he wanted to receive the *venia legendi* as Heinrich Hertz's assistant at Bonn, however, the university demanded that he satisfy their own Latin expert. Hertz asked the faculty to be permitted to test Lenard in all *Gymnasium* subjects, but the senior philologist objected: "'Hertz is a young man; he cannot judge,'"[86] The same as Einstein, Lenard saw little use in learning Latin. The pattern here seems similar to that found in other 'physicalist' foreigners – notably Debye, Ritz, and Frederick Alexander Lindemann – who attended modernist secondary schools.

c. *The Impact of Doctoral Apprenticing*

It will be apparent that, just as events from outside the world of school can have great impact on the formation of a scientist's worldview, so, for many people, the circumstances of advanced training may override the prejudices retailed in secondary school. Elaborating the process of advanced certification in Germany, indeed, has loomed large in historiography. Masters of specialized learning produced disciples to fill university institutes, industrial laboratories, and governmental bureaucracies. University education in Germany, however, admitted of variations, especially in the matter of receiving a doctorate of philosophy. For this reason, it is well to reconsider some features of the German *Dr. phil.*, or Ph.D.[87]

The doctorate in philosophy was awarded by the faculty of a university for an original research essay expected to be published at the author's expense. The requirements for defending a dissertation were defined by the individual faculties. Usually, but not always, the only prerequisites were possession of a *Gymnasium* leaving certificate (the *Abitur*) and about six semesters of study at a German university. In some universities, the student had to sit for state examinations in his field of study. Originality of research, length of dissertation, and residence requirements were all set by the faculties or, in practice, the student's faculty adviser, his 'promoter.' It appears that throughout the nineteenth century, a Ph.D. could have been obtained by almost any determined *Gymnasium* graduate. Deserving foreigners, Jews, and other minorities were seldom if ever refused a Ph.D. Indeed, for those foreigners who had difficulty with German, standards were often relaxed. Questionable practices resulted in occasional scandals, as in the 1870s, when charges of awarding doctorates in absentia were brought against Göttingen, yet this practice continued at various universities into the twentieth century. Lenient standards were to everyone's advantage. Professors received fees for processing candidates, universities received publicity through their alumni, and the young doctor and his wife bore on their greeting card a title reserved for *Kulturträger*, or bearers of civilization. The system allowed professors to supplement their incomes and disseminate their good name as well as to introduce students to scientific research. Yet, unless one were seeking a position as laboratory assistant or unless one thought to write a second thesis in order to become *Privatdozent*, the Ph.D. was of little practical use. It paled beside membership in the officer corps. Government and private employers could require evidence of a *Gymnasium* education, but only implicitly did they ask for a Ph.D. Far more important was mastery of a specialty, an attainment verified by success in state-administered examinations. The examination taken by many future scientists was the one that allowed entry into the lists of secondary-school teachers. It was seen, throughout the nineteenth century, as a difficult hurdle to pass.

A comparative picture is available for Ph.D. production, in physics and a number of other fields, in the years around 1900. In such a picture, physics and the other physical sciences are seen to have attracted students from higher social strata than was the case in other fields. Into the twentieth century, however, the *Gymnasien* ceased to furnish the majority of future physicists and mathematicians. In fact, the *Realanstalten* sent a torrent of students to take university doctorates in the physical sciences by 1913. This change in the secondary-school preparation of physicists is a demographic movement on a large and relatively sudden scale.[88]

Having remarked on the Ph.D. in physics in general terms, I now turn to consider one case of special interest: the students processed through the University of Berlin by Max Planck during the decade before the First World War. It was usual at Berlin for a student to have a principal adviser – the *Doktorvater* or promoter – and a supplemental adviser, an expert whose charge would have been to verify that a particular area of expertise was authentically represented in the dissertation. Beginning in 1904/05, these arrangements are specified in the *Jahres-Verzeichnis* of German doctoral dissertations. According to this source, between 1904/05 and 1913/14 Planck was principal adviser for 16 dissertations and supplemental adviser for 29 more; on the average, he sat on between four and five examining committees per year. In his role as supplemental adviser, he, the theoretician, surveyed for the most part experimental dissertations. His principal faculty collaborator was Heinrich Rubens, who in 1906 succeeded Paul Drude as Berlin's chief experimentalist and who, across his long career, provided essential information to an entire generation of theorists working to extend first Maxwellian electrodynamics and then the quantum theory. In the decade under consideration, Planck coadvised 12 of Rubens's students, and he invited Rubens to sit as coadviser of several of his own, theoretician students. Next in order of collaborative frequency came physical chemist Walther Nernst; Planck coadvised seven of Nernst's students. After Nernst came experimental physicist Arthur Wehnelt, inventor of the interruptor induction-coil that bears his name and discoverer, in 1903/04, of a method for producing slowly moving electrons and soft canal rays; four of Wehnelt's students benefitted from Planck's expertise. Planck also occasionally coadvised students of meteorologists Wilhelm von Bezold and Gustav Hellmann and mineralogist Theodor Liebisch. Planck's relationship with Rubens reflected a continuing committment to interaction with the experimentalists. Planck advised two of Paul Drude's students during Drude's unhappy year in Berlin, and in 1904/05 the theoretician advised two students of Drude's predecessor, Emil Warburg.

A substantial fraction of Planck's research students received secondary-school training in *Realanstalten* (Table I). It is clear that Planck's interests in directing students ran overwhelmingly in the areas of the electron theory and the special theory of relativity. Eleven among

TABLE I

	Kind of Secondary School				
	Gymnasium	*Realgymnasium*	*Oberrealschule*	American	Total
Major adviser	8	6	2	0	16
Minor adviser	14	5	6	4	29

Secondary-school background of Max Planck's research students, 1904/05 to 1913/14, as provided in the *Jahres-Verzeichnis der an den deutschen Universitäten erschienenen Schriften*. Indicated are dissertations for which Planck served as the major adviser or minor adviser, as specified by the order of advisers' names in the *Jahres-Verzeichnis*. *Major adviser*: Günther Falckenberg, Walther Meissner, Kurd von Mosengeil, Hermann Bönke, Ernst Lamla, Walter Schottky, August Gehrts, Abraham Esau, Wilhelm Heil, Adolf Semiller, Hans Witte, Karl Körner, Walther Bothe, Erich Kretschmann, Erich Henschke, Fritz Reiche; *minor adviser*: Egon Alberti, Hans Schneider, Alfred Schülze, Heinrich Bahr, Ernst Genzken, Gustav Hertz, Artur Lehmann, Frederick Alexander Lindemann, Fritz (Alfred) Schulze, Walter Brückmann, Frederick Russell Gorton, Julius Edgar Lilienfeld, Erich Regener, Franz Ritter, Hans Alterthum, Wilhelm Burmeister, Fritz Eckert, Rudolf Lehnhardt, Theodor J. Meyer, Gustav Weitzel, Luise Wolff, Hermann Behnken, Hans Mortiz Cassel, Willy Drägert, Isolde Ganswindt, John Torrence Tate, Arthur Quincy Tool, Hans Vogel, George E. Washburn, Gustav Zickner. Dieter Hermann provides only a partial list of Planck's students in his article, "Max Planck als akademischer Lehrer," in Horst Kant, ed., *Berliner wissenschaftshistorische Kolloquien VIII: Die Entwicklung Berlins als Wissenschaftszentrum (1870–1930); Die Entwicklung der Physik in Berlin* (Berlin, 1984), pp. 55–71.

the fifteen students for whom he served as a major adviser wrote a dissertation – generally a theoretical one – in these areas.

The historian is not recommended to attempt a determinist argument that would distinguish research style merely on the basis of secondary education, especially in such a polyvalent environment as Wilhelmian Germany. The varieties of biographical experience doom such attempts to failure. This point can be elaborated by considering three doctoral dissertations directed by Max Planck in 1912, those of Walter Schottky, Ernst Lamla, and Erich Henschke.[89] Each was an attempt to arrive at a synthetic formulation of the special theory of relativity from a particular point of view. Schottky (who alternately used the terms *Relativtheorie* and *Relativitätstheorie*) sought to synthesize the current state of knowledge on relativistic thermodynamics; Lamla studied the question of compressibility in relativistic hydrodynamics; Henschke rederived recent relativistic results by reconstructing relativistic electrodynamics according to the principle of least action. The three young theoreticians, each of whom must be said to share Planck's approach to relativity, provide indication of a common bibliographic orientation. They began with Einstein's first paper on special relativity and then reconsidered succeeding treatments by Einstein, Minkowski, Abraham, Planck, Born, Ehrenfest, Sommerfeld, Herglotz, Fritz Noether, Tullio Levi-Cività, Gilbert Newton Lewis, and Waldemar von Ignatowski; they also considered classical works by Lorentz, Schwarzschild, Lorentz, Poincaré, Helmholtz, and Emil Cohn. Planck's inspiration is much in evidence in all three texts; he is often thanked for having suggested one or another point of departure.

What attracts our eye from the present perspective, however, is the fact that the three went through different secondary schooling: Schottky, the son of a university professor of mathematics, graduated from the Steglitz *Gymnasium*; Lamla (who did not provide the occupation of his deceased father in the dissertation *Lebenslauf* even though he dedicated the text to his father's memory) graduated from the Friedrichs *Realgymnasium* in Berlin; and Henschke, son of a late fabric manufacturer, graduated from the Cottbus *Oberrealschule*. If their understanding of Einstein's relativity was largely comparable – all, guided by Planck's strong hand, rejected Minkowski's approach – their partiality toward the use of mathematics showed certain clear variations. Schottky, the gymnasiast, sought, almost defiantly, to retain a physical point of view, one mirroring that of Planck and Einstein. He formulated his propositions and motivations by long digressions where no equations entered at all. The second part of his dissertation began with a "thought experiment," surely with the model before him of Einstein's thought experiment on the principle of equivalence. Lamla, the *Realgymnasium* graduate, was more comfortable with mathematical formalism, in his case Hamiltonian integrals; he used his basic equations to reexamine notions of rigidity and rotation which had previously been explored by other theoreticians. Henschke, the *Oberrealschule* graduate, was most at home with mathematics. The first section of his dissertation was devoted exclusively to a review of four-dimensional vector analysis. In later sections he formulated electrodynamics for the vacuum and for moving, material bodies.

The dissertations of Schottky, Lamla, and Henschke illustrate the difficulty of attempting to arrive at a causal connection between fine distinctions in physical discourse and gross educational experience. From general arguments it might be concluded that Henschke's affinity for mathematical elaboration derived from his passage through an

Oberrealschule – certainly Planck would not have been the source of such a proclivity. Yet general arguments cannot have revealed the extent to which *Realanstalt* graduates were unimpressed by Minkowski's approach to relativity. The displacement phenomenon, reasonable in retrospect, does not seem to have been predictable from first principles; it is, after all, an artifice, not unlike Sigmund Freud's notion of the same name. One must not hold such a circumstance against the notion of displacement in the present context: It remains valuable because of the distinction, independently formulated, between Einstein's and Minkowski's ways of viewing relativity.

Displacement is general enough to serve in other historical situations. The notion seems to explain what happened in Germany after 1933, when National Socialist mathematicians were attracted to L.E.J. Brouwer's intuitionist philosophy of mathematics. In Brouwer's view, the primitive notions of mathematics were intuitively posited, and logical development meant finite constructibility of objects to be talked about. Intuitionism seems not inherently more 'intuitive' than formalism, the mathematical doctrine, elaborated by David Hilbert, in reaction to which Brouwer conceived his approach. Resentment against establishment logic and learning led mathematicians sympathetic to the Nazi revolutionaries to look favorably on an alternative approach to the foundations of their discipline.[90] Here, as in the case of gymnasiast Minkowskians, the displaced allegiance cannot be identified as a necessary development.

Although many historians still "engender concepts from disincarnated minds which live their lives beyond time and space," to use Lucien Febvre's phrase of 1928, historian of education Roger Chartier recently emphasized that the past generation has witnessed a fundamental reexamination of literary products. Categories of analysis previously accepted without question have been subjected to intense scrutiny. Chartier identified three of these categories: First there is the traditional distinction between popular and learned activity; then there is the division between producing ideas and consuming them; finally there is the abyss separating hard reality from imagined representations of it.[91]

The preceding pages reflect the *rapprochement* between Chartier's old polar essences. Einstein's relativity has been diffracted through the prism of popular interest in the ongoing scientific revolution. Educational patterns have been probed for their contribution to a susceptibility for one or another approach to the mathematization of nature. The requirements of publication have turned out to shape the form taken by scientific ideas. Relativity in Germany has been rephrased in terms transposed from another historical setting, leading to a portrayal of the physics as, strictly speaking, it never was.

This way of viewing relativity in Germany has placed short- and long-term phenomena under equal scrutiny; it has spoken to a change in historical periods, or epochs, in the small and in the large. The approach would not have appealed, in his later years, to the most important historical actor under consideration – Albert Einstein. The "great revolution" in modern physics, Einstein wrote in 1931, took place with the work of Faraday, Maxwell, and Heinrich Hertz, with Maxwell having "the lion's share in this revolution."[92] The only significant appeal to revolutionary change in Einstein and Leopold Infeld's *Evolution of Physics*, indeed, is to the French *Revolution*; classical physics and quantum physics are, in that book, merely held to "unterscheiden sich radikal."[93] The root of the mature Einstein's views on revolution in science seems to be identical to that of his views on political revolution. He wrote to a Canadian communist in 1935:

> I maintain that, in the long run, the masses cannot be helped unless they achieve political maturity. While you consider the political maturity of the masses a prerequisite to revolution, I suggest that, once the goal of maturity is realized, a revolution will no longer be necessary.[94]

Mature thinkers – physicists and proletarians – would as a matter of course direct themselves to act in a progressive way. The notion of progress here seems to be essentially one dating from the nineteenth century.

In his perceptive essay of 1975 on Einstein and scientific revolutions, Martin Klein has written that Einstein viewed revolutions in science as complete, fundamental, and permanent changes. The twentieth century, he felt, had not witnessed such changes: "When Einstein reacted skeptically to claims that this or that new discovery or new theory had revolutionized physics, he did so in the name of True Revolution."[95] Here, as elsewhere in his work, Einstein resisted what he thought to be passing fashion in the world of physics. In some cases, furthermore, observation could be held to be related to fashion. Beginning at least around the eve of the First World War, Einstein believed that the essential truth of a theory could not be keyed directly to the outcome of an apparent experimental confirmation. He wrote to Infeld in 1949 to emphasize his "strict adherence to logical simplicity and [his] lack of confidence in the merit of ever impressive confirmations of theories, whenever questions of principle are involved."[96] The man whose ideas, more than those of any other figure, revolutionized twentieth-century physics, was chary of grandiose proposals for transforming natural laws. He is reported to have affirmed before around 1935: "I have an unbounded mistrust of nature."[97]

"Les révolutions se font malgré les révolutionnaires," Ernest Labrousse noted in 1948.[98] His is the viewpoint adopted today by most theorists of revolutionary change in science. Scientists are seen as not fully conscious of their social roles, and in their comportment they respond to social imperatives. Scientists may hold that they deal only with reasonable arguments advanced in the world around them, but they not infrequently appear in the light of history as if they were, like Hamlet, taking cues from an unseen ghost of their prejudices and emotions. It is on the level of delineating intangible motivation that one may fruitfully employ an analogy between cognitive and social processes. As Chikara Sasaki and others have emphasized, scientific revolutions are social revolutions within the set of scientists under consideration. Cognitive cleavages can and should be defined not merely in social terms, but in socially dynamic ones. For this, bold analogy is an effective vehicle. The material for reconsidering relativity from the perspective of the French Revolution has long been available to historians. It would be of great interest to see whether, in their analogical structures, other scientific revolutions may resemble the flow of circumstance in the world of politics.

ACKNOWLEDGEMENTS

I am grateful to John Heilbron and Thomas F. Glick for criticism. Kuniko Kawachi Mitrovic assisted me in studying Chikara Sasaki's text. My research has been supported in part by the generosity of the Vice-décanat à la recherche, Faculté des Arts et Sciences, Université de Montréal.

NOTES

[1] Einstein to Habicht, 1905, in Carl Seelig, *Albert Einstein: A Documentary Biography*, trans. Mervyn Savill (London, 1956), p. 74; Martin J. Klein, 'Thermodynamics in Einstein's Thought,' *Science*, **157** (1967), 509–516, on p. 511.

[2] Systematic, technical analyses of relativistic physics and related questions during the first two decades of the twentieth century may be found in a large number of works. An exhaustive listing would serve little purpose, as the 'relativity literature' spans a broad spectrum of epistemological, ideological, and historical approaches. There seems to be little agreement about how to make sense out of the historical record. The reader's attention is nevertheless drawn to the following texts: Jun Ishiwara, 'Bericht über die Relativitätstheorie,' *Jarhbuch der Radioaktivität und Elektronik*, **9** (1912), 560–648; Ludwik Silberstein, *The Theory of Relativity* (London, 1914); Wolfgang Pauli, *Theory of Relativity*, trans. Gerard Field (London, 1958); Jean Chazy, *La Théorie de la relativité et la mécanique céleste*, 2 vols (Paris, 1928–1930); Eugene Guth, 'Contribution to the History of Einstein's Geometry as a Branch of Physics,' in Moshe Carmeli, *et al.*, eds, *Relativity* (New York, 1970), pp. 161–207; Vladimir P. Vizgin's two related volumes, *Relyativistskaya Teoriya Tyagoteniya (Istori i Formirovanie. 1900–1915)* (Moscow, 1981) and *Yedinye Teorii Polia v Pervoi Treti XX Veka* (Moscow, 1985); Arthur I. Miller, *Albert Einstein's Special Theory of Relativity: Emergence (1905) and Early Interpretation* (1905–1911) (Reading, Mass., 1981); Abraham Pais, *Subtle is the Lord: The Science and the Life of Albert Einstein* (New York, 1982); Stanley Goldberg, *Understanding Relativity: Origin and Impact of a Scientific Revolution* (Boston, 1984); Lewis Pyenson, *The Young Einstein: The Advent of Relativity* (Bristol, 1985). Some of the issues that I treat in the following pages have been raised by József Illy in 'Revolutions in a Revolution,' *Studies in the History and Philosophy of Science*, **12** (1981), 173–210, a text that brings to mind Régis Debray's study of armed struggle and political revolution in Latin America; I have been unable to follow all of Illy's arguments in detail. In the present text, I generally do not document primary, published sources that may easily be retrieved from standard works, such as the various editions of *J.C. Poggendorff's biographischliterarisch Handwörterbuch* and the *Dictionary of Scientific Biography*, 16 vols (New York, 1970–1976), edited by Charles C. Gillispie; complete references are provided, however, for unusual sources and archival material.

[3] Thomas S. Kuhn, *The Structure of Scientific Revolutions* (Chicago, 1962; 1970). Some of the philosophical response to Kuhn's work is catalogued in the bibliography to *Scientific Revolutions*, ed. Ian Hacking (Oxford, 1981). An illuminating discussion of the history of the concept of a scientific revolution, and of how scientific revolutions differ from political revolutions, may be found in Lewis Feuer, *Einstein and the Generations of Science* (New York, 1974); the matter is considered at length in I. Bernard Cohen, 'The Eighteenth-Century Origins of the Concept of Scientific Revolution,' *Journal of the History of Ideas*, **37** (1976), 257–288. Neither Goldberg, *Understanding Relativity* (ref. 2) nor Illy, 'Revolutions in a Revolution' (ref. 2) is primarily interested in delineating the structure of the relativity revolution. Among a number of studies that reexamine Kuhn's notion of a scientific revolution from a historical perspective, Karl Hufbauer's recent book is of signal clarity: *The Formation of the German Chemical Community (1720–1795)* (Berkeley, 1982). Noteworthy is the philosophical essay by Paul Scheurer, *Révolutions de la science et permanence du réel* (Paris, 1979).

Since the time that I completed this study, I. Bernard Cohen's large book, *Revolution in Science* (Cambridge, Mass., 1985), has come to hand. The book considers how many distinguished scientists, and also historians of science, have made use of the notion of a scientific revolution. According to Cohen (pp. 28–29), actual scientific revolutions proceed in four stages. First comes the apperception or discovery of a radical new solution to outstanding problems. Then there is a commitment to the new way of doing things. Commitment is followed by a dissemination of the new approach ("a revolution in paper," in publications). Finally one finds a large number of scientists carrying out their own research in the new way. In the present text, I depart substantially from Cohen's application of the four-stage model to the reception of the theories of relativity (pp. 36–37); the relativity revolution in Germany seems more complicated than he indicates (pp. 406–419), especially in the period before 1919.

[4] Max Weber, "'Energetische' Kulturtheorien (1909)," in Weber, *Gesammelte Aufsätze zur Wissenschaftslehre*, ed. J. Winckelmann (Tübingen 1973), pp. 400–426, on p. 424. Weber's article is an essay review of Wilhelm Ostwald's *Energetische Grundlagen der Kulturwissenschaft* (Leipzig, 1909). In Weber's view, Ostwald was guilty of attempting to reify abstractions from natural science, trying to transfer forms of thinking from one to another discipline in an uncritical way, trying to brand events and circumstances with an "energeticist" tag, and transforming the *Weltbild* of a discipline into a *Weltanschauung*; Weber thought that Ernst Mach made many of the same mistakes.

[5] Chikara Sasaki, *The Historical Structure of Scientific Revolutions* (Tokyo, 1985), p. v. The book is in Japanese.

[6] *Ibid.*, pp. v–vi.

[7] *Ibid.*, p. 10.

[8] *Ibid.*, p. 552.

[9] *Ibid.*, p. 49.

[10] *Ibid.*, introduction (pp. 1–56) and chapter 6 (pp. 599–628).

[11] *Ibid.*, p. 279.

[12] Erwin Panofsky, *Renaissance and Renascences in Western Art* (Stockholm, 1960; New York, 1969), pp. 3–4.

[13] I noted and commented on this difference in an earlier publication: 'La Place des sciences exactes en Allemagne à l'époque de Guillaume II,' *Europa: A Journal of Interdisciplinary Studies* (Montreal), **4**, no. 2 (1981), 187–217, on pp. 199–200. Compare Cohen, *Revolution* (ref. 3), pp. 254–261, where the changing language of revolution in Germany is discussed.

[14] Felix Hausdorff, in *Zeitschrift für mathematische und naturwissenschaftliche Unterricht*, **33** (1902), 190–193, on p. 190.

[15] Paul Drude, *Die Theorie in der Physik: Antrittsvorlesung gehalten am 5. December 1894 an der Universität Leipzig* (Leipzig, 1895), p. 15.

[16] Siegfried Aronhold, 'Die reine Mathematik in den Jahren 1884–1899,' in E. Lampe, ed., *Die reine Mathematik in den Jahren* 1884–1899 *nebst Actenstüken zum Leben von Siegfried Aronhold* (Berlin, 1899), pp. 7–31, on p. 31.

[17] Ernst Cassirer, *Leibnitz' System in seinen wissenschaftlichen Grundlagen* (Marburg, 1902), p. viii.

[18] Karl Schwarzschild, "Ueber Himmelsmechanik," *Physikalische Zeitschrift*, **4** (1903), 765–773, on p. 771.

[19] Felix Klein, 'Mathematik, Physik, Astronomie an den deutschen Universitäten in den Jahren 1893–1903,' *Physikalische Zeitschrift*, **5** (1904), 764–775, on pp. 766–767.

[20] Felix Klein, 'Bemerkungen zum mathematischen und physikalischen Unterricht,' *Physikalische Zeitschrift*, **5** (1904), 710–717, on pp. 712, 713.

[21] Paul Staeckel, 'Bericht über die Entwicklung des Unterrichtsbetriebes in der angewandten Mathematik an den deutschen Universitäten,' *Physikalische Zeitschrift*, **3** (1902), 92–97, on p. 92.

[22] Otto Kammerer, 'Ueber den Zusammenhang der Maschinentechnik mit Wissenschaft und Leben,' *Physikalische Zeitschrift*, **1** (1899), 186–190, on p. 189.

[23] Ludwig Boltzmann, 'On the Development of the Methods of Theoretical Physics in Recent Times,' in Boltzmann, *Theoretical Physics and Philosophical Problems*, ed. Brian McGuiness, trans. P. Foulkes (Dordrecht, 1964), pp. 77–100, on p. 82. Boltzmann's text appeared in the *Physikalische Zeitschrift*, **1** (1899/1900), and the extract is located on p. 78.

[24] Ludwig Boltzmann, 'Antrittsvorlesung, gehalten in Wien im Oktober 1902,' *Physikalische Zeitschrift*, 4 (1903), 274–277, on p. 276.

[25] The electromagnetic view of nature is analyzed in a series of brilliant publications by Russell McCormmach: 'H.A. Lorentz and the Electromagnetic View of Nature,' *Isis*, **61** (1970), 457–497; 'Lorentz, Hendrik Antoon,' in the *Dictionary of Scientific Biography* (ref. 2), *s. v.*; 'Einstein, Lorentz, and the Electron Theory,' *Historical Studies in the Physical Sciences*, **2** (1970), 41–87.

[26] Pyenson, *Young Einstein* (ref. 2), pp. 42–43.

[27] Paul von Zech, 'Die Physik vor hundert Jahren und heute,' *Deutsche Revue*, **17**. I (1892), 188–198, on p. 197.

[28] O. Krigar-Menzel and M. von Laue, eds, *Vorlesungen über Elektrodynamik und Theorie des Magnetismus von H.v. Helmholtz* (Leipzig, 1907), pp. 3–4. The lectures were reworked from Helmholtz's notes.

[29] Walter Kaufmann, 'Die Entwicklung des Elektronenbegriffs,' *Physikalische Zeitschrift*, **3** (1901/02), 9–15, on p. 9.

[30] Eduard Riecke, 'Die jetztigen Anschauungen über das Wesen des metallischen Zustandes,' *Physikalische Zeitschrift*, **10** (1909), 508–519, on p. 508. The Naturforscher, or Gesellschaft deutscher Naturforscher und Aerzte, held annual meetings that were attended by a large proportion of Germany's practising scientists.

[31] Wilhelm Wien, 'Zur Elektronentheorie,' *Physikalische Zeitschrift*, **5** (1904), 294–295, on p. 295.

[32] Cornelis Harm Wind, 'Elektronen und Materie,' trans. Alfred Gradenwitz, *Physikalische Zeitschrift*, **6** (1905), 485–494, on p. 493.

[33] Einstein to Solovine, 3 May 1906, in Albert Einstein, *Lettres à Maurice Solovine* (Paris, 1956), pp. 2–7.

[34] Revealing in this regard is Max Born's recollection that Einstein spoke to the Salzburg Naturforscherversammlung in 1909 on "die neueren Umwandlungen, welche unsere Anschauungen über die Natur des Lichtes erfahren haben." Born, 'Physics and Relativity (1955),' in Born, *Physics in My Generation* (New York, 1969), pp. 100–115, on p. 107. Einstein published his paper, however, with the title: 'Ueber die Entwicklung unserer Anschauungen über das Wesen und die Konstitution der Strahlung,' *Physikalische Zeitschrift*, **10** (1909), 817–825, the same title that is provided in the program for Einstein's session which was published in the *Verhandlungen* of the German Physical Society. At the beginning of his paper, Einstein emphasized that in the future "a deep-going *Aenderung* of our views of the essence and of the constitution of light is indispensable." Here Einstein chose not to speak about an *Umwandlung* or *Revolution*. For Born, nevertheless, the import of Einstein's message was in the direction of a revolutionary change.

[35] Max Planck, 'Die Stellung der neueren Physik zur mechanischen Naturanschauung, *Physikalische Zeitschrift*, **11** (1910), 922–932, on p. 923.
[36] Ernst Lamla, *Ueber die Hydrodynamik des Relativitätsprinzips* (diss., Univ. of Berlin, 1912), p. 7.
[37] Hans Witte, *Raum und Zeit im Lichte der neueren Physik* (Brunswick, 1914), quotations on pp. 2, 59, 83, 77.
[38] Johannes Classen, 'Ueber das Relativitätsprinzip in der modernen Physik,' *Zeitschrift für den physikalischen und chemischen Unterricht*, **23** (1910), 257–267, on pp. 257–258.
[39] Otto Berg, 'Das Relativitätsprinzip der Elektrodynamik,' *Abhandlungen der Fries'-schen Schule*, **3** (1910), 333–382, on p. 381.
[40] Max Frischeisen-Köhler, *Wissenschaft und Wirklichkeit* (Leipzig, 1912), p. 324.
[41] Max Bernhard Weinstein, 'Die Relativitätslehre und die Anschauung von der Welt,' *Himmel und Erde*, **26** (1913), 1–14, on p. 1.
[42] Max Bernhard Weinstein, *Die Physik der bewegten Materie und die Relativitätstheorie* (Leipzig, 1913), p. 307.
[43] David Hilbert to Hugo Andres Krüss, 1913. Göttingen, Niedersächsische Staats- und Universtätsbibliothek. Hilbert Nachlass 494.
[44] Henri Poincaré, 'Die neue Mechanik,' trans. P. Schwahn, *Himmel und Erde*, **23** (1910/11), 97–116, on p. 97.
[45] Gustav Mie, 'Ionen und Elektronen,' in *Fortschritte der naturwissenschaftlichen Forschung*, **2** (1911) [ed. E. Abderhalden], 163–192, on p. 163.
[46] Melchior Palágyi, *Die Relativitätstheorie in der modernen Physik; Vortrag gehalten auf dem 85. Naturforschertag in Wien* (Berlin, 1914), quotations on pp. 32, 44–45, and 5, in sequence.
[47] Otto Wiener, 'Entwicklung der Wellenlehre des Lichtes,' in Emil Warburg, ed., *Physik*, (Leipzig, 1915), pp. 517–574, on p. 572. The general editor of the series *Kultur der Gegenwart* was Paul Hinneberg.
[48] The main lines of current thought regarding comparative history may be retrieved from recent articles and debates in the pages of the *American Historical Review*, **85** (1980),753–857, 1055–1166; **87** (1982), 123–143, and also in *Comparative Studies in Society and History*, **22** (1980), 143–221. Featured are contributions by twenty-four scholars. Comparing the relativity revolution in Germany with the French Revolution would not be excluded in principle by Theda Skocpol and Margaret Somers, 'The Uses of Comparative History in Macrosocial Inquiry,' *Comparative Studies in Society and History*, **22** (1980), 174–197.
[49] There has been much huffing and puffing about such analyses of late. For two philosophers making a case against and for this way of doing history: Larry Lauden, 'The Pseudo-Science of Science,' *Philosophy of the Social Sciences*, **11** (1981), 173–198; David Bloor, 'The Strengths of the Strong Program,' *ibid.*, 199–213. Neither Lauden nor Bloor seems much interested in marshalling and evaluating the secondary literature in history and sociology. For a balanced and critical (if not overly optimistic) survey, see: Vojin Milić, 'Sociology of Knowledge and Sociology of Science,' *Social Science Information*, **23** (1984), 213–273.
[50] I have been guided in my understanding of the French Revolution by François Furet and Denis Richet, *La Révolution* (Paris, 1965). A translation and abridgment by Stephen Hardman appears as *The French Revolution* (London, 1970) and includes a summary bibliography. A further list of references would serve no purpose, as each reader will have

favorite surveys. It would be pleasing to think that readers might forgive here, as elsewhere in the present text, the absence of exhaustive referencing. The recent remarks of Dominick LaCapra are not unrelated to this point: 'Is Everyone a *mentalité* Case? Transference and the 'Culture' Concept,' *History and Theory*, **23** (1984), 296–311.

[51] The books of Barrington Moore, Jr., have provided much to my understanding of revolution: *Social Origins of Dictatorship and Democracy: Lord and Peasant in the Making of the Modern World* (Boston, 1966); *Injustice: The Social Bases of Obedience and Revolt* (White Plains, N.Y., 1978). In the diffuse and vast literature on revolutionary etiology, I have found Herbert Marcuse's *Counterrevolution and Revolt* (Boston, 1972) of use. I have come back a number of times to ponder the clear and synoptic presentation in Charles Tilly, *From Mobilization to Revolution* (Reading, Mass., 1978), especially regarding the matter of when revolutionary situations produce revolutionary outcomes (pp. 189–222). A sharp and short statement is found in Eugene Kamenka, 'Revolution: The History of an Idea,' in Kamenka, ed., *A World in Revolution?* (Canberra, 1970), pp. 1–14. The stage theory of Crane Brinton – one that draws heavily on the French Revolution – holds some attractions, but I have not consciously followed it here. Brinton, *The Anatomy of Revolution* (New York, 1938; 1965).

[52] Pyenson, 'La Place' (ref. 13); Paul Forman, John Heilbron, and Spencer Weart, *Physics* circa *1900: Personnel, Funding, and Productivity of the Academic Establishments* (Princeton, 1975) [*Historical Studies in the Physical Sciences*, **5**]; David Cahan, 'The Institutional Revolution in German Physics, 1865–1914,' *Historical Studies in the Physical Sciences*, **15** (1985), 1–65. Cahan considers the rise of the big physical laboratories.

[53] John Heilbron, 'Lectures on the History of Atomic Physics 1900–1922,' in Charles Weiner, ed., *History of Twentieth Century Physics* (New York, 1977), pp. 40–108, on p. 40.

[54] See McCormmach's publications cited in ref. 25.

[55] Robert Marc Friedman, *Vilhelm Bjerknes and the Bergen School of Meteorology, 1918–1923: A Study of the Economic and Military Foundations for the Transformation of Atmospheric Science* (diss., Johns Hopkins Univ., 1978), pp. 17–19.

[56] Lenard publically advocated retention of the ether in 1910; evidence for his continuing belief in it until then is circumstantial and may be keyed to his high regard for Bjerknes's hydrodynamical theories, which is recorded in his letters retained in the Vilhelm Bjerknes papers, Institute for Geophysics, University of Oslo. See Andreas Kleinert and Charlotte Schönbeck, 'Lenard und Einstein: Ihr Briefwechsel und ihr Verhältnis vor der Nauheimer Diskussion von 1920,' *Gesnerus*, **35** (1978), 318–333.

[57] Arthur Korn, 'Ueber die Theorie der universellen Schwingungen mit Anwendungen auf die Theorie der Gravitation und der intramolekularen Kraefte,' in *Atti del IV Congresso internazionale dei matematici*, ed. G. Castelnuovo (Rome, 1909), **3**, 81–88; Jakob Kunz, *Theoretische Physik auf mechanischer Grundlagen* (Stuttgart, 1907).

Kunz continued to consider the ether as a construct worth mentioning when he first moved to the United States: 'On the Photoelectric Effect of Sodium–Potassium and Its Bearing on the Structure of the Ether,' *Physical Review*, **29** (1909), 212–228. In the following year he reported how the principle of relativity supported the quantum theory of radiation and rendered the ether superfluous: 'On the Initial Velocity of Electrons as a Function of the Wave-Length in the Photoelectric Effect,' *Physical Review*, **31** (1910), 536–544, on p. 536. Some of the deeper background for the mechanist reactionaries may be found in: Paul Drude, 'Ueber Fernwirkungen,' *Annalen der Physik*, **62** (1897), i–xlix.

[58] Sommerfeld to Schwarzschild, 8 December 1910. New York, American Institute of Physics, Karl Schwarzschild papers. Einstein's threat to Seeliger was clearly perceived by their contemporaries: Hugo Dingler, 'Ueber das Newtonsche Gravitationsgesetz,' *Zeitschrift für positivistische Philosophie*, **1** (1913), 220–226, on pp. 225–226. For the deep background, see Peter Lebedew, 'Die physikalischen Ursachen der Abweichungen vom Newton'schen Gravitationsgesetze,' *Vierteljahrsschriften der astronomischen Gesellschaft*, **39** (1902), 220–226.

[59] Oppolzer to Schwarzschild, 25 August 1902. New York, American Institute of Physics, Karl Schwarzschild papers.

[60] Mie's early work is clear enough in this regard. For his views around 1905: Hans Witte, 'Ueber den gegenwärtigen Stand der Frage nach einer mechanischen Erklärung der elektrischen Erscheinungen,' *Physikalische Zeitschrift*, **7** (1906), 779–785; in the discussion following the paper (presented at the Stuttgart Naturforscherversammlung), Mie was identified as requiring the ether in a mechanical theory.

[61] Arnold Sommerfeld roundly criticized Stark's giving serious credence to the ether in 1909. Sommerfeld to Stark, 10 December 1909. Berlin, Staatsbibliothek preussischer Kulturbesitz, Nachlass Stark. Early in 1910 Hans Witte thanked Stark for his "benevolent judgment" of Witte's work on the ether. Witte to Stark, 11 January 1910, *ibid.*

[62] The work of the experimentalists is convincingly analyzed in Giovanni Battimelli, 'The Electromagnetic Mass of the Electron: A Case Study of a Non-Crucial Experiment,' *Fundamenta Scientiae*, **2** (1981), 137–150.

[63] Pyenson, *Young Einstein* (ref. 2), pp. 202–203.

[64] Lewis Pyenson, *Cultural Imperialism and Exact Sciences* (New York, 1985), pp. 165–167.

[65] Bruno Strasser and Max Wien, 'Anwendung der Teleobjektivmethode auf den Dopplereffekt von Kanalstrahlen,' *Physikalische Zeitschrift*, **7** (1906), 744–746.

[66] Max Planck, 'Die Kaufmannschen Messungen der Ablenkbarkeit der ß-Strahlen in ihrer Bedeutung für die Dynamik der Elektronen,' *Physikalische Zeitschrift*, **7** (1906), 753–759.

[67] Sommerfeld, in discussion reported after *ibid.*, p. 761.

[68] Sommerfeld's work on the electron theory in the years before 1906 is considered in Pyenson, *Young Einstein* (ref. 2), pp. 120–128.

[69] Erwin Voellmy, *Die allgemeine Bewegung des Bornschen Elektrons; eine integrallose Darstellung von Parallelkurven im dreidimensionalen Raum* (diss., Univ. of Basle, 1917).

[70] Abraham's life and impact in Italy are analyzed in Judith Goodstein, 'The Italian Physicists of Relativity,' *Centaurus*, **26** (1982/83), 241–261. Beginning in 1909, Abraham was an associate professor at the Milan Institute of Technology; he accepted this call following an unhappy number of months as a professor of physics at the University of Illinois in Urbana. See also Giovanni Battimelli and Michelangelo De Maria, 'Max Abraham in Italia,' in F. Bevilacqua and A. Russo, eds, *Atti del III Congresso Nazionale di Storia della Fisica* ([Palermo], 1982), pp. 186–192.

[71] The work of Poincaré, Mie, and Wiechert is well known. Helm made his position clear in : 'Das Relativitätsprinzip in der Aetherhypothese,' *Physikalische Zeitschrift*, **13** (1912), 157–158. Gehrcke queried the relativity of time in 'Bemerkungen über die Grenzen des Relativitätsprinzips,' *Verhandlungen der deutschen physikalischen Gesellschaft*, **13** (1911), 665–669. La Rosa's book, *Der Aether* (Leipzig, 1912), announced his point of view. Höfler, Poske, Gehrcke, and La Rosa are analyzed in Joseph Petzoldt's essay,'Die

Relativitätstheorie in der Physik,' *Zeitschrift für positivistische Philosophie*, **2** (1914), 1–56. Also useful in this regard is Moritz Schlick, 'Die philosophische Bedeutung des Relativitätsprinzips,' *Zeitschrift für Philosophie und philosophische Kritik*, **159** (1915), 129–175.

[72] Gunnar Nordström, 'Träge und schwere Masse in der Relativitätstheorie,' *Annalen der Physik*, **40** (1913), 856–878.

[73] Mie to Hilbert, 13 February 1916, in Pyenson, *Young Einstein* (ref. 2), p. 185.

[74] Pyenson, *Young Einstein* (ref. 2), p. 235.

[75] József Illy, 'Geometrical Objects in Lorentz's General Theory of Relativity of 1916' (Budapest, 1985) [Hungarian Academy of Sciences, Institute of Isotopes, preprint IZIN-6/1985].

[76] Gereon Wolters has recently argued that Mach's putative rejection of relativity, published posthumously in 1921, was in fact constructed by Mach's son, under the influence of the philosopher Hugo Dingler. Gereon Wolters, 'Ernst Mach and the Theory of Relativity,' *Philosophia Naturalis*, **21** (1984), 630–641.

[77] Otto Blumenthal to Karl Schwarzschild, 30 December 1913. New York, American Institute of Physics, Karl Schwarzschild papers. Blumenthal remarked on Schwarzschild's last letter, in which he had indicated that he was trying to verify the gravitational redshift in the sun's spectrum.

[78] Robert Darnton, *The Literary Underground of the Old Regime* (Cambridge, Mass., 1982), p. 37; Pierre Goubert and Daniel Roche, *Les Français et l'Ancien Régime*, 2: Culture et société (Paris, 1984), pp. 219–254.

[79] Consider, for example, the paper published by a certain C. Beckenhaupt of Altenstadt-Weissenburg: 'Ueber die physikalischen Verhältnisse, welche bei dem Relativitätsprinzip und der Vierdimensionalität in Betracht kommen,' *Verhandlungen der Gesellschaft deutscher Naturforscher und Aerzte*, **83** (1911), *Sitzungen*, 105–110. The ether is held to be compatible with Einstein's and Minkowski's views on relativity, and some elementary arithmetic is used to illustrate the bounds on the density of the ether. It is difficult to see how Beckenhaupt's publication could have been taken seriously, even by confirmed mechanists.

[80] The succeeding paragraphs follow Pyenson, *Young Einstein* (ref. 2), pp. 194–214.

[81] Quotations, in sequence: Planck to Wilhelm Wien, 30 November 1909; Planck to Wilhelm Wien, 30 May 1911; Planck to Wilhelm Wien, 9 February 1911; Planck to Wilhelm Wien, 29 June 1913. All located in Berlin, Staatsbibliothek preussischer Kulturbesitz, Nachlass Wien.

[82] Planck to Wilhelm Wien, 25 August 1916, *ibid.*

[83] These issues have been addressed in a number of recent works, although there exists a substantial literature from the early part of the twentieth century. See Lewis Pyenson, *Neohumanism and the Persistence of Pure Mathematics in Wilhelmian Germany* (Philadelphia, 1983) [American Philosophical Society, *Memoirs*, **150**]. The following paragraphs draw on this work.

[84] Pyenson, *Young Einstein* (ref. 2), pp. 162–163.

[85] *Ibid.*, pp. 80–100.

[86] Philipp Lenard, 'Erinnerungen eines Naturforschers der Kaiserreich, Judenherrschaft und Hitler erlebt hat,' manuscript dated September 1943. Berkeley, University of California, Office for History of Science and Technology. Quotations from pp. 13–22, 41–43.

[87] The following paragraph follows Pyenson, *Neohumanism* (ref. 83), pp. 17–18.
[88] Pyenson, *Young Einstein* (ref. 2), pp. 162–163; Lewis Pyenson and Douglas Skopp, 'Educating Physicists in Germany *circa* 1900,' *Social Studies of Science*, **7** (1977), 329–366.
[89] Walter Schottky, *Zur Relativtheoretischen Energetik und Dynamik. Abschnitt I und II* (diss., Univ. of Berlin, 1912); Ernst Lamla, *Ueber die Hydrodynamik des Relativitätsprinzips* (diss., Univ. of Berlin, 1912); Erich Henschke, *Ueber eine Form des Prinzips der kleinsten Wirkung in der Elektrodynamik des Relativitätsprinzips* (diss., Univ. of Berlin, 1912).
[90] Herbert Mehrtens, 'Anschauungswelt versus Papierwelt: Zur historischen Interpretation der Grundlagenkrise der Mathematik,' in Hans Poser and Hans-Werner Schütt, eds, *Ontologie und Wissenschaft: Philosophische und wissenschaftshistorische Untersuchungen zur Frage der Objektkonstitution* (Berlin, 1984), pp. 231–276.
[91] Roger Chartier, 'Intellectual History or Sociocultural History? The French Trajectories,' in Dominick LaCapra and Steven L. Kaplan, eds, *Modern European Intellectual History: Reappraisals and New Perspectives* (Ithaca, 1982), pp. 13–46.
[92] Cited in Pais, *Subtle* (ref. 2), p. 319.
[93] Einstein and Infeld, *Physik als Abenteuer der Erkenntnis* (Leiden, 1938), p. 147. Contrast Infeld, *Albert Einstein: His Work and Influence on Our World* (New York, 1950), where Infeld, freed from Einstein's restraining influence, used the word "revolution" in five of his six chapter titles. Infeld wrote: "What do we mean by a revolution in physics? We mean a sudden clarification of our concepts, the forming of a new picture, an unexpected resolution of contradictions and difficulties." Infeld, *Einstein*, p. 22. *The Evolution of Physics* provides a discussion of Einstein and Infeld's ideas on how science progresses. Compare Cohen, *Revolution* (ref. 3), p. xv.
[94] Otto Nathan and Heinz Norden, eds, *Einstein on Peace* (New York, 1960), p. 269.
[95] Martin J. Klein, 'Einstein on Scientific Revolutions,' *Vistas in Astronomy*, **17** (1975), 113–120. Compare Gerald Holton, 'Einstein's Search for the *Weltbild*,' *Proceedings of the American Philosophical Society*, **125** (1981), 1–15, on pp. 14–15. Also compare Cohen, *Revolution* (ref. 3), pp. 435–445.
[96] Einstein to Infeld, 20 September 1949, in Leopold Infeld, *Why I Left Canada: Reflections on Science and Politics*, trans. Helen Infeld, ed. Lewis Pyenson (Montreal, 1978), p. 142.
[97] Reported by Otto Blumenthal in 'Lebensgeschichte,' in David Hilbert, *Gesammelte Abhandlungen*, **3** (Berlin, 1935), on p. 417.
[98] Ernest Labrousse, '1848–1830–1789: Comment naissent les révolutions,' in *Actes du Congrès historique du centenaire de la Révolution de 1848* (Paris, 1949), pp. 1–20, on p. 1.

Université de Montréal

THOMAS F. GLICK

CULTURAL ISSUES IN THE RECEPTION OF RELATIVITY

1.1. COMPARATIVE RECEPTION STUDIES

The objective of an enterprise of comparative scholarship, such as the present volume, is to set out parallel cases of the reception of a single set of ideas, in order to get an overall picture of the balance between general and culture-specific processes. The unit of analysis is the nation and there is no presumption that any one case is more normative than any other. In textbook treatments of reception, the experience of an innovation or new idea in its society of origin may be taken as a normative phenomenon against which its reception elsewhere must be weighed. In the case of paradigmatic change, the bias is even greater: national scientific communities which made no contribution to the paradigm, although receiving and adopting it, are reckoned as uninteresting or derivative. What is the interest, therefore, of the reception of relativity in a country that made no original contribution to it? There are several lines of response to this kind of question. One is to posit, as Mannheim did, that the differential receptivity of cultural or social groups to specific ideas reveals something significant about social and cultural processes. A second is to suppose that such differentiation illuminates science as a social and cultural phenomenon. A third is to challenge received views of what is normative in science by expanding the number of cases examined and broadening the basis of study to include scientific communities thought "peripheral" to mainstream science. A fourth is to change the focus of study from the internal development of ideas to social and cultural contexts of their development and diffusion. A fifth is to cast a social definition of science as ideology; to do so is to finesse the question of what is normative experience, to rid the research program of ethnocentric bias, and to even out the distinction between invention and innovation, discovery and reception. As Tzvetan Todorov has observed in another context, "the reception of . . . statements is more revealing for the history of ideologies than their production."[1] Of course. The proof is in the kind of enterprise this volume undertakes, because such comparisons and contrasts as the national case studies present ultimately reveal ideological distinctions (or, rather, a range of social, cultural and ideological distinctions of which the latter are most significant). Put another way, Goldberg's "national style" involves different ways of doing science, different ways of thinking about science, the embeddedness of science in different value systems, and the attaching to science of different ideological markers. Even if ideology may not inhere in a given scientific idea as relativity, it appears as part of the baggage that ideas carry across cultural boundaries.

1.2. COMPARATIVISM OR ESSENTIALISM

Is the acculturation of an idea simply a constraint upon some cognitive core, or is the cognitive structure itself transformed as it crosses a cultural frontier? Biezunki's *glissements semantiques* are a universal snare of reception studies. Different groups may use the same terminology but attach vastly different meanings to it. Do the contributors to this volume assume a common unit of analysis ("relativity") or would they admit (as David Hull observes with respect to the diffusion of Darwinism) that conceptual systems change – over time, he says, and, we might here add, over space as well.[2] The reception of relativity has much in common with the prior reception of Darwinism (the similarities and differences were noted at the time), as well as with the contemporary reception of Freudian psychology.[3] But it is also true that the reception of relativity, when compared to Darwinism or Freudianism, was, in both cognitive and social senses, a much more sharply bounded phenomenon. Due to the complexity of the ideas the boundary between scientific and non-scientific comment would appear more easily drawn than was the case in the other two receptions where that line was blurred. It is easier to talk of a "French Freud" or "American Darwin" to indicate sharp transformations of the cultural system as German and English ideas, respectively, were acculturated elsewhere. It is less easy to speak of a Russian or Spanish Einstein. The "national style" in relativity was something less than the cultural reinvention that took place in the reception of Freud, but still something more than what is suggested by unquestioning assumption of an essential conceptual core.

A scientist could accept virtually all the tenets of special relativity and still not be a relativist (Poincaré). Or, we can point to the Spanish engineers Herrera, Pérez del Pulgar and Burgaleta who reckoned themselves relativists while rejecting one of the Einstein's principle tenets, the constancy of the velocity of light. In the comparative approach, therefore, we must assume the probability of different semantic structures, socially or culturally based, in different societies and take these structures at face value. The "national style" must involve a process of fusion between a set of ideas and the value system which constitutes any specific culture.

2.1. THE SOCIAL AND CULTURAL STRUCTURES OF SCIENCE

Clearly the "national style" of reception, or any one society's experience, is tightly linked to the professional structure of its scientific community. In the case of relativity national receptions were shaped by the strength and philosophical disposition of the three receiving disciplines – physics, mathematics and astronomy – and the balance of roles played by each discipline in the reception. Elsewhere, I have presented a model which seeks to appraise and define the ability of disciplinary groups to deal with new ideas when they diffuse across cultural boundaries and characterize "active" and "passive" modes of disciplinary reaction.[4] In an "active" reception members of a discipline have direct access to the new ideas (through personal contacts and ability to remain abreast of current research in the language of the new idea); they have both the maturity and sufficient personnel to debate the new ideas critically and institutional structures sufficient to propagate them authoritatively. At its most "active", a disciplinary group will contribute to the further development of the new paradigm through its own research. Among disciplinary groups not as highly institutionalized, reception may be limited to a "passive" role of serving as a conduit for the diffusion of the new paradigm. Passive disciplinary groups typically have no direct contact with mainstream science, are dependent upon translations in order to follow developments in it, and display both temporal and cognitive lags in assimilating new concepts. Such "active" and "passive" traits can be related to the broader social context through the structure of the educational system, the grade of acceptance of the scientific ethos by the elite, and so forth.

2.2. DISCIPLINARY RECEPTIONS

The interrelationship of differently weighted disciplines sets the tone of scientific reception. Thus, in Spain, only mathematics had sufficient weight as a discipline to interpret relativity for the scientific community at large and the public (the participation of individual physicists notwithstanding). But the inability of Spanish mathematicians to contribute to the further development of the paradigm still marks their reception as "passive". The balance among the three disciplines was similar to Italy where, however, mathematicians contributed actively to relativity's development. Within national communities of physicists the balance between theoreticians and experimentalists was obviously important. In this regard two noteworthy patterns of disciplinary substitution present themselves: first, the substitution of mechanics or mathematical physics (considered mathematical subdisciplines) for theoretical physics – a salient feature of French, Italian and Spanish receptions); second, the substitution of engineering for physics wherever the latter was underdeveloped.

Paty notes that in France there was no theoretical physics in the true sense until the 1930s; before that time mathematicians (those who cultivated mathematical physics) dominated the theoretical discussion of relativity and, since they viewed it in formal terms, this proved an obstacle to its introduction as physical concept. Judith Goodstein shows a similar situation in Italy.[5]

Most of our authors did not follow the engineering debate. Reeves discusses the positive views of Lori (an electrical engineer, like many who supported Einstein), and Sredniawa provides examples of both favorable and hostile Polish engineering opinion. Vizgin and Gorelik allude to the hostility of engineers (and applied physicists) to the counterintuitive conclusions of relativity. Only in Spain was the substitution of engineering for physics apparent, although that was the standard situation in Latin America. In his trips to Spain, Brazil, Argentina and Uruguay, Einstein spoke mainly to engineers who comprised the rank-and-file of those with interest in his ideas. It is interesting to note that Argentina did not have in the 1920s a mathematics community comparable in sophistication to that of Spain. Questions posed to Einstein at the Mathematical Society of Madrid in 1923 struck him as interesting and reasonable, while those addressed to him by Argentinian physicists two years later he found difficult to meet with a straight face.[6] Here we have two societies socially and culturally similar, whose scientific communities were quite differently structured.

Disciplinary traditions are of obvious importance. The tradition of non-Euclidean geometry in Russia, the revolt against purism in geometry in Italy and Spain, provided a congenial atmosphere for relativity, one that was congruent with the traditional concerns of the discipline, amongst mathematicians there.

Scientific ideas cross national borders with relative ease; disciplinary boundaries are far harder to breach. As Giuliano Pancaldi observed with respect to the reception of Darwinism, if scientists must reorganize the way they do research as the price of accepting a new theory, there will be resistance. It is harder, in this view, to free oneself from an old disciplinary tradition than it is to shed old theories.[7] Crelinsten notes, with regard to the reception of relativity by American astronomers, the reception of a revolutionary paradigm has the effect of stimulating a process of self-assessment within the discipline.[8] In Spain, this led both to self-confident statements (by J.M. Plans) that Spanish mathematics was no longer isolated from the European mainstream, as well as defensive confessions (by E. Terradas, in a letter to Levi-Civita) that Spain was simply infertile soil for mathematics.[9]

2.3. PHILOSOPHICAL CLEAVAGES

Philosophical cleavages in national science communities were highly significant, with positivism a touchstone. In Spain, Einstein's Catholic followers were at least situational anti-positivists although they didn't deal with the issue in any explicit way. In Brazil, relativity precipitated a show-down in the Academy of Science between anti-relativity positivists and pro-relativity idealists (mainly mathematicians).[10] In Germany, Planck made support of relativity part of his anti-positivist position.[11] In France, entrenched mechanistic positivism in physics preordained a hostile reception. Biezunski argues that French physicists drew on various philosophical traditions including Comtean positivism (which worked against relativity) and Cartesian mechanicism (which could work either for or against it). In the United States, however, Goldberg shows that special relativity was"positivized" to the extent that it was typically presented as experimentally based.

3.1. COMMONALITY OF SCIENTIFIC AND NON-SCIENTIFIC CIRCLES: LANGUAGE

Students of Darwinism and psychoanalysis accept as given a high level of interchange between the cognitive structures involved and their social contexts. Both participants in the development and reception of relativity and latter-day historians of the phenomenon insists that "popularizations" of relativity were lacking in physical meaning. The Monsieurs Homais of the world had their reasonably informed say on Darwin and Freud, but were barred from any real appreciation of relativity.[12]

In this context, it is interesting that some contributors to this volume stress the blurring of the science/non-science boundary. The general point is that which has been made most eloquently by Darwin scholars: that scientists participate in other solidarities within their societies and transpose into those solidarities their scientific concerns.[13] Reeves and Kaneko see language linking scientific and non-scientific spheres of discourse. By the very nature of relativity's common language, relativists themselves could not help but participate in various aspects of the "popular" discussion. Indeed their very participation ensured that the "semantic slides" really slid across the science/non-science border. Biezunski gives press coverage of soirées where the process actually was observed in progress.[14] Illuminating is his notion that when scientists do not agree on the "language", the general public is free to appropriate what it will. There is a logic here: the public is attracted by controversy, but the ideas/images they pick up during such a period may extend well into the period following the formation of scientific consensus. For example, in the first three decades of this century, when Darwinism was "in eclipse" and all kinds of competing evolutionary theories were under discussion,[15] the American public (and perhaps others) received a Lamarckian message which persists to this day. The same was true of the diffusion of the "new" psychology in the 1920s and 30s when competing Freudian, Jungian and Adlerian models of personality and development made available a profusion of concepts and images for public appropriation, frequently in combinations which the leaders of these schools found contradictory.

3.2. INCOMPREHENSIBILITY

The theme of relativity's incomprehensibility has been developed at length by Biezunski.[16] His view is that the cognitive issue is inextricably linked to Einstein's persona: Einstein did not fulfill any of the commonly-held expectations of what a scientist was supposed to be like. That made *him* incomprehensible which, in turn, provided a convenient popular explanation for the incomprehensibility of his ideas. The fact that he expressed complex ideas using everyday terms like time, space, and so on, further added to the confusion, as many commentators have noted.

Ultimately popular notions of the incomprehensibility of Einstein's theories simply reflected, I believe, what was said of them in the scientific community. On scientific incomprehension I find Jean Eisenstaedt's analysis particularly convincing: Einstein's theory was not empirically based yet experience confirmed it.[17] The disjuncture between abstract formulation and observational proofs accounted both for the sense of awe on the part of those scientists whom Einstein convinced and disgust – incomprehension – of those who were opposed. The "contras" were disgusted because the method by which the theory was generated was unthinkable to traditional experimentalists. We are still far from understanding the processes whereby the general public picked up clichés already established in the scientific community and marshalled them to justify what were, at base, ideological stances favoring or opposing Einstein. According to Paty, most French scientists had not heard of Einstein before 1919. In countries where the same was true (all but Germany?) scientists and the general public would have formed their opinions contemporaneously, in which case one would have to assume a complex interplay between scientific and public opinion in the formation of the dispositions of each towards Einstein and relativity (see 5.1).

But there may be a simpler, more universal explanation. According to Pierre Bourdieu, "that the public is irretrievably doomed to incomprehension [is] . . . so profoundly embedded in the social definition of the intellectual's vocation that it tend[s] to be taken for granted."[18] Incomprehension is what intellectuals expect of the public. Following Bourdieu's logic we could argue that lines of force which manifested themselves in the "intellectual" field of a given country during the relativity debate were those demarcating various groups of Einstein's supporters and detractors among scientifically educated and lay members of the intelligentsia. Such groups manifest both competitive relations and those of functional complementarity as well. Here there is complementarity among, for example, anti-relativity scientists, *gens du monde* displaced as arbiters of high culture, and elements of the "public" beyond the intellectual field proper (see section 7, below).

3.3. EXTRA BAGGAGE

In the process of interchange and appropriation, ideas inevitably pick up connotations and interpretations that have no logical or necessary connection with the base idea: for example, Darwinism and progress or the rhetoric of economic competition. In the same wise, cultural and philosophical relativism antedated relativity but, after 1919, were assumed to have been influenced by it. In some, many, domains of discussion, these connotations virtually displaced the base ideas. This semantic association functioned on many levels and bore approximately the same relationship to relativity as the idea of progress did to Darwinism. In typical appropriations, Einstein's theory was assumed to have provided "scientific proof" for a previously held position. Thus Ortega y Gasset was accused of favoring relativity because it appeared to legitimize his own concept of perspectivism, previously formulated. Opponents of philosophical relativism, such as traditional Catholic clerics assumed Einstein to have supported that position.

Relativism, on the whole , was baggage for the non-scientist. But even for scientists, relativity bore ineluctable connotations that had nothing to do with Einstein's theories per se but which become attached to them in the process of reception. As an idea in diffusion, relativity entered specific receiving cultures associated with the prestige of leading scientists. Clearly to diffuse as an already prestigious idea rather than in a more neutral form, enhances a favorable reception. In a Spanish textbook, for example, relativity was identified (in 1916) as the "Einstein-Planck" theory.[19] That relativity was labeled a product of German science surely eased its reception in Russia and Poland, hindered it in France. For a whole slew of experimentalists, applied scientists and engineers, relativity bore the stigma of mathematicism, and of exaggerated abstraction or formalism which justified its rejection *a priori*.

Paul Johnson has observed that "the impact of relativity was especially powerful because it virtually coincided with the public reception of Freudianism"; both doctrines appeared to the public as bearers of the same message: that "the world was not what it seemed".[20] In the light, the association by American commentators of Einstein's ideas with dream experience (Missner) acquires a specific cultural context. That relativity should, in this way, acquire psychoanalytic baggage attests to the defraction of multiple ideas through the same cultural field contemporaneously.

Many traits associated with Einstein's persona (see 5.3) diffused inseparably from his ideas particularly on the popular level: he was not the image of the standard scientist; his ideas did not conform to scientific standards, etc. His science was Jewish just like Einstein; it was revolutionary and so was Einstein.

Relativity then came to be associated with differing images of science held by different cultural, social or professional groups. In all the non-European countries visited by Einstein in the 1920s (Japan, Argentina, Uruguay, Brazil) and also in Spain, he and his ideas were symbols of modernization. In commenting on the insufficiency of mathematics instruction in the Spanish Naval Academy, a technical institution, a critic of the school noted as proof of his complaint that relativity was not imparted there.[21] This was a standard message: you cannot modernize without achieving mastery in science; the symbol of mastery of science is relativity.

To some, relativity revealed basic attitudes towards science. To some science was not allowed to generate doubt (Biezunski's French positivists); or, to turn the prescription around, science had to generate certainty (traditional Catholics). For those inclined towards relativistic perspectives, on the other hand, science does not discover absolute truths (Reeves), an antipositivist view of science which leads to very different philosophical and political conclusions.

4. INTERRELATEDNESS OF NATIONAL RECEPTIONS

To what extent were national experiences colored by the experiences of other societies? Marshall Missner insists that the American response to Einstein in 1921 was sui generis because it was during this trip that his persona became known to the public. It is then that the main elements of the Einstein myth, including the incomprehensibility of relativity, were established.[22] We would then have to assume that subsequent public discussion of Einstein and relativity in the European press was highly colored by the American experience. This supposition would have to be verified by searching the daily press of different countries. In Spain, to cite a counter-example, Einstein's American tour of 1921 was not covered, but his French visit the following year was. Those members of the Spanish intelligentsia who followed the latter episode, as a result, were supplied with a full range of clichés to confront the Einstein phenomenon in Spain in 1923. The Argentinian leg of his 1925 tour was in a like manner set up by extensive coverage in the Argentinian press of the Spanish tour.

The interrelated responses of various scientific groups is clear: German theoretical physics (and Ehrenfest particularly) in Russia; Italian mathematics (Levi-Civita particularly) in Spain; German physics (Jakob Laub particularly) in Argentina.[23]

5.1. EINSTEIN'S PERSONA: WHAT CAUSED HIS FAME?

The issue is related to incomprehensibility, insofar as incomprehensibility was a key element in his fame. The most categorical hypothesis is that adduced recently by Lewis Elton: Einstein's fame arose from controversies surrounding the verification of his theory and from the myth of its incomprehensibility, both of which originated in the Anglo-Saxon world.[24] For Missner, the American press was *the* instrument that made Einstein a celebrity. These two conclusions are complementary because, Missner continues, relativity was famous first and only subsequently did Einstein himself become a celebrity. What actually made him famous was, in Missner's view, the contingent association of the many elements we have been discussing. The contingency created the illusion of a great press campaign guided by an invisible hand.[25]

There are real problems of historical and analytic value here. Did incomprehension originate among scientists or in the press? I think it could be shown that most of the clichés about relativity originated with anti-relativity scientists before 1919 and diffused quickly after November of that year throughout many domains of discourse. In the Anglo-American world or elsewhere? Probably what was said in the English-language press was the great fount of what was said later on everywhere. But early on a variety of different styles emerged. The French journalistic style, in which (Biezunski shows) snobbism was one of the keynotes, weighed heavily in Spain among the non-scientific intelligentsia. The tonality of the Spanish newspaper account of Einstein's visit and relativity was that of the French literary bourgeoisie, with local inflections. What Spanish mathematicians said was tied into the Italian discussion and some of the popular engineering discussion was attuned to the popular discussion in England.

5.2. HIS TRAVELS.

Einstein's much publicized trips (to the United States and Italy in 1921, France in 1922, Japan and Spain in 1922–23, Brazil and Argentina in

1925) raise a number of issues. What did the lionizing crowds who pressed in on him throughout his travels get from such an exercise? Various professional and political groups sought to associate Einstein's prestige with their own cause or objectives. With regard to the professions, Einstein's presence reinforced the prestige of dominant groups (physicians in Spain and Argentina – note his visits with former and future Nobel laureates Santiago Ramón y Cajal and Bernardo Houssay, respectively[26] – mathematicians in Italy and Spain; engineers in the Iberian nations generally).

The interaction between Einstein and self-serving admirers caused a certain backlash. Thus Einstein is reported in Japan to be jobbing "a cheap article at a night stall" (Kaneko), while his staged performances in Spain were likened to a music-hall act. Sometimes Einstein is presented as an establishment pawn: he is carted by officials from place to place, where he opens his bag of three speeches for another uncomprehending audience.[27]

It should be remarked in passing that Einstein's travels generated a tremendous avalanche of local commentary, mainly in newspapers, constituting a great and generally unmined source of information on Einstein as a personality and public figure and the cultural impact of his ideas. American historians of ideas are unused to the newspaper as a major source of data because newspapers have not played the focal role in American culture as they have in Europe. Although Einstein insisted, when travelling, on not giving newspaper interviews, more than one enterprising reporter managed to break down his resistance and record his thoughts. Much of the information gleaned (in Spain and Latin America for instance) consists of trivial comments spun out for the benefit of local readers (the genius' opinion of the local folk dance, architecture, politics, and so forth) and reflected Einstein's knack for telling people what they wanted to hear without thereby revealing too much about himself. Still, when combined with his travel logs, a picture begins to emerge of how he viewed foreign cultures, together with a running comment on relativity, its scientific acceptance, and what the public could or could not derive from it (see Appendix).

What stands out about his travels, viewed in comparative perspective, is the universality of public response to him. The positive mass response occurred everywhere, whether there was a Jewish community or not (the Argentinian experience was very similar to the Spanish, in spite of Einstein's involvement with the Argentinian Zionist movement), or whether the culture was western or non-western, as in Japan. The Japan trip, unlike the others, was a planned media event in that it was sponsored by a publishing house, but the texture of the visit and his interaction with the public was very similar to his reception elsewhere. Comparative history is frequently only contrastive; it is easier to identify the anomalies in a structure than it is to explain the similarities in the face of strikingly different social and cultural contexts. Inevitably, then, the comparative method leads us away from Einstein; it was not he who accounted for the similarity in public response, but something in the nature of those cultures and the way science was viewed in them.

5.3. DIFFERING IMAGES

A number of authors (Biezunski, Missner) stress the theme of Einstein the destroyer as one of the most prevalent images. This is, of course, a negative image, associated with the supposed destruction by Einstein of the Newtonian worldview or the traditional Western cosmology. The destroyer image persists to this day. The Argentinian generals of the 1980's according to Amos Elon, "associated psychoanalysis with Freud, and Freud with Einstein (who was said to have undermined the Christian sense of time and space), and Einstein with Marx and Communist subversion."[28] In the 1920s, antirelativists, particularly those originating in traditional Catholic circles, included relativity in stock lists of modern horrors which they condemned, including modern art, modernist philosophy and socialism.[29] The generals simply present a stripped down list, the more sinister because of its scarcely concealed anti-Semitism.

What strikes me as unusual is that our authors do not mention the opposite image – Einstein the magus, the master (not the destroyer) of time and space – which was just as widely diffused an image. On an excursion to Toledo in 1923, Einstein asked Ortega y Gasset how an abstract idea like relativity could be of interest to the masses. Ortega thought the mass appeal had to do with the conjunction of a new cosmological theory with the post-World War I loss of faith in European society: "In such a circumstance there appeared your work, in which laws are promulgated for the stars, which obey them. The human masses have always perceived astronomical phenomena as religious. In them, science is conjoined with mythology and the scientific genius who masters them acquires a magical halo. You, Sr. Einstein, are the new magus, the confidant of the stars."[30] This image, which underlies much of the popular and "pop" commentary on Einstein and relativity, was similar in content and identical in conception with the much disseminated American myth concerning Thomas Edison, "the *wizard* of Menlo Park". To invent the electric light, the myth held, Edison had to refabricate the myths of Nature. No one else had been able to accomplish this feat but Edison, because he was a wizard, could perceive secrets of nature hidden to others. The public viewed Edisons's "magic" with awe, tempered with the inference of evil that inhered in wizardry.[31] The same was true of Einstein. The awe that scientists expressed over Einstein's magical formulation fed this aspect of the myth and in turn was reinforced by it. The magical creator was also a destroyer and the two images fused seamlessly in the post-1945 myth regarding Einstein and atomic energy.[32]

6. EINSTEIN AND POLITICS

Einstein had a well-delineated image owing to his notoreity in the anti-war movement. That was unusual for a scientist. Others had political images but they were associated with those of their country. Einstein's was unusual, striking, and international. This image was extended in the 1920s by his espousal of pacifism and his adoption of Zionism.

6.1. Varieties of Political Appropriation

Political appropriation is perhaps the most direct way to examine the symbolic values that different cultural and social groups were ready to vest in Einstein. During the course of Einstein's week in Barcelona in 1923, Catalan nationalists made every attempt to associate the physicist with their cause. Such an enterprise was consistent with Einstein's espousal of the right of small, oppressed cultural groups, such as European Jews, to self-determination. Catalan nationalism was, on the whole, a politically conservative movement associated with the industrial bourgeoisie. (Einstein was successful, notwithstanding, in convincing a Catalanist socialist leader, Rafael Campalans, to drop the word "nationalist" from his program because of the term's reactionary connotations.) That rightist dictators like Mussolini and Primo de Rivera could also make use of Einstein to justify the "relativization" of their opponents' political legitimacy attests both the immensity of his prestige and the universality of anti-positivist philosophies which relativism/relativity was seen to support.

Vizgin and Gorelik show, in the case of the Soviet Union, that there was a split among Marxist physicists over how to read relativity ideologically. Those who rejected relativity in the name of dialectical materialism do not seem so different from others elsewhere who rejected the new concepts in the name of positivism. Other Marxists, however, were able to establish a dialectical relation between the relative and absolute and find in this a confirmation of Marxist philosophy.

Like Darwin, therefore, Einstein's perceived message was susceptible to appropriation by groups of such differing political orientation. This is related to the paradoxical use of simple, familiar words like time, space, relative and absolute to (ostensibly) describe complex concepts. The everyday meanings of these terms were so broad that, taken together, they constituted an "open text" which persons of starkly contrasting ideologies could make over to suit their political preconceptions.[33]

6.2. Xenophobia

Einstein and relativity aroused a xenophobic response in United States. France and Spain, all for different reasons. In the United States, incomprehensibility translated socially into an anti-democratic impulse: in the land of "popular science", relativity was viewed as the imposition of European elitism (Missner). In France, anti-German hostility in the immediate post-war period linked up to pre-war antipathy to German scientific "speculation" (Biezunski). In Spain, anti-relativists complained that any new idea to which a foreign name was attached became accepted uncritically and Spanish critics couldn't get a hearing; it was even argued that *anti*-relativists had to cite foreign authorities to get a hearing.[34] The context of Spanish xenophobia was not political but cultural, an expression of that society's defensiveness regarding its perceived historical inability to do science, and, at the same time, the need to redress the stigma of "Spanish decadence" and the "Black Legend".

6.3. Fear of Revolution

If Einstein's ideas were "revolutionary" it followed that revolutionaries would be inclined to favor (Biezunski), conservatives to oppose (Missner), them on purely political grounds. In Barcelona, Einstein was invited to lecture a group of anarchist unionists who admired him for his pacifism. There he was heard by journalists (mistakenly, as it turned out) to identify himself as a revolutionary, "only in science," not politics. Einstein spent the next week in Madrid denying he ever made such a statement and, in his lecture on general relativity there, added a few lines pointing out that he was not a revolutionary in science, only the "translator" of Newton and Galileo into modern physical language.[35] Still, anarchists perceived him as a revolutionary – a testament to the success a certain kind of popularization had in reaching the uneducated – and anarchist journalists made ample use of relativism to justify their political program.[36] Vizgin and Gorelik assert that the relativity boom of the early 1920s corresponded to the general atmosphere of the revolutionary rebuilding of Soviet society. Where there was fear of revolution, as among conservative sectors in most of the countries considered (in particular, the United States) relativity and its pacifist creator could easily be associated with the negative images of the revolutionary left.

7. THE FIELD OF RECEPTION

Bourdieu is sometimes unclear as to the distinction between an intellectual field (composed of intellectuals only) which excludes the public, and the ostensibly broader cultural field which encompasses the entire culture. And yet Bourdieu envisages a functional relationship between public and intellectual field; once that relationship is admitted, there is no reason to exclude the public from the field. I prefer, therefore, to posit a broader field which is functionally defined for specific ideas: in the case of relativity/Einstein the field includes anyone who discusses it/him: scientists, other intellectuals, professionals, and even the masses. Such a broad field can be justified so long as we satisfy the requirement that functional relationships be demonstrated among the field's components.

In societies where science had to buck an adverse cultural tide, interest by the man-in-the-street could be significant. First, the sale of large runs of popularizing books offering "relativity at the reach of all" provides economic legitimation. Mass interest also provides political legitimation in the form of a visible phenomenon that politicians and administrators can easily read. In this way, a significant value change that affects the public climate for science positively is defracted through the field of reception.

The reception of relativity also mobilized the professions. Such mobilization is especially significant in modernizing countries where the professions mobilize against traditional political and cultural elites. As

Bourdieu observes, each science is tied to one of the major professions or to industry – physics and chemistry to the latter, biology to medicine, mathematics to engineering and architecture, and so forth. Through the professions, therefore, individual scientific disciplines receive various kinds of legitimation and support. That is, when a powerful professional group (medical doctors and engineers being the obvious examples) mobilizes on behalf of itself it may pull a pure scientific discipline after it.

The above generalizations sketch out functional relations between different social and cultural groups in a field of reception. Critical lines of force demarcating the constituent groups may divide scientists (experimental versus theoretical physicists, physicists versus mathematicians), between professional groups (physicists versus engineers), between educational groups (scientists and engineers versus lay intellectuals), or political-cum-philosophical ones (liberal materialists versus traditional vitalists), or social ones (middle class intelligentsia versus proletarian unionists). The latter, although poorly educated, may participate in the educational system through union or church organized outreach programs, such as the Spanish anarchists' night schools, where the diffusion of modern scientific ideas was one of the core components informing the program's rationale.

APPENDIX

Some Interviews with Einstein

1. Spain

1.1. Ricardo Baeza, 'Delante del professor Einstein,' *El Sol* (Madrid), July 33, 1921 [from London]. Einstein's philosophical evolution, relativity (the theory that makes the fewest suppositions should be preferred).

1.2. Interview by Andrés Révész, on train from Barcelona to Madrid, *ABC* (Madrid), March 2, 1923. Spanish science, daily routine, smoking, literature (Don Quixote), socialism.

1.3. A. Fabra Rivas, 'Una visita a Einstein,' *El Sol*, March 27, 1930 [in Germany]. Spanish science and reminiscences of his trip to Spain.

2. Brazil

2.1. Interview with Einstein on the *Cap Polonio* (ship) published in *O Paiz* (Rio), March 22, 1925. Einstein says relativity's greatest merit was not to have enlarged the horizons of science but to have restricted them. The infinite cosmos was an erroneous idea. He observes that relativity's absolute components have been ignored by the public. The velocity of light is an absolute standard. Comment on the Sobral eclipse observations.

2.2. Interview in *O Jornal* (Rio), March 22, 1925. Comments on Rio de Janiero, Portugal, Haldane and the British aristocracy, intellectual solidarity, European culture and science, Rathenau, South American mųsic and dances, Japan.

2.3. A long interview with George Santos, elements of which appear in *O Imparcial* (Rio de Janeiro), May 8, 1925, and *Illustração Brasileira*, June 1925. Einstein talks of literature, classical and popular music, cinema, the international mentality and soccer.

3. Argentina

3.1. Interview in Montevideo, March 24, 1925, published in *La Nación* (Buenos Aires), March 25: America is the land of the future; situation of the Jews; Palestine and Hebrew University; internationality of science; European scientific tradition.

3.2. *La Epoca* (Buenos Aires), March 25, 1925: his lectures: people have exaggerated the revolutionary nature of relativity; state of science (astronomy has to be made over and chemistry retouched; Lorentz and others have already shaken up physics).

3.3. Interview in *La Nación*, March 26, 1925, about the nature of his upcoming lectures: nature of ether as presented by Abraham and Föppl (mechanical properties); not true that relativity is inaccessible to the intuition (notion of relative space is more intuitive than that of absolute space).

3.4. Laro Fernández Arias, 'Einstein y su teoría según él mismo,' *Cara y Caretas* (Buenos Aires), April 4, 1925. At home of Bruno Wassermann. Difficulty of popularizing relativity; science is like music (you have to know the notation); the advantage of exact science is that it is like an international language; world politics.

NOTES

[1] See Todorov, *The Conquest of America* (New York, Harper & Row, 1985), p. 54, where he draws attention to the "receivability" of texts in specific socio-cultural settings.

[2] David L. Hull, 'Darwinism as a Historical Entity: A Historiographic Proposal,' in David Kohn, ed., *The Darwinian Heritage* (Princeton, Princeton University Press, 1985). pp. 773–812.

[3] On Darwinism, see Thomas F. Glick, *The Comparative Reception of Darwinism* (Austin, University of Texas Press, 1974), containing case studies for England, the United States, Germany, France, Russia, the Netherlands, Spain, Mexico, and the Islamic World; and articles by P. Bowler (Britain and the United States), P. Corsi and P. Weindling (Germany, France, and Italy), and F. Scudo and M. Acanfora (Russia) in Kohn, *The Darwinian Heritage*. On psychoanalysis, see the comparative *aperçus* in Jack J. Spector, *The Aesthetics of Freud* (New York, Praeger, 1972), pp. 187–209; and O. Mannoni, *Freud* (New York, Vintage, 1974), pp. 166–193. On the receptions of Darwin and Freud compared, see David Shakow and David Rapaport, *The Influence of Freud on American Psychology* (New York, International Universities Press, 1964) ch. 2.

[4] Thomas F. Glick, 'La transferencia de las revoluciones científicas a través de las fronteras culturales', *Ciencia y Desarrollo* (Mexico), Jan.–Feb. 1987, pp. 77–89.

[5] Judith R. Goodstein, 'The Italian Mathematicians of Relativity,' *Centaurus*, **26** (1983), 241–261.
[6] Thomas F. Glick, *Einstein in Spain* (Princeton, Princeton University Press, 1988), chapter 4; *La Razón* (Buenos Aires), April 17, 1925; South American Travel log, Einstein Papers, Princeton, April 16, 1925.
[7] Giuliano Pancaldi, *Darwin in Italia* (Bologna, Il Mulino, 1983), p. 13.
[8] Jeffrey Crelinsten, "William Wallace Campbell and the 'Einstein problem': An Observational Astronomer Confronts the Theory of Relativity," *Historical Studies in the Physical Sciences*, **14** (1983), 88.
[9] J.M. Plans, 'Las matemáticas en España en los últimos cincuenta años,' *Ibérica*, **25** (1926), 172–174, on p. 174: "Mathematics has made a great leap in Spain; the lack of phase with respect to other countries has almost been overcome." Terradas to Levi-Civita, March 22, 1922, Levi-Civita Papers, Accademia dei Lincei, copy deposited in the California Institute of Technology Archives: "What a pity that mathematics is so little developed among us!"
[10] Simon Schwartzman, *Formação da comunidade científica no Brasil* (São Paulo, Editora Nacional, 1979), pp. 109–115.
[11] J.L. Heilbron, *The Dilemmas of an Upright Man: Max Planck as Spokesman for German Science* (Berkeley, University of California Press, 1986), pp. 138–139.
[12] See Glick, *Einstein in Spain*, chapter 7. M. Homais, the pharmacist in *Madame Bovary*, appeared in the popular discussion of relativity in Spain as a symbol of the popularization of science among provincial pharmacists, physicians and intellectuals generally.
[13] Two authors who effectively interweave social and cognitive issues in the history of evolutionary biology are Robert M. Young, *Darwin's Metaphor: Nature's Place in Victorian Culture* (Cambridge, Cambridge University Press, 1985), and Adrian Desmond, *Archetypes and Ancestors: Paleontology in Victorian London, 1850–1875* (Chicago, University of Chicago Press, 1982).
[14] Michel Biezunski, *La diffusion de la théorie de la relativité en France*, unpub. doctoral diss., University of Paris, 1981, p. 207, where a journalist describes an informal discussion between Einstein and some French scientists. The scientists play the simultaneous role of *gens du monde* and pose very different kinds of questions.
[15] See Peter Bowler, *The Eclipse of Darwinism* (Baltimore, Johns Hopkins University Press, 1983).
[16] In *Diffusion de la relativité en France*, chapters 2 and 3.
[17] Jean Eisenstaedt, 'La relativité générale à l'étiage: 1925–1955,' *Archive for History of Exact Sciences* **35** (1986), 115–185.
[18] Pierre Bourdieu, "Intellectual Fields and Creative Project," in Michael D. Young, ed., *Knowledge and Control* (London, Collier-Macmillan, 1971), pp. 161–188, on p. 165.
[19] J. Mañas y Bonví, *Optica aplicada* (Barcelona, 1915), Suplemento a la página 427, p. i: 'Hypothesis of Einstein and Planck: The ether does not exist.'
[20] Paul Johnson, *Modern Times: The World from the Twenties to the Eighties* (New York, Harper and Row, 1983), pp. 5, 11.
[21] The mathematics program of the Naval Academy in 1922 was deemed inadequate to the mastery of "the new and already popularized doctrine of Einstein, through inability to form a clear idea of the four dimensions of space-time," A. Azarola, "El estudio de las Matemáticas en la Escuela Naval Militar," *Revista General de la Marina*, **86** (1920), 441–453.

[22] Marshall Missner, 'Why Einstein Became Famous in America,' *Social Studies of Science*, **15** (1985), 267–291.
[23] On Laub in Argentina, see Lewis Pyenson, *The Young Einstein* (Bristol/Boston, Adam Hilger, 1985), pp. 231–232.
[24] Lewis Elton, 'Einstein, General Relativity, and the German Press, 1919–1920,' *Isis*, **77** (1986), 95–103, on p. 103.
[25] In 'Why Einstein Became Famous in America' (n. 22, above) and an earlier unpublished version of the same article.
[26] On Cajal, Glick, *Einstein in Spain*, chapter 3. On Houssay, who acted as Einstein's host on a visit to a private medical laboratory near Buenos Aires, *La Prensa*, April 7. On his travels, Einstein sought out the best (or best-known) scientists, regardless of discipline.
[27] Images from his Spain tour; Tomás Gómez de Nicolás, 'La relatividad de los valores: Alegrémonos de no ser sabios,' *El Imparcial* (Madrid), March 10, 1923.
[28] Amos Elon, 'Letter from Argentina,' *The New Yorker*, July 21, 1986, p. 75, possibly reflecting a statement attributed to an ideologue of the military regime quoted by Anthony Lewis, in a review of Jacobo Timmerman, *Prisoner Without a Name*, *New York Times Book Review*, May 10, 1981: "Argentina has three main enemies: Karl Marx, because he tried to destroy the Christian concept of society; Sigmund Freud, because he tried to destroy the Christian concept of the family; and Albert Einstein because he tried to destroy the Christian concept of time and space."
[29] Horacio Bentábol, an antirelativist Spanish engineer, lists as examples of modern philosophical disorientation: relativity, non-Euclidean geometry, the Russian Revolution, Cubism, and religious disarray; *Observaciones contradictorias a la Teoría de la Relatividad del profesor Alberto Einstein* (Madrid, 1925), p.86.
[30] José Ortega y Gasset, 'Con Einstein en Toledo,' in *El tema del nuestro tiempo*, 18th ed. (Madrid, Revista de Occidente, 1976), pp. 195–202, on pp. 196–197.
[31] Wynn Wacchorst, *Thomas Alva Edison: An American Myth* (Cambridge, Mass., MIT Press, 1981), pp. 23, 25, 30.
[32] The post-1945 myth of Einstein as genius and destroyer is discussed by Alan J. Friedman and Carol C. Donley, *Einstein as Myth and Muse* (Cambridge, Cambridge University Press, 1985), pp. 171–180. Friedman and Donley do not perceive the magus element in the Einstein myth, even when it occurs in their own evidence (e.g. in the fascinating statements by two psychotics, pp. 190–192, who admired Einstein for his ability to conjure the powers of nature).
[33] On the *Origin of Species* as an "open text", see Gillian Beer, 'Darwin's Reading and the Fictions of Development,' in David Kohn, ed., *The Darwinian Heritage* (Princeton, Princeton University Press, 1985), pp. 543–588, on p. 574.
[34] Bentábol, *Observaciones contradictorias* [n. 29, above], p. 23, explaining why he felt compelled to cite Henri Bouasse in support of his own antirelativity conclusions. The same point was made by José Escofet, 'Crónicas catalanas: Einstein y los matemáticos,' *Las Provincias* (Valencia), March 18, 1923. Escofet's argument, typical of the displaced *gens du monde* (he was editor of *La Vanguardia*, a daily representing the interests of the Catalan bourgeoisie), was that neither supporters nor opponents of Einstein were believed by the Spanish public unless they cited foreign sources. This view was erroneous; the Spanish relativists were believed because of their immense academic prestige.
[35] On Einstein and the Catalan nationalists and anarchosyndicalists, see Glick, *Einstein in Spain*, chapter 3. His Madrid lectures are reproduced in *ibid*, Appendix III.

[36] Francisco Pellicer, 'Revolución científica y revolución económica,' *Redención* (Alcoy), March 22, 1923. The argument here is that physical relativity is the harbinger of a moral relativity which hold that political and economic institutions are relative, not absolute.

Boston University

In Defense of Ether: The British Response to Einstein's Special Theory of Relativity, 1905–1911

BY STANLEY GOLDBERG*

1. INTRODUCTION

At the end of the nineteenth century, J. T. Merz, in his monumental *History of European Thought in the Nineteenth Century*,[1] expounded the thesis that the scientific spirit, which had begun in France, had spread to Germany and then across the Channel to Britain and had now become solidified and unified. He said that "science at last had become international: isolated and secluded centres of thought have become more and more rare. . . . National peculiarities still exist, but are mainly to be sought in the remoter and more hidden recesses of thought. . . . We can now speak of European Thought, when at one time we should have had to distinguish between French, German, and English Thought."[2]

Of course one can cite features of the scientific enterprise that are shared across national boundaries, such as canons of laboratory procedure, and the need for making researches public. But would one have to search out the "remoter and more hidden recesses of thought" to find significant national differences in science? My skepticism on

* Department of Physics, Antioch College, Yellow Springs, Ohio 45387.

1. J. T. Merz, *A History of European Thought in the Nineteenth Century*, 4 vols. (Chicago, 1904–1912).

2. *Ibid.*, *1*, 16–20.

this point was a major motivation for this study. I believe that national differences have persisted in science after Merz's time, and that they have involved more than linguistic difficulties. An analogy may be drawn to music: while the notes are a universal language, the interpretation of the notes is a highly cultural and personal affair between the conductor, musicians, and audience. In physical science the notes become the mathematical notation, which in every case must be interpreted.

Gerald Holton has analyzed the process of scientific inquiry and has suggested that there is more in it than empirical observation and analytic process. He insists on a third component, which he calls "themata": "The strong hold that certain themes have on the mind of the scientist helps to explain his commitment to some point of view that may run exactly counter to all accepted doctrine and to the clear evidence of the senses."[3] Are national scientific characteristics an important component in shaping the themata Holton has spoken of? It was this question that led me to study the response to Einstein's theory of relativity in four countries: France, Germany, Britain, and the United States.[4] I chose to study the period between 1905, the year of Einstein's first paper on relativity, and 1911, the year of the first Solvay Conference, when the thoughts of many early contributors to the theory of relativity turned to the quantum theory. During this time, over ninety-nine percent of the literature on Einstein's theory emanated from these four countries.[5] The present paper is a portion of this larger study, an examination of the British response to relativity. This was a unique response in at least two respects: it was far more uniform than the responses in other countries; and British scientists were more outspoken on the subject than were their Continental colleagues. Before turning to a detailed consideration of the British response, it will be helpful to sketch the context in which Einstein's theory was first advanced, and to evaluate briefly how it was received elsewhere.

3. G. Holton, "The Thematic Imagination in Science" in Holton, ed., *Science and Culture* (Boston, 1965), pp. 88–108, quotation on p. 100.

4. S. Goldberg, *Early Response to Einstein's Special Theory of Relativity, 1905–1912: A Case Study in National Differences* (unpubl. Ph.D. diss., Harvard University, 1969).

5. M. Lecat, *Bibliographie de la relativité* (Brussels, 1924), pp. 201–202.

2. THE THEORY OF RELATIVITY

There has been much confusion and difference of opinion over the originality of Einstein's contribution to the theory of relativity. On the one hand, it has been argued that the theory was largely the creation of H. A. Lorentz and H. Poincaré, and that Einstein's 1905 formulation added little that was new. This is the line that Whittaker took,[6] and more recently Keswami put forward the same view.[7] On the other hand, Holton and I have argued that Einstein's theory of relativity has little more than the formalism in common with the electron theory of Lorentz and Poincaré.[8]

There are three innovative features of Einstein's theory that set it apart from the work of his predecessors. First, the theory is derived from two postulates, that of relativity, and that of the invariance of velocity of light. Second, the theory is one of measurement, and in this sense, it is a kinematical, not a dynamical, theory; it says nothing about the structure of electrons or of matter in general, but only of the way events are measured in different inertial frames of reference. Finally, the theory does not depend in any way on the existence of a medium for the propagation of electromagnetic waves.

In the introduction to his first paper on special relativity, Einstein poses a paradox: when a conductor and a magnet are moved relatively to one another, the current induced in the conductor does not depend on which of the two is moved. Yet, according to the theory, the motion of a magnet gives rise to an electric field, while the motion of a conductor does not. Coupling this paradox with the

6. E. T. Whittaker, *A History of the Theories of Aether and Electricity*, 2 vols. (New York, 1960), 2, 27–77, esp. p. 44.

7. G. H. Keswami, "Origin and Concept of Relativity," *Brit. J. Phil. Sci., 15* (1965), 286–306.

8. G. Holton, "On the Origins of the Special Theory of Relativity," *Amer. J. Phys., 28* (1960), 627–636; "Influences on Einstein's Early Work in Relativity Theory," *The American Scholar, 37* (1968), 60–79. S. Goldberg, "Henri Poincaré and Einstein's Theory of Relativity," *Amer. J. Phys., 35* (1967), 933–944; "The Silence of Poincaré and Einstein's Relativity," *Brit. J. Hist. Sci.*, in press; "The Lorentz Theory of Electrons and Einstein's Relativity," *Amer. J. Phys.*, in press.

G. Holton has recently analyzed the foundations of relativity in his valuable "Einstein, Michelson, and the 'Crucial' Experiment," *Isis, 60* (1969), 133–197. See especially pp. 160–167. My analysis was written prior to seeing Holton's paper; however, there is substantial agreement in our points of view. I take this opportunity to acknowledge a special debt of gratitude to Holton, since my analysis owes much to conversations we have had over the past ten years.

futility of attempts to measure the motion of the earth through the ether, Einstein was led to suggest that the "phenomena of electrodynamics as well as mechanics possess no properties corresponding to the idea of absolute rest."[9] Without further consideration he stated his intention to "raise this conjecture . . . to the state of a postulate,"[10] the principle of relativity. Einstein never questioned this postulate or appealed to experience to make it seem more reasonable. This contrasts with the attitude of Poincaré, who saw the relativity principle as a generalization from experience, one that required only one contrary instance to render it inoperative.[11]

It was not until after Einstein's 1905 paper that Lorentz considered the principle of relativity as a basic concept rather than as a result of the electron theory. Whereas Lorentz' theory of 1892 predicted no first-order effects due to motion through the ether, it did predict second-order effects; that is, phenomena depending on the square of the ratio of the velocity of the apparatus through the ether to the velocity of light in free space. Second-order experiments, such as the Michelson-Morley experiment, however, failed to yield the expected results. The equations of the electromagnetic field appeared to be covariant in practice while the theory predicted otherwise. In the face of this puzzle, Lorentz was content to work out the transformation equations between inertial frames of reference that rendered the equations of the electromagnetic field covariant. By an accomplished piece of inductive detective work, he found, in 1904, results that were formally identical to those that Einstein published a year later.[12] However, in Einstein's paper the transformation equations are *derived* from considerations based on his two postulates.

Einstein's second postulate was that "light is always propagated in empty space with a definite velocity *c* which is independent of the

9. A. Einstein, "On the Electrodynamics of Moving Bodies," *Ann. d. Phys., 17* (1905), 891–917, trans. W. Perret and G. B. Jeffrey in Einstein, *et al., The Principle of Relativity* (New York, 1923), p. 37.

10. *Ibid.*, p. 38.

11. H. Poincaré, "Sur la dynamique de l'électron," *Rev. gen. sci. pures et appliquées, 19* (1908), 386–342. See S. Goldberg, "Henri Poincaré and . . . Relativity." Poincaré's continuing doubts about the principle were brought into sharp focus by Kaufmann's 1906 claim that his experiments on the specific charge of high-speed electrons were not in accord with either the Einstein or the Lorentz theories. For his part, Einstein maintained a significant silence.

12. H. A. Lorentz, "Electromagnetic Phenomena in a System Moving with Any Velocity Less than That of Light," *Proc. Acad. Sci. Amst., 6* (1904), 809, repr. in A. Einstein, *et al., The Principle of Relativity*, pp. 11–34.

state of motion of the emitting body."[13] In his "Autobiographical Notes," written approximately forty years after the publication of his first paper on relativity, Einstein reminisced about the source of his second postulate: "After ten years of reflection . . . a [general] principle resulted from a paradox upon which I had already hit at the age of sixteen: If I pursue a beam of light with the velocity *c* (velocity of light in a vacuum), I should observe such a beam of light as a spatially oscillatory electromagnetic field at rest. However, there seems to be no such thing, whether on the basis of appearance or according to Maxwell's equations."[14]

Again Einstein had leaped from a suggestive thought experiment to a heuristic postulate. In 1916 he published what is perhaps still the most charming and the clearest, if the most terse, popularization of the theory of relativity. In it he asks how the postulate of the invariance of the velocity of light could be confirmed experimentally. He proceeds to show that since a one-way measurement of the velocity of light is impossible, it is necessary to make a stipulation: "that light requires the same time to traverse the path . . . [AM] as the path. . . . [BM, M being the midpoint of the line AB] is in reality neither *a supposition nor a hypothesis* about the physical nature of light, but a *stipulation* which I can make of my own free will."[15] Such a stipulation is necessary in any inertial frame of reference if one is to arrive at an intuitive definition of simultaneity. And such a stipulation is equivalent to the postulate of the invariance of the velocity of light. It *follows* that observers in different inertial frames disagree on judgements of simultaneity. Though such results deduced from the postulates are counterintuitive, there is no contradiction between the postulates.

As early as 1898, Poincaré had also considered the problem of synchronizing clocks using light signals; and he had stated that the invariance of the velocity of light was a fact about which there could be no question. However, in his view, it was an empirical fact which could be disproved by experiment. In any event, Poincaré never saw the need for making the invariance of the velocity of light

13. A. Einstein, "On the Electrodynamics of Moving Bodies," *op. cit.* (note 9), p. 38.

14. A. Einstein, "Autobiographical Notes" in P. A. Schillp, *Albert Einstein: Philosopher-Scientist* (New York, 1959), p. 53.

15. A. Einstein, *Relativity, The Special, The General Theory* (Chicago, 1951), p. 18.

a postulate and, indeed, in his analysis of the process of synchronizing clocks, he showed clearly that he believed that the invariance was due to a strange compensation, an effect of the Lorentz-FitzGerald contraction. With Lorentz, he would have agreed that there was only one frame of reference in which the velocity of light was really constant—the frame of the ether. In all other frames, it only appeared to be constant. Lorentz shared Poincaré's view on this matter.

As Einstein pointed out in his first paper on relativity, electrodynamic theory is based on "the kinematics of the rigid body, since the assertions of any such theory have to do with the relationships between rigid bodies (systems of coordinates), clocks, and electromagnetic processes. Insufficient consideration of this circumstance lies at the root of the difficulties which the electrodynamics of moving bodies at present encounters."[16] His paper is accordingly prefaced by a kinematical part in which the postulates are presented. Here Einstein deduces the relativity of simultaneity, the Lorentz transformation equations, the "Lorentz-FitzGerald" contraction, time dilation, and the composition rule for velocities. In the sequel dynamical part of the paper, these results are applied to specific circumstances in electrodynamics.

For example, with regard to the contraction of lengths in the direction of motion, Einstein may be interpreted as follows: when one makes a measurement of length, one must either have the object in his frame of reference or, failing that, make simultaneous observations of the ends of the object (or some other equivalent procedure). But since observers in different frames of reference will disagree about judgements of simultaneity, it follows that they will disagree as well with regard to judgements of length. The disagreement is not due to any dynamical change in the object; indeed, the construction and nature of the object are not even considered. The disagreement is an artifact of the measuring process itself and may be traced ultimately to Einstein's two postulates. Einstein maintained this kinematical view in his subsequent work. Thus when Paul Ehrenfest pleaded in 1907 that Einstein supply more information about his "theory of electrons" so that a decision could be made on certain paradoxes of electron rigidity Ehrenfest had pointed out, Einstein's

16. A. Einstein, "On the Electrodynamics of Moving Bodies," *op. cit.* (note 9), p. 38.

reply was simply to deny that he had a theory of electrons. Rather, he said, his was a theory concerned with the relationship between rigid rods, perfect clocks, and light signals.[17]

On the other hand, the theory of Lorentz and Poincaré was not restricted to the problem of measuring events by systems in relative motion. It was their hope to be able to subsume all physical phenomena under the assumptions of one theory. Thus within the electron theory, one attempted to explain such seemingly diverse phenomena as heat conductivity, electrical conductivity, energy radiation, reflection, refraction, and dispersion of light, elasticity, and strength of materials. Newtonian forces were not taken as fundamental; they were to be understood in terms of the interactions of electrons with other electrons through the absolutely fixed, space-filling ether. The problem of the electrodynamics of bodies in motion became important only because it was a troublesome area, an area which resisted, until Lorentz' 1904 paper, every assault.

Lorentz assumed the contraction hypothesis first as a way of explaining the Michelson-Morley and other second-order experiments. He later independently postulated the same kind of contraction for electrons. Prior to 1905 he never derived the contraction effect from first principles via his own transformation equations. To Lorentz the contraction was a real phenomenon in the sense that things actually became shorter in the direction of motion as viewed from different inertial frames of reference. Furthermore, in Lorentz' view, not only was the contraction independent of judgements of simultaneity, but discrepancies in such judgements were not significant and were themselves only disagreements on paper. The time dilation which ensues from the transformation equations was only an "aid to calculation."

The Lorentz theory was a dynamical theory throughout. Lorentz even tried to account for contraction by hypothesizing an alteration in the electrical forces between electrons in materials in motion. And when Abraham challenged the deformable electron of Lorentz' on the grounds that such an electron would be unstable, Poincaré

17. A. Einstein, "Bemerkungen zu der Notiz von Herrn Paul Ehrenfest: 'Die Translation deformierbarer Elektronen und der Flächensatz,' " *Ann. d. Phys.*, *23* (1907), 206–208. See M. J. Klein, "Thermodynamics in Einstein's Thought," *Science*, *157* (1967), 509–516, esp. pp. 515–516, and S. Goldberg, "Henri Poincaré and Einstein's Theory of Relativity," *op. cit.* (note 8), 938.

suggested that the stability of the electron was maintained by an external force exerted by the ether.[18] Such considerations are completely absent from the Einstein theory.

In his 1905 paper on relativity, Einstein makes only one mention of the concept of the ether—not, as is often assumed, to say that the ether does not exist, but rather to say that in working out the theory, "the introduction of a 'luminiferous ether' will prove to be superfluous. . . ."[19] In the Lorentz theory, while the ether does not play an active role (disregarding the "Poincaré pressure" which maintained the deformable electron), it is vital. True time, according to Lorentz, is the time measured in the frame of reference at rest in the ether. And it is only in the ether frame that the velocity of light is really the constant *c*. Lorentz did eventually recognize that one could indeed arrive at the same formal results as he had without using the concept of an ether, but he always preferred a theory which utilized the concept. Then, too, it must be recalled that the theory of electrons required the ether to play the role of mediator and transmitter of electromagnetic fields; the fields of electrons interacted with the fields of other electrons in the ether. Lorentz would have agreed with Poincaré: "Beyond the electrons and the ether, there is nothing."[20]

3. THE RESPONSE TO EINSTEIN'S THEORY OF RELATIVITY IN FRANCE, GERMANY, AND THE UNITED STATES

Outside of scientists at German-speaking universities, the response to Einstein's theory was anything but immediate. In fact, in France, there was no response at all, and one can hardly find any mention even of Einstein's name in the context of problems associated with the theory of relativity prior to Einstein's 1910 visit to France.[21] During the 1905–1911 period, by far the most outstanding figure in France actively working on the problem of the electrodynamics of

18. For discussions of the points raised here, see G. Holton's and S. Goldberg's articles cited in note 8, and also S. Goldberg, "The Abraham Theory of the Electron: Symbiosis between Theory and Experiment," *Archive for History of Exact Sciences*, in press.
19. A Einstein, "On the Electrodynamics of Moving Bodies," *op. cit.* (note 9), p. 38.
20. H. Poincaré, *Science and Method* (New York, Dover, n.d.), p. 209.
21. See A. Einstein, "Le principe de la relativité et ses conséquences dans la physique moderne," *Arch. sci. phys. et nat., 29* (1910), 5–28.

moving bodies was Henri Poincaré. Poincaré never once referred to Einstein in this context. Since we have no private papers of Poincaré to guide us, this puzzle may never be solved with certainty. Recently I have analyzed Poincaré's attitudes towards electrodynamics and towards theories in general[22] and have suggested that he may have held the view that Einstein's work was only a small and rather insignificant part of a much larger theory that Lorentz and he had developed. Apparently Poincaré's influence was responsible for the fact that almost no one else in France had much to say about Einstein's theory prior to Poincaré's death in 1911. Even as late as 1955, the complaint was heard that the syllabus for the *certificat de mécanique rationnelle* contained not one lesson on the kinematics of high velocities.[23]

The response to the theory of relativity in Germany stands in sharp contrast to that in France. All during the period under investigation, Einstein's theory was discussed, criticized, elaborated upon, and defended. While I intend to make a detailed analysis of the response in Germany another time,[24] I want to mention several features of that response here. It was only in Germany that the theory was elaborated upon: physicists such as Max Planck, Max Born, Max von Laue, and Jakob Laub published lengthy studies of various implications of the theory, including generalized mechanics, relativistic thermodynamics, the concept of rigidity, and relativistic optics.[25]

Many German physicists opposed Einstein's theory, but it is only in Germany that its opponents understood it. Thus Max Abraham and Wilhelm Kaufmann both opposed the theory on the grounds that experiments on the specific charge of high-speed electrons gave results which were not in accord with the theory; yet they exhibited a firm grasp of the principles on which Einstein based his analysis.

22. S. Goldberg, "The Silence of Poincaré . . . ," *op. cit.* (note 8).

23. H. Arzelies, *La Cinématique Relativiste* (Paris, 1955), p. vii.

24. I am preparing four papers on the German response to Einstein's theory. The first is on the role of experiment in judgements about the theory; the second is on the debate on the nature of rigid bodies; the third is on the development of generalized relativistic mechanics; and the fourth is on relativistic thermodynamics.

25. Two contemporary reviews of the theory of relativity serve as excellent guides to the literature: A. Einstein, "Ueber das Relativitätsprinzip und die aus demselben gezogene Folgerungen," *Jahrb. Rad. u. Elek., 4* (1907), 411–462; J. Laub, "Ueber die experimentellen Grundlagen des Relativitätsprinzips," *Jahrb. Rad. u. Elek.,* 7 (1910), 405–463.

They explicitly rejected those principles.[26] Fritz Hasenöhrl also rejected the theory, even though he recognized that, using it, Planck had given a very satisfactory rendition of generalized mechanics and the physics of moving black bodies. Hasenöhrl preferred, he said, Lorentz' theory because it explicitly introduced the concept of the ether.[27]

Regardless of their attitudes towards the theory, German physicists took the theory seriously. This may sound like a strange comment to make as a characteristics of a national response, but the German response stands in sharp contrast to the response in other countries in just this particular. It was the seriousness of the German response, in my view, which ultimately led to the acceptance of relativity, for it insured that the theory would be examined, criticized, and elaborated upon. Once that process began, and the heuristic power of the theory was revealed, it was only a matter of time before it came into general use.

When one turns to the response to relativity in the United States, one finds little evidence that the theory was taken seriously or that any commentator had anything but a dim notion of Einstein's program. The first to deal with the theory itself were G. N. Lewis and R. C. Tolman who, in 1909, published a highly original exposition of the theory.[28] But, like most American scientists, Lewis and Tolman emphasized the practical aspects of the theory. The difference between them and other American scientists was this: they considered the theory to be a practical and useful one with experimentally proven postulates, while most of the others did not. Indeed, the tendency in America was to ridicule the theory of relativity as being totally impractical and absurd. For example, in his Presidential Address to the American Association for the Advancement of Science in 1911, W. F. Magie was moved to proclaim that "I do not believe that there is any man now living who can assert with truth that he can conceive of time which is a function of velocity or is willing to go to the stake for the conviction that his 'now' is another man's 'future' or still another man's 'past.' "[29] Magie concluded that

26. S. Goldberg, *Early Response . . . , op. cit.* (note 4), Chap. 2.
27. *Ibid.*
28. G. N. Lewis and R. C. Tolman, "The Principle of Relativity and Non-Newtonian Mechanics," *Phil. Mag.*, *18* (1909), 510–523.
29. W. F. Magie, "Primary Concepts of Physics," *Science*, *35* (1912), 281–293; passage cited is on p. 292.

Einstein's theory could not be fundamental, since a fundamental theory must be intelligible to all people. And there were few scientists in America willing or able to dispute him.

4. THE BRITISH RESPONSE TO EINSTEIN'S THEORY

If there is a word to characterize British physics in the nineteenth century it is "ether." "Ether mechanics," as I will term the investigation of the properties of the ether, was not a study restricted to Britain. However, its effect on British physics in the early twentieth century was far more profound that it was on the physics of any other country in Europe. And when one examines the sources of ether theory *construction* in the nineteenth century, the preponderance of British authors is easily discernible.[30] One can trace the concept of the ether in Britain at least as far back as the time of Newton. However, the year 1800 is a convenient bench mark. In that year Thomas Young published his first paper on light as a wave phenomenon. British concern with the existence and nature of the ether grew steadily[31] until, at the end of the century, there were those who believed that the time was close at hand when the true structure of the etherial medium would be revealed to man.[32]

In a sense, it is misleading to speak of a reaction to the theory of relativity in Britain. More accurately, the British were reacting to what they perceived to be an attack on the ether; for, in fact, many British scientists were quite ignorant of the details of the theory of relativity. As late as 1923, N. R. Campbell could write of the average British physicist that he was "still ignorant of Einstein's work and not very much interested in it. Physicists of great ability, who would

30. See the first volume of Whittaker, *op. cit.* (note 6).

31. The role that the ether played in nineteenth-century British physics has been analyzed by several authors. See E. T. Whittaker, *A History of the Theories of Aether and Electricity, op. cit.* (note 6); J. T. Merz, *A History of European Thought in the Nineteenth Century, op. cit.* (note 1); C. C. Gillispie, *The Edge of Objectivity* (Princeton, 1959); Mary Hesse, *Forces and Fields* (New York, 1959). Work on the ether in Britain was concentrated in what has become known as the "Cambridge School," a group of mathematical physicists who studied and worked at Cambridge, and who included Green, Stokes, and Kelvin.

32. The most confident hopes for the imminent solution of the problem of the ether were expressed by Oliver Lodge in his *Modern Views of Electricity,* 1st ed. (London, 1889), p. ix. "[What is the Aether?] is *the* question of the physical world at the present time. But it is not unanswerable. It is in my belief, not far from being answered" (emphasis in original). This statement was repeated in the 2nd edition in 1892, but was deleted from the 3rd edition in 1907.

be ashamed to admit that any other branch of physics is beyond their powers, will confess cheerfully to a complete inability to understand relativity."[33] Although there were other factors involved, such as the direction Rutherford and his experimental school gave to early twentieth-century British physics, the comparative ignorance of and disinterest in relativity on the part of British physicists was primarily due to the unique character of their theoretical commitment to the ether.

Pervasiveness of the Concept of Ether in Britain

Perhaps no figure of nineteenth-century physics contributed more to ether mechanics than William Thomson, Lord Kelvin.[34] Section A, Mathematical Physics, at the British Association for the Advancement of Science in 1907 witnessed Kelvin's last public appearance before his death. At this meeting, he read a paper on the motions of the ether. The view he put forward (his views of the ether were in constant flux) was that the ether must be an "elastic, compressible, non-gravitational solid."[35] He cited a paper by Oliver Lodge[36] which had concluded that "every cubic millimeter of the universal ether of space must possess the equivalent of a thousand tons and every part must be squirming with the velocity of light.[37] Lodge based this conclusion on the fact that the presence of transverse waves in the interior of a fluid can only be explained using gyrostatic principles, and on the fact that the internal circulatory motion of a fluid must be comparable with the speed with which such waves are propagated, which, in this case, was the velocity of light.

The significance of Kelvin's and Lodge's views is not that they are based on the assumption of an ether—many physicists in Europe still clung to the belief in the ether in 1907. Rather, what is significant

33. N. R. Campbell, *Relativity* (Cambridge, 1923), p. v.

34. On the influence of Kelvin in this regard, see S. P. Thomson, *The Life of William Thomson, Baron Kelvin of Largs*, 2 vols. (London, 1910); E. T. Whittaker, *op. cit.* (note 6), *1*, Chap. 9, "Models of the Aether"; J. T. Merz, *op. cit.* (note 1), *2*, Chap. 2.

35. Anon., "Mathematics and Physics at the British Association, 1907," *Nature*, *76* (1907), 457–462, p. 457.

36. Oliver Lodge (1851–1940) studied at the Royal College of Science and at University College, London. In 1881 he became Professor of Physics at Liverpool, and in 1900 he was appointed first principal of the new university at Birmingham.

37. Anon., "Mathematics and Physics at the British Association, 1907," *op. cit.* (note 35), p. 459.

about these views—views that were shared by many leading British physicists—is that they regard the ether as a mechanical object. The fact that the ether must exhibit properties unlike any substance within human experience had driven Lorentz, for example, to divest the ether of all mechanical properties. In fact, to Lorentz the ether ultimately represented nothing more than the frame of reference in which absolute time was to be measured, a frame that was, by his own admission, undetectable in principle.[38]

It is easy to trace the source of Kelvin's 1907 mechanical ideas on the ether; he had spent a major portion of his professional life building a host of mechanical models which might represent the ether. Nor is it hard to find the source of Lodge's conception of the ether. His major research interests were in the propagation of electromagnetic waves; and if one did away with the ether, in his view, it would be impossible to account for the fact of wireless transmission.[39] But the ether meant much more than that to him. His biographers said that the "concept of the ether of space dominated his whole life. In his youth as a physicist he was well in the forefront of the science of his day, following the steps of Faraday and Maxwell by concentrating his attention on the properties of the medium surrounding electrified bodies."[40] Late in life, in the preface to a book in which he summed up his philosophical outlook, he wrote that the "Ether of Space has been my life study. . . . I always meant some day to write a scientific treatise about the Ether of Space; but when in my old age I came to write this book, I found that the Ether pervaded all my ideas, both of this world and the next. I could no longer keep my treatise within the proposed scientific confines; it escaped in every direction and now I find has grown into a comprehensive statement of my philosophy."[41]

38. H. A. Lorentz, "Alte und Neue Fragen der Physik," *Phys. Zs.*, *11* (1910), 1234–1257. See S. Goldberg, "The Lorentz Theory of Electrons . . . ," *op. cit.* (note 8).

39. O. Lodge, *Relativity: A Very Elementary Exposition*, 3rd ed. (London, 1926). The first edition was published in 1925.

40. R. A. Gregory and A. Ferguson, "Oliver Lodge," *Obituary Notices of Fellows of the Royal Society*, *3* (1941).

41. O. Lodge, *My Philosophy: Reporting My Views on the Many Functions of the Ether of Space* (London, 1933), Preface. Lodge had a great interest in psychical phenomena, conducting experiments on extrasensory perception and consulting with some of the major mediums of his day on their work. He thought that besides the role that the ether played in the physical world, the ether was the agent responsible for the transference of thoughts in extrasensory phenomena (A. Ferguson, "Oliver J. Lodge," *DNB 1931–40*).

From an early date Lodge's clear ideas on the ether played an integral part in his work. Subsequent editions in 1892 and 1907 of his *Modern Views of Electricity* reveal that his etherial conception of electricity did not change. In his 1907 preface he remarked that "it is noteworthy that few actual *corrections* have had to be made; showing that new discoveries are of a supplementary rather than a revolutionary character. The doctrine expounded in this book is the etherial theory of electricity. Crudely, one may say that as heat is a form of energy, or mode of motion, so electricity is a form of ether or a mode of etherial manifestation."[42] He thought that nonphysicists "may possibly be surprised to see the intimate way in which the ether is now spoken of by physicists and the assuredness with which it is experimented on. They may be inclined to imagine that it is still a hypothetical medium whose existence is a matter of opinion. Such is not the case. The existence of an ether can legitimately be denied in the same terms as the existence of matter can be denied, but only so. . . . The evidence for ether is as strong and direct as the evidence for air."[43] In an appendix to *Modern Views of Electricity,* based on a lecture given in 1882, Lodge elaborated upon his conception of the ether. He noted that the idea of an ether was not a new one. The idea, he said, arose as soon as it became clear just how enormous was the extent of interstellar space. As to what was in this space, metaphysics had given two different answers. Some thinkers believed a vacuum could not possibly exist, while others believed that emptiness was necessary to provide room for motion. If metaphysics was to have any weight or validity for Lodge, it had to provide an unconscious appeal to common sense.

> If a highly developed mind or set of minds, find a doctrine about some comparatively simple and fundamental matter absolutely unthinkable, it is evidence and is accepted as good evidence that the unthinkable state of things is one that has no existence. . . .
>
> . . . .
>
> Now if there is one thing with which the human race has been more conversant from time immemorial than an ether and concerning which

42. O. Lodge, *Modern Views of Electricity,* 3rd ed., pp. v–vi. Emphasis in original. Whether or not Lodge was referring to the theory of relativity here is not clear, since there is no reference to the theory in this edition.

43. O. Lodge, *Modern Views of Electricity,* 1st ed., pp. vii–viii. These statements also appear in the prefaces to the second and third editions.

> more experience has been unconsciously accumulated than about almost anything else that can be mentioned, it is *the action of one body on another*. . . . Every activity of every kind that we are conscious of may be taken as an illustration of the action of one body on another. Now I wish to appeal to this mass of experience, and to ask, is not the direct action of one body on another across empty space, and with no means of communication whatever, is not this absolutely unthinkable? I think that whenever one body acts by obvious contact, we are satisfied and have a feeling that the phenomenon is simple and intelligible; but that whenever one body apparently acts on another at a distance, we are irresistibly impelled to look for the connecting medium.[44]

"The medium," Lodge claimed, "is now accepted as a necessity by all modern physicists."[45] He characterized the "modern view" of the ether as "one continuous substance filling all space; which can vibrate as light; which can be sheared into positive and negative electricity; which in whirls constitutes matter; and which transmits by continuity and not by impact, every action and reaction of which matter is capable."[46] Though Lodge first expressed these ideas in 1882, he repeated them in 1892, 1907, and, indeed, to the end of his life.[47] Lodge's belief in the ether was not simply an unthinking adherence to British tradition. It stemmed from a simple and straightforward metaphysics, an inability to conceive of two bodies affecting each other without some connection between the two.

In a new chapter[48] added to the 1907 edition of *Modern Views of Electricity,* Lodge again detailed his opinions on the constitution of the ether. He said it must be continuous[49] and frictionless, and at the

44. O. Lodge, "The Ether and Its Functions," delivered at the London Institution, 28 Dec. 1882; see *Modern Views of Electricity,* 1st ed. pp. 327–358, quotation on pp. 328–332.

45. *Ibid.*, p. 338.

46. *Ibid.*, p. 358.

47. O. Lodge, *My Philosophy, op. cit.* (note 41), *passim.*

48. This chapter was also published as a paper, "Modern Views of the Aether," *Nature, 75* (1907), 519–522; and it served as the basis for Lodge's 1907 address to the British Association.

49. The property of continuity is one on which Lodge had much to say. In his Presidential Address to the British Association in 1913, entitled "Continuity," Lodge commented on some recent trends in physics:

> Indeed, a whole system of non-Newtonian mechanics has been devised having as its foundation the recently discovered changes which must occur in bodies moving at speeds nearly comparable with that of light. . . . But I do not consider it so revolutionary as to overturn Newtonian mechanics. After all, a variation of mass is familiar enough and it would be a great mistake to say that Newton's second law breaks down

same time it must be rigid and perfectly elastic. Following Kelvin, Heaviside, FitzGerald, and Larmor, Lodge assumed a magnetic field to be a circulation of etherial fluid. The direction of flow was along the lines of magnetic induction, and the magnetic intensity at any point was proportional to the speed of flow. Assuming that the circulation of ether around the equator of an electron was equal to the velocity of the electron's forward motion and that the mechanical energy and the magnetic energy of the field were equal, Lodge arrived at a relationship between known electromagnetic quantities and the unknown density of the ether. From this he calculated the density to be of the order of magnitude of 10^{12} grams per cubic centimeter.[50] All of this supports the view that, to Lodge, the ether was a real substance capable of carrying electromagnetic energy, of transmitting all known forces, and of serving as the Cartesianlike source of charge and of matter. In Lodge's hands, the ether had a substantiation which one was not likely to find in France or Germany.

To a modern scientist, reading Lodge's work is similar to falling to the bottom of Lewis Carroll's rabbit hole. But it would be too easy to dismiss Lodge's work as the fantastic creation of an overimaginative individual. I think it is important that Lodge's assessment of the place and importance of the ether be taken quite seriously as a barometer of the thinking of many of his British colleagues, and as a key to the kinds of issues that were considered important in physics in Britain near the turn of the century. For while Lodge's passion for the ether was more vocal and more obvious than was that of most of his colleagues, he is more typical than aberrant. For example, while not everyone agreed with Lodge's calculations concerning the density of the ether, the calculations were taken seriously

---

> merely because mass is not constant. . . . In fact, variable masses are the commonest, for friction may abrade any moving body to a microscopic extent (*Brit. Assoc. Rep., 1907*, pp. 13–14).
>
> . . . .
>
> I urge that we remain with or go back to, Newton. I see no reason against retaining all Newton's laws discarding nothing, but supplementing them in the light of further knowledge" (*ibid.*, p. 15).

Lodge believed that the main trend of the controversy over relativity was over continuity vs. discontinuity (*ibid.*, p. 22), and he declared himself an "upholder of ultimate continuity and a fervent believer in the Ether of Space" (*ibid.*, p. 32).

50. O. Lodge, "Modern Views of the Aether," *op. cit.* (note 48).

and did arouse considerable comment in both Britain and America. C. W. Richardson, professor of physics at Princeton University, thought that the flow of ether should be identified with the Poynting vector. This, he argued, would seem most natural.[51] C. V. Burton, on the other hand, eschewed any program that identified the flow of ether either with the magnetic or the electric fields or with the Poynting vector. Rather, Burton argued, the ether was "a-phenomenal," by which he meant that it was "without influence on the senses of the observer or upon any instrumental test or measurement which he can make."[52]

For Ebenezer Cunningham,[53] the ether "is in fact not a medium with an objective reality but a mental image which is only unique under certain conditions."[54] Just what those conditions were, Cunningham did not specify. But he was willing to ascribe to the view that each frame of reference has its own ether, each ether having a uniform velocity corresponding to the motion of its frame with respect to absolute space.[55] In other words, there would have to be an ether corresponding to every degree of uniform motion. In this way, an observer in any frame of reference would be unable to detect the ether associated with his frame. But what of the ethers associated with all the other frames of reference? Cunningham had no comment. Presumably, since he urged that the ether be taken as a "mental image," he was not bothered by such a question.

Cunningham drew inspiration for this position from Joseph Larmor,[56] though Larmor himself could not concur with any theory that ascribed "a uniform motion to the whole of the ether because there is no conceivable means of producing or altering such a motion."[57] Larmor believed that "an infinitely extended aether postulates absolute motion as a fact in the only real sense of that term, namely, motion relative to the remote quiescent regions of the aether: since

51. O. W. Richardson, "Structure of the Aether," *Nature, 76* (1907), 78.
52. C. V. Burton, "The Structure of the Aether," *Nature, 76* (1907), 150–151.
53. Ebenezer Cunningham (1881–) attended St. John's College, Cambridge, where he graduated senior wrangler in 1902. He was Smith's Prizeman in 1904. He was lecturer in mathematics at the University of Liverpool between 1904 and 1907; he later taught at University College, London. He returned to St. John's College in 1911, where he remained in active work until 1946.
54. E. Cunningham, *The Principle of Relativity* (Cambridge, 1914), Chap. 15.
55. E. Cunningham, "Structure of the Aether," *Nature, 76* (1907), 222.
56. *Ibid.*
57. J. Larmor, "The Aether and Absolute Motion," *Nature, 76* (1907), 269–270.

that determination is made, arguments from relativity of motion must lapse."[58] Larmor,[59] whose influence on British physics was great, rejected any substantial view of the ether. He expressed himself on this subject most forcefully in his book *Aether and Matter*,[60] where he said that an understanding of the general dynamical and physical relationships of matter must rest on the assumption that matter consists of atomic aggregates immersed in the ether. But "all that is known (or perhaps need be known) of the aether itself may be formulated as a scheme of differential equations deriving properties of a continuum in space, which it would be gratuitous to further explain by a complication of structure."[61] Eddington reported that Larmor believed the ether was not a material medium, and that he saw no point in trying to explain its properties as though it were. Larmor's ether was only to be defined by mathematical equations, and in this regard it resembled the ether of Lorentz. Larmor, like Lodge, believed that the ether was something other than a material medium. But Lodge believed that it was at least a detectable entity; his mysticism, his penchant for physical models,[62] and his experience with electromagnetic radiation led him to the belief that the ether, while not material, was substantial. Larmor, however, trained from the start as a theoretician, embraced a mathematical, abstract view of the ether.

J. J. Thomson, like Lodge, was an experimentalist,[63] and his views

58. *Ibid.*

59. Joseph Larmor (1871–1942) attended St. John's College, Cambridge, where he graduated first wrangler in 1880. He succeeded G. G. Stokes as Lucasian Professor at Cambridge.

60. J. Larmor, *Aether and Matter: A Development of the Dynamical Relations of the Aether to Material Systems on the Basis of the Atomic Constitution of Matter* (Cambridge, 1900). This book, which was based on several earlier papers, won the Adams Prize in 1900.

61. *Ibid.*, p. 78.

62. Lodge devoted a good deal of time to constructing models similar to those of Clerk Maxwell. His model for electricity was an elaborate gear and pinion affair. "If we think of electricity in the several molecules of the insulating medium connected like so many cogwheels gearing into one another and also into those of the metal, it is easy to picture a sideways spread of rotation brought about by the current just as a moving rack will rotate a set of pinions gearing into it and into each other." (O. Lodge, *Modern Views of Electricity*, 1st ed., p. 177. The same model is presented in all later editions.)

63. J. J. Thomson (1856–1940) was Cavendish Professor of Experimental Physics at Cambridge from 1894 to 1919. According to Rayleigh, Thomson's whole life was devoted to experimental physics. (Rayleigh, "Thomson, J. J.," *Obituary Notices of Fellows of the Royal Society, 3* [1941].)

on the ether were similar to those of Lodge. He expressed these views most clearly in his Adamson Lecture at the University of Manchester in 1907.[64] His concern here was with the degree of generality of Newton's third law. He noted that a system in which the third law did not apply could not be imitated by any mechanical model, and that there did seem to be situations in which the third law was not valid. Consider, he said, two electrical bodies A and B in rapid motion. The electromagnetic forces these two bodies exert on each other are not equal and opposite, unless the two bodies are moving with the same speed in the same direction. As a result, the momentum of the system composed of A and B does not remain constant. If one were to conclude from this that electrified bodies are not subject to the third law, it would mean "giving up the hope of regarding electrical phenomena as arising from the properties of matter in motion." But, Thomson noted with relief, such a conclusion was not necessary. The solution lay in hypothesizing an "invisible system," connected to the system containing A and B, which possesses mass and momentum; and any momentum gained or lost by the system containing A and B could be ascribed to the invisible system.[65] He identified the invisible system with the ether. Thomson might be interpreted as suggesting that the ether was simply a device to save and extend the applicability of Newton's third law; but the ether meant more to him than this. His Presidential Address to the British Association at Winnipeg in 1909 leaves no doubt about his position: "The aether is not a fantastic creation of the speculative philosopher; it is as essential to us as the air we breathe." It is the "seat of electrical and magnetic forces," and it is the "bank in which we may deposit energy and withdraw it at our convenience."[66]

I have presented a sampling of opinion in Britain during the first decade of this century on the nature of the ether. This opinion was not a response to Einstein's theory, since most British physicists had no knowledge of the theory at this time. It would be the easiest thing in the world to sweep early twentieth-century British considerations of the ether aside. They were not heuristic. They did not lead any-

64. J. J. Thomson, *On the Light Thrown by Recent Investigations on Electricity and on the Relation between Matter and Ether; The Adamson Lecture Delivered at the University on November 4, 1907* (Manchester, 1908).

65. *Ibid.*, pp. 7–8.

66. J. J. Thomson, "Presidential Address to the British Association at Winnipeg, 1909," *The Electrician, 63* (1909), 776–779, quotation on p. 778.

where, but rather represented something of a last breath for a theory that was nearly moribund. But they were also unique. In Germany, physicists, even those physicists committed to the ether, were engaged in a searching analysis of Einstein's theory. While one or two leading German physicists can be cited who ignored the theory and concentrated on ether mechanics, this was not typical, nor was their work given prominent display or attention.[67] In France, there was little attention paid to Einstein's work, but then, too, there was little active concern with the nature of the ether. In America, though there was also visible concern with the ether, physicists followed British work more than striking out in independent directions.

British Model Building and the Ether

In J. T. Merz's view the differences in scientific attitude and methodology between countries of Western Europe were more pronounced at the beginning than at the end of the nineteenth century. At the end of the century the "intercourse of the different nations," he said, had "done much to destroy these national peculiarities."[68] However, when one compares British and German physics in 1905, there is little evidence that national peculiarities had been destroyed. The comments of the French philosopher-physicist Pierre Duhem, writing on British physics in 1906, are illuminating on this point.[69] The French, he said, are astonished by the way the English always accompany the exposition of their theories with models. For instance, the "French or German physicist conceives, in the space separating two conductors, abstract lines of force having no thickness or real existence; the English physicist materializes these lines and thickens them to the dimensions of a tube which he will fill with vulcanized rubber." Lodge epitomized for Duhem the Englishman's commitment to the purely mechanical explanation of physical phenomena.

67. See W. Kass, "Die Natur des Äthers," *Das Weltall, 8* (1908), 134–136; P. Lenard, "Äther und Materie," *Sitzgb. Heidelberg. Ak. Wiss.* (1910), 37 ff.; W. Ritz. "Du rôle de l'éther en physique," *Riviste di Scienza, 3* (1908), 260–274; H. Witte, Ätherfrage," *Jahrb. Rad. u. Elek., 7* (1910), 205–261.
68. J. T. Merz, *A History of European Thought, op. cit.* (note 1), *1*, 252.
69. P. Duhem, *The Aim and Structure of Physical Theory* (New York, 1962), pp. 69–72.

When we open Lodge's *Modern Views of Electricity,* we find "nothing but strings which move around pulleys, which roll around drums, which go through pearl beads, which carry weights; and tubes which pump water while others swell and contract; toothed wheels which are geared to one another and engage hooks. We thought we were entering the tranquil and neatly ordered abode of reason, but we find ourselves in a factory." Duhem observed that "understanding a physical phenomenon is, therefore, for the physicists of the English school, the same thing as designing a model imitating the phenomenon; whence the nature of material things is to be understood by imagining a mechanism whose performance will represent and simulate the properties of bodies. The English school is completely committed to the purely mechanical explanation of physical phenomena."

To Duhem, the construction of mechanical models was not a sufficient characteristic in itself to set off British physics from the physics of the Continent. It was rather the particular form that model building in Britain assumed that was distinct. In continental Europe, he declared, "the sense for abstraction may lapse, but it never falls asleep completely,"[70] as it had done in England.

> When an English physicist seeks to construct a model appropriate to a group of physical laws, he is not embarrassed by any cosmological principle, and is not constrained by any logical necessity. He does not aim to deduce his model from a philosophical system nor even to put it into accord with such a system. He has only one object: to create a visible and palpable image of the abstract laws that his mind cannot grasp without the aid of this model. . . . The English physicist does not therefore ask any metaphysics to furnish the elements with which he can design his mechanisms. He does not aim to know what the irreducible properties of the ultimate elements of matter are.[71]

While much of what Duhem says is valid, he is wrong on two counts. First of all, there were Englishmen, such as Larmor, who were perfectly willing to let the concept of the ether rest on nothing more than the mathematical equations that describe the electromagnetic field. Second, it is simply incorrect to say that the British penchant for model building overrode or ignored all cosmological principles

70. *Ibid.*, p. 73.
71. *Ibid.*, p. 74.

and that it did not stem from any uniform metaphysics. The model building itself can be construed as representing a metaphysics. The metaphysics of Oliver Lodge is quite plain: he could not imagine a world in which action at a distance could be operative. It would be quite natural for him to rely on some medium or system of particles to intervene between bodies interacting electrodynamically. J. J. Thomson wanted to preserve the conservation of momentum in all physical interactions and to leave open the possibility of a mechanical explanation of all phenomena; this, again, is certainly a metaphysics.

The case of J. J. Thomson and the vortex atom offers a counter-example to Duhem's further claim that the English scientist was prone to adopt inconsistent models for different phenomena.[72] The study of vortex motion, especially in connection with the motion of the ether, was a favorite subject in Britain in the second half of the nineteenth century. J. J. Thomson submitted an essay for the Adams Prize in 1882 on the subject of the interaction of two vortex rings. He was attracted to the subject by Kelvin's suggestion that matter might consist in vortex rings in an ideal fluid.[73] The "spartan simplicity" of this suggestion appealed to him, as did its promise to lead to a "theory more fundamental and definite than any that had been advanced before." Though Kelvin's enthusiasm for the explanatory power of the vortex ring faded, J. J. Thomson became more enthusiastic with the passing of time.[74] "I regard," the latter once commented, "the vortex-atom explanation as the goal [sic] at which to aim."[75] In his 1907 *Corpuscular Theory of Matter,* he wrote

72. There is some truth in Duhem's claim. Kelvin was probably the greatest offender in the building of inconsistent models. There are plenty of such examples in his 1884 Baltimore lectures. It was in these lectures that he made his famous admission that "I never satisfy myself until I can make a mechanical model of a thing. If I can make a mechanical model, I understand it" (*Lectures on Molecular Dynamics and the Wave Theory of Light* [Baltimore, 1884]).

73. J. J. Thomson, *Recollections and Reflections* (London, 1936), pp. 94–95.

74. Kelvin finally abandoned the model: "I am afraid it is not possible to explain all the properties of matter by the vortex-atom theory alone. . . . We may expect that the time will come when we shall understand the nature of an atom. With great regret I abandon the idea that a mere configuration of motion suffices" (quoted in J. T. Merz, *A History of European Thought, op. cit.* [note 1], 2, 182). On the other hand, J. J. Thomson said: "With reference to the vortex-atom theory, I do not know of any phenomenon which is manifestly incapable of being explained by it; and personally, I generally endeavour (often without success) to picture to myself some kind of vortex-ring mechanism to account for the phenomenon with which I am dealing" (*ibid.*, p. 183).

75. *Ibid.*

that he considered the vortex theory to be "fundamental."[76] The point is that in all of his work Thomson was faithful to the vortex model. Thomson, like Lodge, eventually accepted the results of relativity, but not the spirit. His commitment to the ether remained unspoiled; he could account for the results of relativity his way. Maxwell's equations and the electrical theory of matter, without the aid of relativity, had explained the contraction of moving bodies and the variation of mass with velocity. He thought that it was "reasonable to regard Maxwell's equations as the fundamental principle rather than that of relativity, and also to regard the ether as the seat of mass, momentum, and energy of matter."[77]

Duhem's thesis must be qualified in yet another way. He implies that British model building was part of a commitment to a completely mechanical view of nature. One can certainly find this commitment in Kelvin who, failing to find an adequate mechanical model of electromagnetic phenomena, eventually despaired of being able to understand electromagnetism. But by the turn of the century British physicists were beginning to abandon their frustrating attempts to provide a mechanical model of electromagnetism. Rather than trying to show that all electrical phenomena were fundamentally mechanical, the British now undertook to show that all mechanical phenomena were fundamentally electrical. Larmor's *Aether and Matter* was written from this point of view. J. J. Thomson, while clinging to the vortex theory of the atom, was of the opinion that the experimental results of electrodynamics could be accounted for on the basis of the electrical theory of matter. He was also of the opinion that the entire mass of the electron was electrical,[78] and that, since the entire mass of the atom is due to positive and negative charges, the entire mass of the atom was electrical.[79]

The change in allegiance from a mechanical to an electrical view of the universe was partly due to a feeling of failure and discouragement. It also owed much to the influence of the work of Lorentz,

76. J. J. Thomson, *The Corpuscular Theory of Matter* (London, 1907), pp. 1–2.
77. J. J. Thomson, *Recollections and Reflections, op. cit.* (note 73), pp. 432–443.
78. J. J. Thomson, *The Corpuscular Theory of Matter, op. cit.* (note 76), pp. 301–302.
79. *Ibid.*, pp. 1–2. Thomson's adherence to an electrical view of matter must be qualified in at least one regard. While holding that the entire mass of the atom should be considered as electrical, he did cling to the hope that electrical phenomena could be understood in terms of the properties of bodies in motion.

work which many British physicists saw as paralleling that of Larmor.[80] Lorentz' attempts at constructing first- and second-order theories of moving bodies were widely interpreted, with justification, as assuming an electrical view of matter. In the eyes of most British physicists, Lorentz' attempts had been successful, and they saw this as tending to confirm the electrical view of matter. To them the Lorentz theory was not *ad hoc;* rather, it represented a type of theory they were quite familiar with, a type that involved the tailoring of a model to fit the data.

Lodge viewed the null result of the Michelson-Morley experiment as confirmation of the contraction effect, which Lorentz and FitzGerald independently hypothesized to account for that experiment. It was while Lodge was discussing the difficulties of the Michelson experiment with FitzGerald that FitzGerald first hit upon the contraction idea. Lodge later recalled their conversation: FitzGerald suggested that " 'perhaps the stone slab is affected by the motion.' I rejoined that it was a 45 degree shear that was needed. To which he replied, 'Well that's all right—a simple distortion.' *And very soon he said, 'And I believe it occurs and that the Michelson experiment demonstrates it.'* And is such a hypothesis gratuitous? Not at all: *in the light of the electrical theory of matter such an effect ought to occur.*"[81]

When J. J. Thomson said that the fact that momentum was not conserved in the interactions of moving charges *proved* that the invisible universe of the ether really existed, he was expressing a limited view of physical explanation similar to that of Larmor, FitzGerald, and other British physicists. Thomson argued that theories were not intended to explain anything in an ultimate sense: "From the point of view of the physicist, a theory of matter is a policy rather than a creed; its object is to connect or coordinate apparently diverse phenomena, and above all to suggest, stimulate and direct experiment."[82] But if that is the case, why were British physicists unwilling to give an ear to Einstein's theory, a theory which could certainly "coordinate apparently diverse phenomena"?

80. In a recent personal communication with the author, Cunningham referred to the Lorentz transformations as "commonly known as the Larmor-Lorentz transformations."

81. O. Lodge, "Continuity," *op. cit.* (note 49), pp. 56–57. Emphasis added. See also O. Lodge, *My Philosophy, op. cit.* (note 41), p. 69.

82. Thomson, *The Corpuscular Theory of Matter, op. cit.* (note 76), pp. 1–2.

The answer is that the theory could not be accepted in Britain until it could be made consonant with the existing British ideas about the ether. First, British physicists had to implicitly deny relativity the status of a theory; second, they had to make it a subordinate principle, modifying the electron theory of matter; and third, they had to redefine the ether so as to render it undetectable in principle and describable by equations agreeing with those of the theory of relativity.

The Contributions of Ebenezer Cunningham

Lodge's observation in 1907[83] that new theories were supplementary to the electron theory of matter rather than revolutionary provided the stepping-stone to the introduction of relativity theory in Britain. Ebenezer Cunningham was the first British physicist to give serious consideration to relativity. He wrote about it first in 1907 in a paper that dealt with Abraham's objection to the Lorentz electron.[84] Abraham had argued that there was no external source of energy available for deforming the Lorentz electron and that, furthermore, the deformable electron would be unstable.[85] Cunningham demonstrated that Abraham's conclusions were true only because Abraham had imposed his own rigid electron[86] on Lorentz' equations for the deformable electron. He claimed that if one allowed a change in dimensions in the direction of motion, both energy and momentum would be conserved and the electron would be stable.

Cunningham said that "Maxwell's equations represent equally well the sequence of electromagnetic phenomena relative to a set of axes moving relative to the aether as relative to a set of axes fixed in the aether."[87] This was his statement of the principle of relativity, from which he deduced "how a light wave traveling outward in all directions with a velocity *c* relative to one observer A, may at the

83. Lodge, "Continuity," *op. cit.* (note 49).
84. E. Cunningham, "On the Electromagnetic Mass of a Moving Electron," *Philosophical Magazine, 14* (1907), 538–547.
85. M. Abraham, *Theorie der Elektrizität* (Leipzig, 1905), 2, 205 ff.
86. See M. Abraham, "Prinzipien der Dynamik des Elektrons," *Annalen der Physik, 10* (1903), 105–179.
87. E. Cunningham, "On the Electromagnetic Mass of a Moving Electron," *op. cit.* (note 84), p. 538.

same time be traveling outward in all directions with the same velocity relative to an observer B moving relative to A with velocity *v*."[88] By the technique of equating the expressions for two spheres of light, he derived the Lorentz transformations. When A. H. Bucherer challenged this derivation on the grounds that he was "not aware that such a requirement [invariance of the velocity of light] . . . [was] necessary to explain any known fact of observation,"[89] Cunningham replied[90] that he was not making the claim that the invariance of the velocity of light was a fact of experience; rather, he took it to be a statement derivable from the principle of relativity. His interpretation of relativity stemmed from his conviction that the ether was not a substantive entity and that an ether had to be defined for each inertial frame of reference. In each such frame the ether would be at rest, and the velocity of light would be the same in all directions.

Cunningham's eclectic attitude can be further illustrated by two papers he published in 1909. One of these, "The Motional Effects on the Maxwell Aether-Stress,"[91] represents an effort in pure ether mechanics. The task that Cunningham set himself in this paper was to bridge the gap between Maxwell's ether-stress and Newton's third law. In the second paper Cunningham came back to the problem of relativity. He wrote that the absence of detectable effects of the earth's motion relative to the ether had been "fully accounted for by Lorentz and Einstein *provided* the hypothesis of electromagnetism as the ultimate basis of mater can be accepted so that the only available means of estimating the distance between two points is the measuring of the time of propagation of the effects between the bodies, such propagation taking place in accordance with the equa-

88. *Ibid.*

89. A. H. Bucherer, "On the Principle of Relativity," *Philosophical Magazine, 15* (1907), 318. Bucherer had developed his own theory of the electron which contained a transformation equation for the mass of a moving electron and for the dimensions of the electron. The theory relied on the assumption that the volume of the electron was an invariant to all observers, and the contraction occurred in such a way as to maintain a constant volume. Subsequently Bucherer relinquished his theory in favor of Einstein's as a result of experiments he performed on the specific charge of the electron.

90. E. Cunningham, "On the Principle of Relativity and the Electromagnetic Mass of the Electron: A Reply to Dr. A. H. Bucherer," *Philosophical Magazine, 16* (1908), 423–428.

91. E. Cunningham, "The Motional Effects on the Maxwell Aether-Stress," *Proceedings of the Royal Society*, Ser. A, *83* (1909), 109–119.

tions of the electron theory."[92] Cunningham saw the "foundations" of the theory of relativity as the Lorentz transformation.[93]

In 1911 Cunningham was asked by the Cambridge University Press to write a book on the theory of relativity.[94] The book came out in 1914, and in it he gave his view of the place of relativity in physics at the time. He believed that *"there is a real place for it as a hypothesis supplementary to and independent of electrical theory* owing to the limitations to which that theory is subject"[95] (emphasis added). And he believed that it

> holds the same place in the physical thought of today, that the [Newtonian] principle of dynamical relativity held in the time when the laws of dynamics were considered as ultimate and all embracing. It consists in the general hypothesis, based on a certain amount of experimental evidence that *the problem of determining in a physical sense the absolute velocity of a body is one that can no more be solved uniquely by the help of optical and electrical phenomena, than it could by means of dynamical observations.* If we speak of a "fixed aether" as the background of electrical activity, it is the hypothesis that *the velocity of any piece of matter relative to the aether is unknowable.* To put it more precisely, it states that we *neither have nor expect to have any experimental evidence of the uniqueness of the framework which we call "the aether" but that if there is one, there is an infinite number of such frames of reference, any one of which has, relative to another, a uniform velocity of translating without rotation, the velocity being of arbitrary magnitude. . . .* [Relativity] is an *empirical* principle, suggested by an observed group of facts, namely the failure of experimental devices for determining the velocity of the earth relative to the luminiferous aether, and would make it a *criterion* of theories of matter that they should give an account of this failure, and it suggests modifications where the theory is insufficient to do so. But like all physical principles, it is to be probed by further experience.[96]

In one fell swoop, Cunningham had reduced the "theory of relativity" to the first postulate, "the principle of relativity." He had

92. E. Cunningham, "The Principle of Relativity in Electrodynamics and an Extension Thereof," *Proceedings of the Mathematics Society of London, 8* (1909), 77–98, quotation on p. 77.
93. *Ibid.*, p. 78.
94. E. Cunningham, personal communication.
95. E. Cunningham, *The Principle of Relativity* (Cambridge, 1914), p. v.
96. *Ibid.*, pp. 7–8.

made it supplementary to the electron theory of matter, and he had reasserted his position of the undetectability of the ether. He viewed the second postulate of relativity as a consequence of the first. The first postulate itself was, in his view, a generalization from experience about the ether—not a "metaphysical dogma." His consignment of the theory of relativity to the status of a principle supplementary to the electron theory was intimately related to his view of that principle as an empirical generalization. After reviewing some experiments on the motion of the earth through the ether, he explained that "it is because the experimental evidence extends into regions where existing electrical theory is insufficient, that the principle of relativity becomes of importance as a supplementary and independent hypothesis."[97]

Cunningham differentiated between the work of Lorentz, Larmor, and Einstein on the one hand, and the work of nineteenth-century mechanists on the other, terming the latter work "experimental relativity."[98] By this he meant that the failure to find evidence of absolute motion relative to the ether had incidentally led to a revision of the concepts of space and time.[99]

> It may be said that the arguments differ *only* in the point of view, but for the moment it is the new point of view that is vital. The older point of view was the outcome of the strivings of the mechanical school of physicists after an objective mechanical elastic ether and their adoption of a metaphysical concept of space and time. But when mechanics is placed in a derivative place and electrical theory in the fundamental position, when we are prepared to admit that many of the properties of matter are, when analysed, found to be compounded of electrical action, it becomes very necessary to consider with an open mind the possibility that some of what seem the most obvious modes of thought may require revision, and among other things to avoid the vicious circle of ideas involved in likening the aether to a species of matter with molar properties which are an idealization of the properties of matter as we imperfectly know them.[100]

There was no doubt in Cunningham's mind that the mechanists' search for what he called an "objective aether" had failed. But he

97. *Ibid.*, p. 41. Emphasis in original.
98. *Ibid.*, p. 45.
99. *Ibid.*
100. *Ibid.*, p. 50.

did not want to abandon the ether as a concept. He recognized that the ether would have to be "subject to the Einstein transformation of space and time when the frame of reference is altered."[101] But while he was cautious about the ether, he devoted a whole chapter of his book to "Relativity and an Objective Aether." He argued that he could make the principle of relativity and the concept of an ether compatible if he were allowed to make the ether *a priori* undetectable. This is the sense in which he conceded to the ether an "objective reality." Thus the program that Cunningham had outlined as early as 1907 was culminated.[102] Despite the efforts of Cunningham to make a reconciliation between the theory of relativity and the ether, most British physicists simply ignored relativity and the work of Einstein and maintained an unquestioning commitment to an ether-based physics.[103] In the entire period between 1905 and 1911, only one physicist seems to have questioned the validity of the concept of the ether; that physicist was Norman R. Campbell.

The Contributions of Norman R. Campbell

Campbell[104] challenged the ether in a series of three papers beginning in 1910.[105] The concept of the ether, he said, seemed unsatisfac-

101. *Ibid.*, p. 52.

102. See E. Cunningham, *Relativity and the Electron Theory* (London, 1915). This book is an abbreviated version of *The Principle of Relativity*. In the 1915 publication Cunningham repeated his claim that the theory of relativity was supplementary to the electron theory. But he noted with some force the *ad hoc* nature of the Lorentz contraction (*ibid.*, p. 25).

103. During the period 1905–1911, publications on ether mechanics continued unabated. Typical titles are C. V. Burton, "The Sun's Motion with Respect to the Aether," *Philosophical Magazine, 19* (1910), 417–423, and J. J. Thomson, "Light Thrown by Recent Investigations on Electricity on the Relation between Matter and Ether," *Smithsonian Reports, 1909*, pp. 233–244.

104. N. R. Campbell (1880–1949) was trained at Trinity College, Cambridge, becoming a fellow there in 1904. He worked in the Cavendish Laboratory under J. J. Thomson. In 1913 he was awarded the Cavendish research fellowship at Leeds, where Bragg was then Cavendish Professor. In 1919 he joined the staff of the research laboratories of General Electric; he was instrumental in establishing the freedom of the research worker in an industrial laboratory. He was deeply interested in philosophy and wrote several tracts on the philosophy of physics, including his classic *Physics: The Elements*, which has been reprinted as *The Foundations of Science* (New York, 1957).

105. N. R. Campbell, "The Aether," *Philosophical Magazine, 19* (1910), 181–191; "The Common Sense of Relativity," *Philosophical Magazine, 21* (1911), 502–517; "Relativity and the Conservation of Mass," *Philosophical Magazine, 21* (1911), 626–630.

tory in modern physics, even though some authors still considered it of supreme importance. He noted that one of the founders of the atomic theory of radiation, J. J. Thomson, had recently devoted his entire Presidential Address at the British Association to the properties of the ether. Campbell suggested that perhaps it was the shyness of scientists in discussing the "essential foundations" of their studies that explained their reluctance to attack or defend the concept of the ether. He did not share their reluctance, and he produced an example to demonstrate the untenability of the ether and the need for relativity.

> Consider the case of two or more electrically charged bodies moving with different uniform velocities relative to some observer. Round each body is distributed electrical energy localized in the aether; the positions of the portions of the aether which contain stated amounts of energy (belonging to one and the same body), relative to each other or to the charged nucleus, are not changed by the motion. It seems obvious and simple to identify points in the aether . . . by the amounts of energy contained in them. Then the velocity of the aether relative to the observer would be different according as one or the other of the charged bodies was considered [since the charged bodies are moving relative to each other] and would be in each case the same as the velocity of the corresponding charged body relative to the observer.
>
> Such, I think, is the simple and obvious view leading to the principle of relativity.[106]

But such an example did not necessarily lead one to the theory of relativity. Faced with the kind of contradiction Campbell proposed, Cunningham had simply redefined the ether in such a way that the contradiction would make no difference. Campbell himself recognized that one could do this simply by refusing to identify the ether as the locus of electromagnetic energy. But in that case, the ether became undetectable and therefore physically meaningless.

Cunningham said that the ether apologists had tried a new tack: "They now said that the 'difference between the velocities relative to the aether of any two bodies was equal to their relative velocities, but that the velocity relative to the aether of any body was uncertain to the extent of a constant.' And then they tried to show that this constant was far beyond the range of experiment right now." This

106. N. R. Campbell, "The Aether," *ibid.*, pp. 184–185.

seemed utterly unconvincing to him. He conjectured that the "future historian of physics will be astounded that the vast majority of physicists should accept a system of such bewildering complexity and precarious validity rather than abandon ideas which seem to have their sole origin in the use of the word 'aether' and reject those to which so many lanes of thought point insistently." He saw no point in trying to save the ether concept, since it "has never been the source of anything but fallacy and confusion of thought"; it should be relegated to the "dust-heap where 'phlogiston' and 'caloric' are now mouldering."[107] His position was that the concept of the ether should be given up since there was no way to test its existence. He insisted that all the apparent paradoxes of relativity will disappear "if attention is kept rigidly fixed upon quantities which are actually observed."[108]

Campbell's assertions about the meaninglessness of the concept of the ether were unique in Britain at this time. They are similar to the strictures of Duhem. Like Duhem, Campbell believed that metaphysics and science were completely separate: "one of the chief characteristics which distinguishes science from metaphysics, and the feature which makes men of science so averse from the latter, is that in science, but not in metaphysics, it is possible to obtain universal assent for conclusions, and to present results which do not lose their value because when they are presented, they are so obvious as to be indubitable."[109] But to claim that science is without metaphysics, or to claim indubitability of scientific results, was to express a metaphysics. It was this positivistic strain in Campbell that made him blind to the value that the ether concept had had in physics. It also prevented him from understanding the basis of the British physicists' commitment to the ether.

According to Campbell, the reason for the lack of interest in relativity theory and for the persistence of the concept of the ether was that the man in the laboratory "has been offered profound mathematical investigations, which are intensely important and interesting but tend to obscure the fundamental points at issue in the mind of one who thinks physically rather than mathematically. And on the other hand, he has been offered collections of apparently paradoxical

107. *Ibid.*, p. 190.
108. N. R. Campbell, "The Common Sense of Relativity," *ibid.*, p. 515.
109. N. R. Campbell, *The Foundations of Science, op. cit.* (note 104), p. 10.

conclusions deduced from the principle which are sometimes elegant and entertaining, but more often fallacious."[110] Campbell was mistaken. What stopped the British physicists was not primarily the mathematics or the paradoxical conclusions, which, in any case, were present in Lorentz' and Larmor's theories too, but the foundations and ether-denying implications of relativity. Campbell's "measurement metaphysics" was so intense that he preferred to use it to deduce reasons for the British physicists' difficulties with the theory rather than look for the actual reasons.

In spite of Campbell's emphasis on measurement and performable experiments, and his objection that the concept of the ether was meaningless, he was not averse to the use of models. In reply to the objections to relativity theory that it had no physical meaning and that it "destroyed utterly the old theory of light based on an elastic aether and puts nothing in its place,"[111] he maintained that a physical theory of light could be produced which was consistent with the principle of relativity. Campbell was of the opinion that a model would be an essential part of the construction of such a theory.[112]

5. CONCLUSIONS

It needs emphasizing that there was little work done in Britain on relativity in 1905–1911. The British contribution to the relativity literature represents less than ten percent of the total in this period.[113] Cunningham notes that he "was surprised to be asked to speak to the British Association in 1910" on the subject of relativity, because he "did not think that there was much interest in the matter at that time."[114] Indeed, even those British scientists who, in one way or another, became aware of the theory seemed to have difficulty in understanding it. For example, in 1909 the astronomer F. C. Searle wrote to Einstein to thank him for sending a paper on rela-

110. N. R. Campbell, "The Common Sense of Relativity," *op. cit.* (note 105), p. 502.

111. *Ibid.*, p. 515.

112. *Ibid.* See also N. R. Campbell, *The Foundations of Science, op. cit.* (note 104), p. 130. For a discussion of the conflict over the use of models between Duhem and Campbell, see Mary Hesse, *Models and Analogies in Science* (London, 1963).

113. This estimate is based on M. Lecat, *Bibliographie de la relativité, op. cit.* (note 5).

114. Personal communication.

tivity: "I have not been able so far to gain any really clear idea as to the principles involved or as to their meaning and those to whom I have spoken in England about the subject seem to have the same feeling."[115] Searle thought that it would be a good idea if Einstein were to write a short account of relativity for an English journal to help the English gain a better grasp of the ideas. When, in 1911, the British Association finally devoted a session to the "Principle of Relativity,"[116] there was little sign of any change in British attitudes toward relativity. The *Nature* reporter at the meeting noted C. V. Burton's "satisfaction that no one had confessed a disbelief in the aether."[117] Through 1911 the British rejection of relativity was founded on a tenacious adherence to an ether-based physics.

While some British physicists like Cunningham argued that experiment would be the final arbiter in deciding the worth of the theory of relativity, the bulk of those physicists concerned with the problems of electrodynamics were not concerned with experimental issues. It is simply incorrect to state, as Campbell did, that the major difficulty the British physicist encountered in dealing with relativity was with the consequences of the theory. In fact the British physicist was unwilling to accept the premises and assumptions on which the theory was based. But since Campbell did not believe that scientific matters involved metaphysics, or more accurately, since he believed that these were two separate and separable areas and that he and his colleagues were doing "science," there was no reason for him to look for assumptions or presuppositions as such. But the presuppositions were there. For example, many physicists shared Lodge's inability to conceive of a theory which did not account for "apparent" action at a distance in terms of pushes and pulls.

Toward an Understanding of the British Response to Relativity

There are, no doubt, many features of British culture and society that could aid in an understanding of the nearly uniform British

115. Letter from F. C. Searle to Albert Einstein, 20 May 1909, Einstein Archives, Princeton University. I wish to thank Dr. Otto Nathan, executor of the Albert Einstein Estate, for permission to quote from this letter.

116. Anon., "Mathematics and Physics at the British Association, 1911," *Nature*, *87* (1911), 498–502.

117. *Ibid.*, p. 500.

neglect of Einstein's theory and the concomitant insistence on retaining the concept of the ether. The feature I wish to bring out here—and one I feel is instrumental—is the structure of British education in physics. A British theoretical physicist interested in electrodynamics was trained to do ether mechanics; it is what he had to learn, and it is what he knew best.

Most British physicists, experimental and theoretical, were trained at Cambridge University. The Mathematical Tripos at Cambridge was the most competitive arena for mathematicians and mathematical physicists in Britain. The importance for students of developing skills in solving mechanical problems of the ether is suggested by questions appearing in the 1901 Tripos:

> Obtain the energy function of an isotropic elastic medium and assuming that waves of dilation are propagated through the medium with an indefinitely great velocity and that the difference between different media is one of density only, find the intensities of the reflected and refracted waves when plane waves are incident on a plane interface separating two media.

> Waves of light are incident on a face of a uniaxial crystal cut perpendicularly to its axis. Find on MacCullagh's theory the intensities of reflected and refracted waves (1) when they are polarized in the plane of incidence, (2) when they are polarized perpendicularly to it.[118]

Presumably, one would have needed to commit MacCullagh's theory to memory. This theory, which had been created in 1839, required an elastic-solid ether that exhibited the property of "rotational elasticity." No known elastic solid had ever exhibited this property, and there had been considerable early skepticism about the theory for this reason. The 1901 Tripos also placed heavy emphasis on fluid mechanics, a subject closely related to the concerns of ether physicists. A typical question was: "Give an account of Helmholtz' and Lord Kelvin's theory of vortex motion." A few more sample questions on elastic-solid theory and fluid mechanics from the 1903 and 1905 examinations point to the ether-mechanics competency expected of Cambridge graduates:

> Investigate Helmholtz' expression for the velocity of a fluid in vortex motion in terms of a vector potential.

118. *Mathematical Tripos, Part II, Thursday, May 30, 1901* (Cambridge, 1901).

> Assuming that the spin at any point within a circular vortex ring of finite cross-section is proportional to the distance of the point from the axis of the ring, prove that the vector potential at any point P is directed at right angles to the axial plane through P, and is equal in magnitude to the value at P of the gravitational potential of the ring, supposed to have at any point Q a density proportional to the projection on the axial plane through P of the perpendicular from Q to the axis of the ring.
>
> Deduce expressions for the stream-function in the case of Hill's spherical vortex, and express the velocity of translation in terms of the strength of the vortex.[119]
>
> Discuss generally evidence for the conclusion that the velocity of the earth's motion does not sensibly affect observable optical phenomena. Show however that if two independent sources of light could interfere, a simple experiment would at once reveal the earth's motion in space.[120]
>
> The potential energy of an elastic medium is a homogeneous quadratic function of the components of strain; prove that, if the directions of transverse vibrations are always in the wave front, the wave surface is Fresnel's.[121]

Such questions may be found in almost any of the Tripos examinations through 1910. It is important to note as well that these were timed examinations and that the student had to be extremely dexterous with the mathematical manipulations.

J. J. Thomson, who studied ("coached") with Routh at Cambridge, recalled that

> Routh's system certainly succeeded in the object for which it was designed, that of training men to take high places in the tripos; for in the thirty-three years from 1855 to 1888 in which it was in force, he had 27 Senior Wranglers and he taught 24 in 24 consecutive years. . . . Until quite near the time when he gave up "coaching," candidates for the Mathematical Tripos were expected to be acquainted with the whole of the wide range of pure and applied mathematics included in the examination. . . . [Routh] was Senior Wrangler in the year when Clerk Maxwell was second. Perhaps no other man has ever exerted so much influ-

119. *Mathematical Tripos, Part II, Saturday, June 6, 1903* (Cambridge, 1903).
120. *Ibid.*
121. *Mathematical Tripos, Part II, Wednesday, May 31, 1905* (Cambridge, 1905).

> ence on the teaching of mathematics; for about half a century the vast majority of professors of mathematics in English, Scotch, Welsh, and Colonial universities, and also the teachers of mathematics in the larger schools, had been pupils of his and to a very large extent adopted his methods. In the textbooks of the time old pupils of Routh would be continually meeting with passages which they recognized as echoes of what they had heard in his classroom or seen in his manuscripts. . . . Routh like Maxwell studied mathematics under Hopkins, the great "coach" at that time, who had taught Stokes and William Thomson and scored 17 Senior Wranglers before he retired.[122]

There existed then a natural filter for processing mathematicians and mathematical physicists in Britain. It is understandable why so many of them acted with such uniformity when confronted with the theory of relativity. They had studied under Hopkins, or under one of his pupils, such as Routh. They had been trained to master the partial differential equations of waves moving through "froths, jellies and vortices." It is understandable also why a person like Lodge, who had not been trained at Cambridge, might have had a similar outlook, since English theoretical physics was dominated by Cambridge. (What is difficult to understand is how Campbell, who was trained at Cambridge, was able to escape the prevailing ether orientation of British physics.)

Cunningham states that the Tripos he took in 1902 hardly dealt with electrodynamics, and that Maxwell's work had not yet filtered down to the point of being used as a textbook. With regard to the theory of relativity, he recalls that it was he who introduced systematic teaching of it in 1911. He says that this time was "a period of transition, which culminated in 1910 when the order of merit in the Tripos was abolished and teaching in all subjects took a much more theoretical turn, while the Cavendish Laboratory moved on into atomic theory. . . . Modern physics was in its infancy. I had to find for myself the work of Lorentz, Planck, etc. after I had graduated. Eddington, Jeans and others were still to come."[123] Cunningham's remarks have a familiar ring. Others[124] have noted the length of

122. J. J. Thomson, *Recollections and Reflections, op. cit.* (note 73), pp. 38–40.
123. Personal communication.
124. See E. Ashby, *Technology and the Academics: An Essay on Universities and the Scientific Revolution* (London, 1959), Chap. 2; J. T. Merz, *op. cit.* (note 1), *1*, 233 ff.

time required for scientific innovations to filter down to the classrooms and examinations of Cambridge—from Leibniz' "d's" through Young's wave theory of light to Einstein's relativity.

ACKNOWLEDGMENTS

I am indebted to the following people for comments and criticisms of earlier drafts of this paper: Professor Paul Forman, University of Rochester; Professor L. K. Nash, Harvard University; Professor E. L. Yates, University of Zambia. I am particularly grateful to Professor Russell McCormmach, University of Pennsylvania, who commented extensively on the manuscript, and to Professor Gerald Holton, Harvard University, under whose guidance and direction this study was first conceived and carried out.

*ANDREW WARWICK**

CAMBRIDGE MATHEMATICS AND CAVENDISH PHYSICS: CUNNINGHAM, CAMPBELL AND EINSTEIN'S RELATIVITY 1905–1911

PART I: THE USES OF THEORY

'To give a new concept' can only mean to introduce a new employment of a concept, a new practice.

Ludwig Wittgenstein[1]

1. Introduction

THE THEORY of relativity is widely regarded today as one of the greatest intellectual products of the twentieth century. It is fundamental to the practice of modern physics and has revolutionized our understanding of the concepts of space and time. As an intellectual achievement, its origin and development both warrant, and have received, close examination by historians of physics.[2] Yet it is not simply the monumental *status* of relativity theory that has generated the extensive literature devoted to its early international reception.[3]

*History of Science and Technology Group, Sherfield Building, Imperial College, London SW7 2AZ, U.K.

Received 28 *September* 1991; *in final form* 7 *April* 1992.

[1]L. Wittgenstein, *Remarks on the Foundations of Mathematics*, eds G. H. Wright, R. Rhees and G. E. M. Anscome, translated by G. E. M. Anscome (Oxford: Blackwell, 1967), p. 195e.

[2]The literature on the early history of relativity is massive. For a useful overview and bibliography see D. Cassidy, 'Understanding the History of Special Relativity', *Historical Studies in the Physical and Biological Sciences* **16** (1986), 177–195.

[3]For studies of the international reception of relativity theory see M. Biezunski, 'La diffusion de la théorie de la relativité en France' (unpublished Ph.D. Dissertation, University of Paris, 1981); C. Cattani and M. De Maria, 'Max Abraham and the Reception of Relativity in Italy', in D. Howard and J. Stachel (eds), *Einstein Studies, Volume 1: Einstein and the History of General Relativity* (Boston: Birkhäuser, 1989), pp. 160–174; J. Crelinsten, 'Reception of Einstein's General Relativity Among American Astronomers, 1910–1930' (unpublished Dissertation, University of Montreal, 1982); T. F. Glick (ed.), *The Comparative Reception of Relativity* (Dordrecht: Reidel, 1987), contains ten national case studies; S. Goldberg, 'Early Response to Einstein's Special Theory of Relativity, 1905–1912: A Case Study in National Differences' (unpublished Ph.D. dissertation, Harvard University, 1969); J. Goodstein, 'The Italian Mathematicians of Relativity', *Centaurus* **26** (1983), 241–261; L. Pyenson, 'The Göttingen Reception of Einstein's General Theory of Relativity', (unpublished Ph.D. Dissertation, Johns Hopkins University, 1973). On the British reception of relativity theory see S. Goldberg, 'In Defense of Ether: The British Response to Einstein's Special Theory of Relativity, 1905–1911', *Historical Studies in the Physical Sciences* **2** (1970), 89–124; J. M. Sanchez-Ron, 'The Reception of Special Relativity in Great Britain', in Glick (ed.), *ibid.*, pp. 27–58; A. C. Warwick, 'Einstein's Theory of Relativity and British Physics in the Early Twentieth Century', *Ideas and Production* 7 (1987), 82–95; A. C. Warwick, 'The

Of equal importance to these reception studies is the fact that the founding principles of the so-called *special* theory of relativity can be traced to a single paper, 'On the Electrodynamics of Moving Bodies', published by a single author, Albert Einstein, in September of 1905.[4] The identification of Einstein's paper as the unique harbinger of a new, well-defined, and radical physical theory, has prompted several historians to look at early responses to the theory in national context. The range of these responses, it has been argued, provides a useful means of comparing national styles of physics during the first decade of the twentieth century.[5]

In this essay I discuss the role of early commentaries on Einstein's work in the assessment of different styles of physics. Taking Cambridge University as my example, I shall develop three lines of argument. First, I shall argue that although important innovations in physics can be associated with Einstein's first paper on relativity, it is misleading to view these innovations as the *cause* of early international responses to his work. Rather than viewing commentaries on Einstein's paper as straightforward *responses* to the theory of relativity, I shall attempt to explain *why* some physicists in Cambridge identified Einstein's work as relevant to their own research. Second, since Cambridge commentators attributed a range of *different* meanings to Einstein's work, I shall argue that (at least in the British case) the nation state is not the most appropriate unit by which to characterize styles of physics. Third, I shall argue that the different uses to which Einstein's work was put in Cambridge can be explained by relating them to different schools of institutionalized physics within the University.

The essay is divided into two parts.[6] In this first part I develop the methodological implications of points one and two above, and illustrate my remarks with reference to the work of the Cambridge mathematical physicist Ebenezer Cunningham. Cunningham is widely remembered as the man who introduced relativity theory into Britain and acted as its chief spokesman prior to World War I. I explain Cunningham's interest in Einstein's work in terms of his training in the Cambridge Mathematical Tripos and conclude that his work

Electrodynamics of Moving Bodies and the Principle of Relativity in British Physics 1894–1919' (unpublished Ph.D. Dissertation, Cambridge University, 1989).

[4]A. Einstein, 'On the Electrodynamics of Moving Bodies', in J. Stachel (ed.), *The Collected Papers of Albert Einstein* (Princeton: Princeton University Press, 1989), vol. 2, pp. 140–171, English translation by A. Beck and P. Havas. Einstein's unique authorship of the theory was first emphasized by Gerald Holton; see 'On the Origins of the Special Theory of Relativity' in his *Thematic Origins of Scientific Thought: Kepler to Einstein* (Cambridge Mass.: Harvard University Press, 1973), p. 165.

[5]See Goldberg 'In Defense of Ether...', *op. cit.*, note 3, p. 90; and Glick, 'Cultural Issues in the Reception of Relativity', in his (ed.) *The Comparative Reception of Relativity*, *op. cit.*, note 3, pp. 381–400, p. 381.

[6]The second part will appear in the next issue of this journal.

represented a direct contribution to Cambridge electrodynamics and not to the development of relativity theory.

In the second part of the essay I add the important comparative dimension to my study by discussing the uses that were made of Einstein's work by physicists at the Cavendish Laboratory in Cambridge. In this second part I focus primarily, though not exclusively, on the work of the experimentalist Norman Campbell, and show how his reading of Einstein's work reflected both his training in the Cavendish Laboratory (as opposed to the Mathematical Tripos) and his antipathy to the kind of physics practiced by Cunningham and other Cambridge mathematicians. When treated comparatively, these case studies reveal an important division that arose in late-Victorian and Edwardian Cambridge between traditional mathematical physics and the emergent experimental tradition of the Cavendish Laboratory.

2. A Cambridge Response to Relativity?

I noted in the introduction that it is misleading to depict the earliest commentaries on Einstein's 1905 paper as responses to a *theory of relativity*. In this section I shall argue that by viewing these commentaries as active re-interpretations of his text, rather than as responses to a commonly perceived theory of relativity, the text becomes much more powerful as a tool for the comparative study of different traditions in physics.[7] As we shall see, however, the traditions revealed by my analysis do not represent 'national' styles in physics, but point to different networks of collaboration and competition that are neither typical of, nor (necessarily) bounded by, the nation state.

Historian Stanley Goldberg, who undertook the first study of the early international reception of relativity, began by noting the three features of Einstein's 'innovative theory' that set it apart from other contemporary studies of electrodynamics.[8] According to Goldberg, the three features were: first, that the theory was derived from two postulates; second, that it was a *kinematical* rather than *dynamical* theory; and, third, that it was not dependent upon the

[7]I am assuming here (as have all other historians working on the British reception of relativity) that Einstein's 1905 text (*op. cit.*, note 4) was the one known to British readers before 1911. It seems very likely, however, that some British readers (especially Campbell and Searle) actually used the survey article on relativity written by Einstein at the end of 1907; see 'On the Relativity Principle and the Conclusions Drawn From It', in *The Collected Papers of Albert Einstein*, *op. cit.*, note 4, vol. 2, pp. 252–311. Where evidence exists that the 1907 text was being employed I shall draw attention to the fact.

[8]Goldberg followed Holton in identifying the innovative features unique to Einstein's work; see Goldberg 'In Defense of Ether...', *op. cit.*, note 3, p. 91.

existence of an electromagnetic ether.[9] It is, for Goldberg, these features that establish Einstein as the unique author of the special theory of relativity.[10] Of most relevance to our present concerns, however, is that it is precisely the *novelty* of these features and the *clarity* with which they were expressed by Einstein, that has made special relativity such a seemingly appropriate topic for reception studies. Those who read Einstein's paper in the *Annalen der Physik* — then the leading German journal of theoretical physics — would surely have noticed the novel content and profound implications of the new theory. On Goldberg's reading, Einstein's paper seems almost calculated to raise a response.

The reception accorded to Einstein's work in Cambridge, however, belies this expectation in three ways. First, very few physicists in Cambridge saw fit to publish *any* response to Einstein's new theory. We can note right away, therefore, that those who *were* prompted to respond, identify themselves not as typical of their compatriots, but rather as unusual. Second, the meaning ascribed to Einstein's work by different physicists, even within one university, differed dramatically. It follows that the nation state is not, in this case, going to be an appropriate unit by which to characterize communities of physicists. Third, of the tiny handful of men in Cambridge who did comment on Einstein's work, most failed to notice *all three* of the allegedly novel features identified by Goldberg. Moreover, this apparent oversight cannot be ascribed to professional incompetence or to cursory reading of the paper. As we shall see, physicists in Cambridge *attributed* meaning to Einstein's work that accorded with their *own* research interests.

What emerges from my remarks so far, is that despite the novel features of Einstein's 1905 paper, there is little meaningful sense in which early Cambridge commentaries can be construed as responses to a 'theory of relativity'. Most Cambridge physicists simply did not recognize the innovative features which, according to Goldberg, constituted the new theory. Furthermore, the term 'response' is itself rather misleading because it seems to attribute Einstein's work with *active* capacity to command the attention of the reader. I stated above that it was not simply the high status of relativity theory in modern physics that prompted the extensive literature on its early reception, but that status *has* shaped the historiography of early twentieth-century physics in a way that impinges directly upon reception studies. Historians have projected the modern status of relativity theory back onto Einstein's work of 1905, so

[9] *Ibid.*, p. 91. I agree with Goldberg that there were innovative features in the sense that Einstein would probably have identified them himself as crucial in 1905.

[10] Einstein himself neither presented, nor saw, his 1905 paper as embodying a revolutionary new theory. He adopted the term 'relativitätstheorie' rather reluctantly, only after it had been used by others; see 'Einstein on the Theory of Relativity', in J. Stachel (ed.), *The Collected Papers of Albert Einstein*, *op. cit.*, note 4, pp. 253–274, p. 254; and L. Pyenson, 'The Relativity Revolution in Germany', in Glick (ed.), *The Comparative Reception of Relativity*, *op. cit.*, note 3, pp. 59–111, p. 59.

that its importance to the physics of the first decade of the twentieth-century has become self-evident.[11] Within this *relativity-centred* historiography, it has become unnecessary to provide a historical *explanation* of why any of Einstein's contemporaries chose to comment on his early publications. Despite the extreme paucity of such comments, their existence is treated as unexceptional.[12]

This failure to *account* for interest in Einstein's work further highlights the problematic, even paradoxical, nature of the notion of an international response to relativity theory: it is tacitly assumed in most reception studies that a few physicists somehow grasped (and *responded* to) the profound nature of Einstein's work, even though they clearly failed to recognize any of the features by which it would come to count as profound.[13] I shall argue that the importance of Einstein's work was not *self-evident* to any of his readers in Cambridge.[14] Indeed, we shall see in the second part of this essay that on the only occasion that Einstein actually sent a copy of a paper on relativity to a British physicist, G. F. C. Searle, the response was one of confusion and bafflement.

Rather than taking it for granted that physicists would automatically and self-evidently have been interested in Einstein's work, I shall offer a more symmetrical account by seeking to *explain* why a tiny handful of Cambridge commentators were prompted to make the remarks that they did. Put slightly differently, I shall claim that these commentators should not be regarded as mere passive respondents to the appearance of a new and novel theory, but as working physicists who, for whatever reason, *actively* identified the work of a then little-known physicist as relevant to their own research. The importance of this seemingly small shift of emphasis is considerable. First, by abandoning the relativity-centred historiography, we need no longer define the work of those who commentated on Einstein's paper in terms of relativity theory itself.

[11]Joseph Illy goes as far as to suggest that recent historical reconstructions have turned the history of physics into the history of relativity; see 'Revolutions in a Revolution', *Studies in History and Philosophy of Science* 12 (1981), 175–210, 206.

[12]Some notable attempts have been made to explain Planck's early interest in Einstein's work of relativity; see S. Goldberg, 'Max Planck's Philosophy of Nature and his Elaboration of the Special Theory of Relativity', *Historical Studies in the Physical Sciences* 7 (1976), 125–160; J. Heilbron, *The Dilemmas of an Upright Man: Max Planck as Spokesman for German Science* (Berkeley: University of California Press, 1986), pp. 28–32; L. Pyenson, 'Physical Sense in Relativity: Max Planck Edits the Annalen der Physik, 1906–1918', in his *The Young Einstein: The Advent of Relativity* (Bristol and Boston: Adam Hilger, 1985), pp. 194–214.

[13]This problem is evident, for example, in Sanchez-Ron's account of Cunningham's work. He claims that, in 1907, Cunningham had 'knowledge of the existence of' Einstein's 'special theory of relativity' but no 'knowledge of the meaning of' Einstein's theory; see 'The Reception of Special Relativity in Great Britain', *op. cit.*, note 3, p. 40.

[14]The initial invisibility of Einstein's 1905 paper contrasts sharply with the wave of interest prompted by his publication of the general theory of relativity in 1915. Einstein's fame as the author of the (by 1915) well-known special theory of relativity, together with the novelty of a radically new theory of gravitation, led to its rapid recognition and unproblematic attribution to Einstein; see Warwick, 'The Electrodynamics of Moving Bodies . . .', *op. cit.*, note 3, ch. 7.

The burden of historical explanation is thereby shifted from accounting for the failure of most commentators to understand what Einstein really meant, to accounting for what it was about the research interests of the commentators themselves that made Einstein's work seem relevant. As this last remark implies, I shall not attribute a unique or essential meaning to Einstein's text, and will not, therefore, speak of Cambridge commentators as having 'misread' or 'misunderstood' the text.[15]

Second, since the *meaning* of Einstein's text clearly changed according to the tradition within which it was being incorporated, the stability of a particular interpretation must originate and reside beyond the text itself. I shall argue that individual interpretations derive from the locally taken-for-granted research practices within which an interpretation holds currency. The local traditions made up from such practices are usually evoked only insofar as they are perceived to act as obstacles to the adoption of the essential 'Einsteinian' interpretation of the text.[16] I shall argue that the reconstruction of individual interpretations (including Einstein's own interpretation) must be couched in terms of these local practices and, indeed, is historically meaningless without them. An important implication of this argument is that the successful transmission of a new physical theory cannot be achieved simply by the distribution of a text that allegedly embodies the theory. The taken-for-granted practices employed by the author himself must also be transmitted if the meaning is to be preserved. In the case studies that follow we shall see that very little transfer of these skills took place and that neither Cunningham nor Campbell became a successful practitioner of relativity theory. The processes by which such practices *are* actually transmitted remain to be investigated by historians of physics.

Third, by looking comparatively at the *different* research traditions within which Einstein's work was made meaningful by physicists from the same country, institution or other chosen unit of analysis, we can begin to disaggregate communities that were previously assumed to be homogeneous. Cambridge is an excellent case in point. The blatant differences between the interpretations of Einstein's work produced by Cunningham and Campbell have been noted in previous studies of the British reception of relativity, but

[15]For an excellent discussion of how interpretive communities *produce* the meaning of texts (in the case of literary criticism) see S. Fish, *Is There a Text in This Class?: The Authority of Interpretive Communities* (Cambridge, Mass.: Harvard University Press, 1980), especially pp. 338–355. Fish argues persuasively that there is no esential meaning in a text. It is, rather, the discipline of literary criticism that sanctions the range of interpretive strategies that are to be considered legitimate at any given time.

[16]Goldberg, for example, argues that relativity theory could not be accepted in Britain 'until it could be made consonant with the existing British ideas about the ether'; see S. Goldberg, 'In Defense of Ether . . .', *op. cit.*, note 3, p. 113.

left completely unexplained.[17] However, when one begins to investigate the research projects within which Einstein's work was being exploited by these men, it quickly becomes apparent that they belonged to quite different (actually antagonistic) communities with different disciplinary allegiances. In failing to follow up these differences of interpretation, the historian therefore neglects an important dimension of comparative reception studies (the identification of different traditions where only one was formerly believed to exist) and leaves the interpretations themselves unexplained.

3. Local Practice and Theoretical Technology

The switch from relativity-centred to more local accounts of theoretical work in physics is not, in itself, sufficient to open up the comparative reception of relativity to detailed historical analysis. This localist turn also requires new ways of describing research practices in mathematical physics. The work done by mathematical physicists is traditionally characterized in terms of abstract theories whose essence can be stated in purely propositional or conceptual terms.[18] This characterization of theoretical work is sometimes unhelpful to the historian, however, because it captures neither the real-time experience of learning and developing a particular theoretical approach, nor the enormous skill required to apply that approach to practical problems. In short, the history of idealized theories makes no place for the local cultural resources that generate and sustain individual theoretical enterprises. It is these missing components that need be restored if the uses to which Einstein's work was put in Cambridge (or anywhere else) are to be explained in local, historical, terms.

A promising strategy for reinstating the cultural-historical component of theoretical work is provided by revisionist studies in the history of experiment. Historians of science have recently fashioned new approaches to the history of the experimental sciences by drawing upon micro-sociological studies of contemporary laboratory practice.[19] These sociological studies have stressed that experimenters are not merely passive observers of a pre-determined natural world because the techniques and instruments that they employ in producing natural phenomena derive from a 'man-made environment of skills

[17]See Sanchez-Ron, 'The Reception of Special Relativity in Great Britain', *op. cit.*, note 3, pp. 37–41; and S. Goldberg, 'In Defense of Ether', *op. cit.*, note 3, p. 124; Goldberg states that he is unable to understand how Campbell 'was able to escape the prevailing ether orientation of British physics'.

[18]For an interesting critique of the notion of 'abstract theory', see B. Latour, *Science in Action: How to Follow Scientists and Engineers through Society* (Milton Keynes: Open University Press, 1987), pp. 332–347.

[19]For a useful discussion and select bibliography of this material, see D. Gooding, T. Pinch and S. Schaffer (eds), *The Uses of Experiment* (Cambridge: Cambridge University Press, 1989), pp. 1–27, pp. 461–467.

and material artifacts'.[20] According to these studies, the products of experiment can themselves be considered as cultural artifacts, whose significance is determined by the communal and consensual nature of the scientific enterprise. From the historian's perspective, it is the identification of the *cultural contingency* of experiment that renders experimental knowledge accessible to broader social-historical analysis. If experiments are derivative of the instruments and skilled techniques possessed by a particular culture at a particular time, then knowledge claims based upon those experiments can be related to broader issues within that culture.[21]

But although this new history has sought culture-specific explanations of *experimental* practice, it has had much less to say about the culture-specific nature of *theoretical* practice.[22] Where experimental practices in physics are now widely viewed as locally produced and communally validated, theorizing in physics continues to be treated, largely by default, as a transcendent and solitary activity which cannot usefully be analysed in terms of local cultural resources.[23] This asymmetry between historical accounts of experimental and theoretical work is, however, both undesirable and unnecessary. The same micro-sociological studies that inspired the historical investigation of experimental practice, also provide a framework for the historicization of *theoretical* practice.

Bloor and Collins, for example, have both developed models of theoretical practice that characterize theoreticians as skilled artisans rather than as contemplative philosophers.[24] According to these *enculturational* models, scientists learn their trade not by being taught the logical application of strict theoretical principles (the *algorithmical* model of learning), but by being taught to tackle a range of problems by reference to a number of canonical solutions. It is as part of this process of enculturation that physicists learn how to 'go on' when confronted by new and unfamiliar problems.[25] Likewise at the research

[20]*Ibid.*, p. 10.

[21]For excellent studies of this kind, see B. Barnes and S. Shapin (eds), *Natural Order: Historical Studies of Scientific Culture* (Beverly Hills, Calif.: Sage, 1979); and S. Shapin and S. Schaffer, *Leviathan and the Air Pump: Hobbes, Boyle, and the Experimental Life* (Princeton: Princeton University Press, 1985).

[22]I do not mean to imply here that experimenters do not use theory in their work, merely that I am concerned with specialist theoreticians who have little or no direct contact with experimental work.

[23]For an excellent discussion on the pervasive notion in western culture that the most authentic intellectual agents are the most solitary, see S. Shapin, '"The Mind Is Its Own Place": Science and Solitude in Seventeenth-Century England', *Science in Context* 4 (1990), 191–218.

[24]See D. Bloor, *Wittgenstein: A Social Theory of Knowledge* (London: Macmillan, 1983); and H. M. Collins, *Changing Order: Replication and Induction in Scientific Practice* (London: Sage, 1985).

[25]Anyone who has been a student of physics will appreciate that merely being taught the basic principles of a theory does not allow one to become a competent practitioner of that theory. The student must also be shown examples of how the theory is used to set up and solve typical (canonical) problems so that he or she knows how to proceed when faced with a new problem. Indeed, the method of going almost directly to worked examples when learning physics is extremely effective (as evidenced by the lasting popularity of *Schaum's Outline Series* for example).

level, the enculturational model of science focuses upon the taken-for-granted, or tacit, nature of much of the practice of theoretical work.[26] The skills that are common to *all* members of a community will seldom be referred to explicitly, yet it is precisely this pool of common expertise that lends currency to physical theories. The success of a theoretical school will therefore depend, in large measure, upon its ability to build and maintain a consensus of action amongst its members where, to quote Collins, 'the locus of knowledge is not the written word or symbol but the community of expert practitioners'.[27]

Recognition of what might be called the skill- or practice-ladenness of theory has two important implications for the historian of mathematical physics.[28] First, in seeking to understand mathematical physics as skilled work, we must differentiate between the idealized conceptual schema of a general theory, and the piecemeal steps actually followed by physicists in solving particular problems. It is only by common acknowledgement of the legitimacy of such steps that other members of the community will recognize the proposed solutions as correct.[29] I shall employ the term *theoretical technology* to describe pieces of theoretical work that are not constitutive of a general theory, but which are used to solve particular problems and which are taken for granted by members of a local community.[30] This term will provide a very useful alternative to the word 'theory' on occasions when I wish to emphasize the local origins and specificity of application of individual pieces of theoretical work.

The second implication of the practice-ladenness of theory is that the taken-for-granted status of individual theoretical technologies will be culture-specific and, therefore, explicable in terms of local cultural resources. I began this essay

[26]On the 'tacit' nature of experimental work see Collins, *Changing Order*, *op. cit.*, note 24, p. 56. Fish also notes the tacit nature of the rules governing what is to count as an acceptable interpretive framework in the discipline of literary theory. 'Nowhere is this set of acceptable ways written down', he writes, 'but it is a part of everyone's knowledge of what it means to be operating within the literary institution as it is now constituted'; see *Is There a Text in This Class?*, *op. cit.*, note 15, pp. 342–343.

[27]Collins, *Changing Order*, *op. cit.*, note 24, p. 159.

[28]The terms 'skill-ladenness' and 'practice-ladenness' of observation were introduced (by analogy with the philosophers' theory-ladenness) by Gooding and Pinch respectively to capture the human agency in the production of natural phenomena; see D. Gooding, 'Empiricism in Practice: Teleology, Economy, and Observation in Faraday's Physics', *Isis* **73** (1982), 46–67; and T. Pinch, 'Towards an Analysis of Scientific Observation: The Externality and Evidential Significance of Observational Reports in Physics', *Social Studies of Science* **15** (1985), 3–36, 14.

[29]For an excellent account of a dispute over the legitimacy of novel mathematical techniques in physics, see B. J. Hunt, 'Rigorous Discipline: Oliver Heaviside Versus the Mathematicians', in P. R. Dear (ed.), *The Literary Structure of Scientific Argument: Historical Studies* (Philadelphia: University of Philadelphia Press, 1991), pp. 72–95.

[30]My use of the term *technology* to describe the taken-for-granted assumptions and calculational procedures employed by theoretical physicists has much in common with the *material*, *literary*, and *social* technologies discussed by Shapin and Schaffer in *Leviathan and the Air Pump*, op. cit., note 21, pp. 25–79. In particular I would like to borrow their emphasis (p. 25) that such technologies are 'knowledge-producing tools'.

by referring to the almost globally acknowledged status of the theory of relativity, but this international consensus of opinion is a *result* (rather than a cause) of the emergence of an international community of physicists during the twentieth century. This community is able to retain a global consensus of action (and opinion) amongst its members because they are trained in the same basic practices in physics, contribute to (and read) the same international journals, freely exchange personnel between major international centres, hold regular international conferences and conventions, and share major experimental resources.[31]

Prior to the emergence of this global network, however, local communities of physicists differed very markedly in the research projects and practices that they took for granted. These differences are important to the historian because they provide a range of case studies in how local communities of theoreticians are structured and how they rely upon broader resources for their existence. Just as the products of experimental work can be seen as cultural artifacts of the instruments and skills from which they are constituted, so the products of theoretical work can be viewed as the cultural artifacts of the theoretical practices learned and articulated by theoreticians. In the study that follows I shall argue that the mathematical techniques employed by Cunningham in his most original research derived directly from his training in the Cambridge Mathematical Tripos — a training that was unique to Cambridge during this period. As we shall see, Cunningham certainly understood his work as a contribution to a more general theory (the Electronic Theory of Matter) and took care not to contradict any of the basic principles of that theory, but neither the detailed structure of his argument nor the use that he made of Einstein's work can be explained without reference to the individual research tools (the theoretical technology) that he and his colleagues took for granted.

4. Cunningham's Training in Cambridge

Cunningham's research in electrodynamics has attracted little informed comment from historians of science. It has been variously characterized as an *anticipation* of relativity theory, a direct *contribution* to relativity theory and, most recently, as an attempt to *modify* relativity theory in order to prevent it

[31]Latour has given an account of how *experimental* work carried out in the 'intensive-care' environment of the laboratory can be made to count in the harsh environment of the outside world; see B. Latour, 'Give Me a Laboratory and I will Raise the World', in K. Knorr-Cetina and M. Mulkay (eds), *Science Observed: Perspectives on the Social Study of Science* (London and Beverly Hills: Sage, 1983), pp. 141–170; and *Science in Action*, *op. cit.*, note 18, pp. 247–257. See also Collins, *Changing Order*, *op. cit.*, note 24, pp. 51–78. Similar studies have not been made of the successful transmission of theoretical work.

from undermining the concept of the electromagnetic ether.[32] Notice that in each case his work is defined not in terms of Cambridge mathematical physics, but in terms of the emergent theory of relativity to which his work is seen, retrospectively, as a response. By abandoning this relativity-centred approach and paying careful attention to the research in which Cunningham was engaged when he began to exploit Einstein's paper, I shall argue that his work was a direct contribution to Cambridge electrodynamics.

Cunningham went up to Cambridge in 1899 having won a major scholarship to read mathematics at St John's College.[33] At St John's he was taught electromagnetic theory by Joseph Larmor, and was coached for the Mathematical Tripos by one of the most famous Cambridge mathematics coaches, R. R. Webb.[34] Cambridge was the most important centre for the study of mathematics and mathematical physics in Victorian Britain, and being placed highly in the annual Tripos examinations was a mark of considerable intellectual distinction. The candidates who scored the highest marks were known as 'wranglers', their names being published in order of merit. The student who scored the highest marks was known as the 'Senior Wrangler', the student immediately below him 'Second Wrangler', and so on. Preparation for those who hoped to become high wranglers was intense, their studies being directed not by the professors or lecturers of the University — with whom they had little contact — but by a small number of private tutors, or 'coaches'. The coaching system was unique to Cambridge, and looking back on the Tripos during the nineteenth century W. W. Rouse Ball, himself Second Wrangler in 1874, commented upon its importance as follows:

> A remarkable feature of this period was the fact that until the closing years of the century, nearly every student read with a private tutor, and that two or three private tutors obtained in succession almost a monopoly of the direction of studies of advanced honours students. Thus William Hopkins ... had among his pupils 175 wranglers of whom 17 were senior. ... So again Edward John Routh ... had

[32]Goldberg, for example, portrays Cunningham as an eclectic who tried to bring about a reconciliation between the 'ether-denying' theory of relativity and 'pure-ether' mechanics; see 'In Defense of Ether', *op. cit.*, note 3, pp. 113–117. McCrea, on the other hand, argues that Cunningham's early work shows that the 'body' of physicists (including the British), had 'already arrived at all the essentials of the theory of special relativity without reliance upon Einstein'; see W. McCrea, 'Ebenezer Cunningham' (Obituary Notice), *Bulletin of the London Mathematical Society* **10** (1978), 116–126, 124.

[33]For the biographical details of Cunningham's life see E. Cunningham, *Ebenezer* (unpublished autobiographical typescript dictated by Cunningham in 1969, copy available in the library of St. John's College, Cambridge). See also McCrea, 'Ebenezer Cunningham', *op. cit.*, note 32 which contains a list of Cunningham's publications.

[34]The name 'Mathematical Tripos' refers to the Senate House Examination in mathematics at Cambridge University, the title being officially adopted in the early nineteenth century. For a brief history of the Tripos, see W. W. Rouse Ball, 'The Cambridge School of Mathematics', *The Mathematical Gazette* **6** (1912), 311–323.

between 600 and 700 pupils, most of whom became wranglers, 27 being senior in the Tripos and 41 Smith's prizemen.[35]

Under Hopkins and Routh alone, virtually all of the best known representatives of Cambridge mathematical physics — known during the late nineteenth and early twentieth centuries as the 'Cambridge School' — were trained.[36] Rouse Ball also gave an interesting account of the techniques that were employed (in this case by Routh) in attempting to ensure the success of their students in the Tripos examination. Routh apparently gave:

> ... catechetical lectures three times a week to classes of eight or ten men of approximately equal knowledge and ability. The work to be done between two lectures was heavy, and included the solution of some eight or nine fairly hard examples on the subject of the lectures. Examination papers were also constantly set (bookwork and riders), while there was a weekly paper of problems set to all pupils alike.... The effectiveness of teaching of this kind was dependent on intimate personal intercourse, and the importance of this cannot be overrated.[37]

Apart from the obvious success of this system in training students to score highly in the Tripos examination, it also led to a remarkable uniformity of technical achievement by the most able students educated in mathematics at Cambridge. The expertise of this small group of successful tutors resided in their mastery of a series of mathematical techniques that were employed in the solution of canonical Tripos problems, and which were passed on to successive generations of wranglers during the 'catechetical lectures' and by 'intimate personal intercourse'.[38] The employment of this method of teaching goes some considerable way towards explaining the existence of a well-defined Cambridge School of mathematics during the second half of the nineteenth century.[39] It clearly provided a highly conservative skill-base for the retention and transmission of the theoretical technology employed by Cambridge wranglers.

Under Webb's guidance Cunningham mastered the well-worn skills of the Cambridge applied mathematician which, as we shall see, he later put to use in

[35]*Ibid.*, p. 320.

[36]On the importance of the Mathematical Tripos in nineteenth-century British mathematical physics, see P. M. Harman (ed.), *Wranglers and Physicists* (Manchester: Manchester University Press, 1985), 1–11.

[37]Rouse Ball, 'The Cambridge School of Mathematics', *op. cit.*, note 34, p. 321.

[38]There were many critics of this system of training, especially amongst those who believed that experimental physics should be taught; see D. B. Wilson, 'Experimentalists Among the Mathematicians: Physics in the Cambridge Natural Sciences Tripos, 1851–1900', *Historical Studies in the Physical Sciences* **12** (1982), 325–371.

[39]Goldberg also draws attention to the importance of the Mathematical Tripos in the training of wranglers; see 'In Defense of Ether', *op. cit.*, note 3, p. 124. He claims, however, that this explains why they 'acted with such uniformity when confronted with the theory of relativity'. Since Cunningham is the only wrangler whose 'response' to relativity Goldberg actually discusses, I assume that the above remark is intended to explain why no other wranglers responded at all. This is an excellent example of relativity-centred historiography where the local tradition is evoked simply to explain why the 'theory of relativity' was ignored or misunderstood.

his research in electrodynamics.[40] Cunningham was Senior Wrangler in 1902 and took Part II of the Mathematical Tripos the following year. He then began research in mathematics for which he was awarded a Smith's prize in 1904.[41] Alongside his research in mathematics, Cunningham also found time to read Joseph Larmor's recently published book, *Aether and Matter*, in which Larmor gave a comprehensive account of his *Electronic Theory of Matter* (hereafter ETM).[42]

According to the ETM, all matter was composed solely of positive and negative electrons moving in a sea of ether. The forces acting between the electrons were purely electromagnetic, so that the motions of the electrons were dictated by Maxwell's electromagnetic field equations. The ETM predicted that electronically constructed matter would contract as it moved through the ether with velocity v by a factor of α (where $\alpha = \sqrt{(1 - v^2/c^2)}$), exactly the amount postulated by George Fitzgerald in 1889 to account for the null result of the Michelson–Morley ether drift experiment.[43] The correspondence between the prediction of his theory and FitzGerald's hypothesis led Larmor to argue that, far from being problematic, the null result now furnished firm experimental evidence that the ETM was correct.[44]

The aspect of the ETM that quickly caught Cunningham's attention was the use that Larmor had made of what would later be called the Lorentz transformations. As these transformations are central to the discussion that follows, I shall give a brief account of the way in which they were employed by Larmor. Larmor believed electric and magnetic fields to be the macroscopic manifestations of micro-physical processes taking place in an underlying dynamical ether. In order to find out how electromagnetic fields related to the physical state of the ether, however, it would be necessary to view electromagnetic processes from the stationary-ether frame of reference. But since, in

[40]R. R. Webb (1850–1936) dominated the training of wranglers (following Routh's retirement in 1888) from 1889 until the turn of the century: see *The Eagle* (St John's College, Cambridge) **49** (1936), 272–273.

[41]The 'Smith's Prize' was offered each year for the two best essays in mathematics or natural philosophy by students in their first year of research. For the history of the prize and a list of the winners, see J. R. Tanner (ed.), *Historical Register of the University of Cambridge To the Year 1910* (Cambridge, 1910), pp. 299–301.

[42]J. Larmor, *Aether and Matter* (Cambridge: Cambridge University Press, 1900). Cunningham's professional research interests in mathematics during this period mainly concerned the classification and solution of differential equations. Typical examples of his early publications are: 'An Extension of Borel's Exponential Method of Summation of Divergent Series Applied to Linear Differential Equations', *Proceedings of the London Mathematical Society* **3** (1905), 157–169; 'On the Normal Series Satisfying Linear Differential Equations', *Philosophical Transactions of the Royal Society* **A 205** (1905), 1–35; 'On Linear Differential Equations of Rank Unity', *Proceedings of the London Mathematical Society* **4** (1907), 374–383.

[43]On FitzGerald's reasons for proposing the contraction see B. J. Hunt, 'The Origins of the Fitzgerald Contraction', *British Journal for the History of Science* **21** (1988), 67–76.

[44]Larmor's claim is discussed in detail in A. C. Warwick, 'On the Role of the Fitzgerald–Lorentz Contraction Hypothesis in the Development of Joseph Larmor's Electronic Theory of Matter', *Archive for History of Exact Sciences* **43** (1991), 29–91, 56–60.

practice, all measurements of electromagnetic processes were made on the surface of the earth — which Larmor assumed to be moving through the ether with a velocity of several miles per second — he needed to find a method of correlating these measured values with those that would be made in the stationary ether frame of reference.

Larmor found the required correlation by assuming that Maxwell's electromagnetic field equations must hold true (at least to second order) for all observers moving uniformly through the ether. Having made this assumption, he sought transformations that would preserve the form of Maxwell's equations when transforming between moving frames of reference. The transformations that he found, and published in *Aether and Matter*, were those that were subsequently given by Lorentz in 1904 and Einstein in 1905.[45] It should be noted, however, that Larmor did not treat all frames of reference symmetrically. As far as he was concerned, the stationary ether frame of reference was privileged — only in this frame could clocks be properly synchronized. It followed from this definition of synchronization, that moving matter would always be physically contracted in its direction of motion through the ether by a factor of α.[46]

As an expert in the solution of differential equations, Cunningham investigated the transformation technique introduced by Larmor and soon made what was, for him, a very important discovery. When Larmor had first employed the Lorentz transformations in *Aether and Matter* he had noted, erroneously, that they transformed Maxwell's equations *only to second order*.[47] Cunningham quickly spotted that they actually transformed Maxwell's equations exactly, and at once informed Larmor of this important finding. Larmor responded by remarking that he had since noticed this property himself, but that he considered it unimportant because it was not possible to ascertain *experimentally* whether motion through the ether was actually detectable above second order.[48] The following summer (1904) H. A. Lorentz drew attention in print to the exact nature of the transformations — which thereafter bore his name — and incorporated them into his theory of the electrodynamics of moving bodies.[49]

[45] *Ibid.*, 60–69. The same transformations were given by Lorentz in 'Electromagnetic Phenomena in a System Moving with any Velocity Less than that of Light', *Proceedings of the Royal Academy of Science of Amsterdam* 6.II (1903/4), 809–831; and Einstein in 'On the Electrodynamics of Moving Bodies', *op. cit.*, note 4, p. 151.

[46] Larmor's method of synchronization is discussed in Warwick 'On the Role of the Lorentz–FitzGerald Contraction...', *op. cit.*, note 44, p. 65.

[47] *Ibid.*, Appendix D. In this Appendix Larmor's reasons for reaching this erroneous conclusion are discussed in detail.

[48] This incident is recorded by Cunningham, *Ebenezer*, *op. cit.*, note 33, p. 29; and in E. Cunningham, 'On the Electromagnetic Mass of a Moving Electron', *Philosophical Magazine* 14 (1907), 538–547, 539.

[49] Lorentz's article (*op. cit.*, note 45) first appeared in English in July 1904.

Cunningham left Cambridge in October of 1904 to take up a junior lectureship in mathematics at Liverpool University, where he remained for the next three years. Sometime during this period he came across Einstein's now famous paper 'On the Electrodynamics of Moving Bodies' and began the research for which he is now remembered. It was also during this period that he began to use the term 'principle of relativity' but, as we shall shortly see, his understanding of the principle differed markedly from Einstein's understanding.

5. The Principle of Relativity and the ETM

Cunningham first made explicit reference to the principle of relativity in 1907 in a reply to a critique of Lorentz's electron theory by Max Abraham. Lorentz had made it a fundamental principle of his electron theory that moving electrons contracted by a factor of α in their direction of motion through the ether.[50] Abraham, who supported a rigid electron theory, claimed that Lorentz's deformable electron was untenable because it led to an ambiguous definition of the electromagnetic momentum of the electron.[51] The central point of Cunningham's reply to Abraham was that Abraham's own discussion was flawed because he had not made use of the *principle of relativity* in analysing the electron's motion. In re-working Abraham's analysis, Cunningham showed that Lorentz's deformable electron *was* consistent with a unique definition of electromagnetic momentum and, crucially, revealed his own understanding of the principle of relativity.

Cunningham began his paper by explaining that it had been proved *analytically* that the Lorentz transformations transformed Maxwell's equations exactly, provided that the electric and magnetic fields measured by observers in states of relative motion were related in a certain well-defined way.[52] He then defended Lorentz's deformable electron by asserting that:

> . . . Lorentz's hypothesis of the reduction of the dimensions of a body when it moves relatively to an observer is reduced by this geometrical correspondence to the assumption that in the variables associated with the axes moving with it its shape is unaltered — an assumption suggested by the fact that the electromagnetic equations associated with those variables are independent of the motion through the ether, and by the attempt to form a purely electromagnetic theory of nature.[53]

[50]On Lorentz's notion of a contracting electron see A. I. Miller, *Albert Einstein's Special Theory of Relativity: Emergence (1905) and Early Interpretation (1905–1911)* (Reading, Mass.: Addison-Wesley, 1981), pp. 72–73.

[51]*Ibid.*, pp. 77–79.

[52]The relationship between the fields measured in systems in states of relative motion was first given by Larmor; see Warwick, 'On the Role of the Fitzgerald-Lorentz Contraction Hypothesis...', *op. cit.*, note 44, p. 66.

[53]Cunningham, 'On the Electromagnetic Mass of a Moving Electron', *op. cit.*, note 48, p. 540.

Cunningham's justification of Lorentz's deformable electron postulate thus contained two steps. He first claimed that it was unnecessary separately to *postulate* the contraction of moving electrons because the contraction followed as a geometrical consequence of the Lorentz transformations. If these transformations were applied, then an electron would appear spherical in its own rest frame, but contracted by a factor of α when moving with velocity v. This first step in Cunningham's argument appears compatible with Einstein's definition of the principle of relativity, but the second step reveals his differences with Einstein.

The crucial second step in Cunningham's argument was the assertion that the correctness of the Lorentz transformations had to be *justified* on theoretical grounds. They were justified *mathematically* because they (unlike the Galilean transformations) could preserve the form of Maxwell's electromagnetic field equations for all uniformly moving frames of reference; and they were justified *physically* because the contraction of moving matter implied by the transformations was predicted by the ETM. For Cunningham, therefore, the term 'principle of relativity' was merely a shorthand statement of the fact that the Lorentz transformations provided a powerful mathematical tool for dealing with certain problems in the electrodynamics of moving bodies.

Cunningham's position on the status of the Lorentz transformations was thus more akin to Larmor's position than to Einstein's. Cunningham noted, for example, that the 'transformations in question' had been given by Einstein but added that there were, 'in substance', the same as those given by Larmor in *Aether and Matter*.[54] Larmor and Cunningham nevertheless differed crucially over the precise physical status of the transformations: Cunningham was prepared to accept that the *exact* mathematical correlation between the electromagnetic variables of different systems implied by the transformations was a physical fact (so that motion through the ether could never be detected), whereas Larmor insisted on restricting the transformations to the limit of experimental accuracy.[55]

But despite this important difference with Larmor, Cunningham remained a long way from adopting Einstein's interpretation of the principle of relativity, and did not even recognize Einstein's second postulate (the principle of the constancy of the velocity of light). This last point is clarified by Cunningham's comments during a brief dispute that followed the publication of his reply to

[54]*Ibid.*, p. 539.

[55]On Larmor's reasons for retaining this restriction see Warwick 'On the Role of the Lorentz–FitzGerald Contraction...', *op. cit.*, note 44, pp. 68–69. On the differences in attitude towards electrodynamics between wranglers trained in the 1870s (such as Larmor) and those trained around the turn of the century (such as Cunningham), see A. C. Warwick, 'The Sturdy Protestants of Science: Larmor, Trouton and the Earth's Motion through the Ether', in J. Z. Buchwald (ed.), *Table-Top Experiments* (Chicago: University of Chicago Press, 1993, forthcoming).

Abraham. Towards the end of this reply, Cunningham had remarked, in passing, that the principle of relativity that he (Cunningham) had employed was, in essence, identical to a similar principle that had been stated elsewhere by the German physicist Alfred Bucherer.[56] Cunningham claimed, further, that the employment of *any* such principle was bound to lead to the Lorentz transformations (which Bucherer had not used), because it was necessary to *explain* how the velocity of light could be the same for all observers in relative states of motion.[57] This comment prompted a rapid response from Bucherer who had come to no such conclusion. Bucherer replied that he was not aware of any 'known fact of observation' that *required* one to consider the velocity of light to be the same for all observers.[58] Cunningham replied as follows:

> . . . I did not wish to assert that it was required by any known fact of observation but took it to be *involved in the statement of the principle* . . . if the Maxwell equations are assumed to hold when referred, as occasion requires, to various frames of reference moving relatively to one another, the deduction cannot be escaped that the velocity of propagation of a spherical wave will be found to be exactly the same, whatever the frame of reference.[59]

As is well known, Einstein grounded his kinematical arguments upon two *independent* principles (the principle of relativity and the principle of the constancy of the velocity of light), the second of which uncoupled his arguments from electromagnetic theory. Cunningham, by contrast, assumed the principle of relativity to be a statement of the fact that Maxwell's equations transformed exactly under the Lorentz transformations. If one also assumed, as did Cunningham, that Maxwell's equations were equally applicable in every frame of reference, then it followed that the velocity of light must itself appear the same in all frames of reference. Thus Einstein's second *postulate* was, for Cunningham, simply a consequence of the principle of relativity. Indeed, the principle of relativity itself was not really a fundamental *postulate* for Cunningham because, as we have seen, it rested ultimately upon the ETM and the mathematical properties of the Lorentz transformations.

[56]A. H. Bucherer, 'On a New Principle of Relativity in Electromagnetism', *Philosophical Magazine* **13** (1907), 413–420. Bucherer attributes the principle of relativity to Lorentz and does not mention Einstein. This paper appears to have been submitted first to the *Annalen der Physik*, but was rejected by the editor Max Planck. For a discussion of this episode see Pyenson, *The Young Einstein*, *op. cit.*, note 12, pp. 201–202.

[57]Cunningham, 'On the Electromagnetic Mass...', *op. cit.*, note 48, p. 544.

[58]A. H. Bucherer, 'On the Principle of Relativity and on the Electromagnetic Mass of the Electron. A Reply to Mr. Cunningham', *Philosophical Magazine* **15** (1908), 316–318.

[59]E. Cunningham, 'On the Principle of Relativity and on the Electromagnetic Mass of the Electron. A Reply to Dr. A. H. Bucherer', *Philosophical Magazine* **16** (1908), 423–428, 423; my italics.

6. Reading Einstein

Having sketched Cunningham's interpretation of Einstein's principle of relativity, I now return to the problem of why he first identified Einstein's work as relevant to his own research interests. A complete answer to this problem will have to wait until the following section in which I explain how Einstein's work partly prompted Cunningham's own original contribution to electrodynamics. In this section I shall suggest how Cunningham came to the understanding of the principle of relativity described above.

It is not difficult to appreciate why Cunningham, unlike the vast majority of his British colleagues, would at least have taken the trouble to read Einstein's paper, even though it was by a little-known author and written in German. We have seen that, by 1904, Cunningham was committed to the ETM and had discovered for himself that Maxwell's field equations transformed exactly under the Lorentz transformations. Furthermore, the exact nature of the transformations, the significance of which Larmor had denied, had subsequently been exploited by Lorentz in his electromagnetic theory of 1904. Following this last development, Cunningham took a keen interest in all publications on the electrodynamics of moving electrical systems and would certainly have been attracted by a short paper explicitly entitled 'On the Electrodynamics of Moving Bodies'.[60]

There were, I suggest, two closely related aspects of Einstein's paper that would have caught Cunningham's interest. The first of these concerned Einstein's development of the *mathematics* of the transformation technique introduced by Larmor and Lorentz. Larmor, it will be recalled, had set out to explain why second-order electromagnetic effects of the Earth's motion could not be detected experimentally. He solved this problem by finding transformations (the Lorentz transformations) that retained the form of Maxwell's equations up to second order. When Cunningham subsequently found that the transformations given by Larmor actually preserved the form of Maxwell's equations to *all* orders he (unlike Larmor) believed this to imply that no effects of the Earth's motion through the ether could ever be detected experimentally.

For Cunningham, Einstein's paper of 1905 was primarily an insightful discussion of the *mathematical* consequences of ascribing truly equivalent status to the space–time and electromagnetic variables of all uniformly moving reference systems. Einstein's analysis showed, for example, that if the space and time variables of all moving systems were granted equivalent status, then a

[60]The references in Cunningham's publications reveal that he was very familiar with Anglo-German work on electrodynamics. He does not appear to have known of Poincaré's work.

new formula for the addition of velocities was implied.[61] Once the new velocity addition formula was adopted, moreover, the Lorentz transformations formed a mathematical group (successive transformations being replaceable by a single transformation of the same form).[62] Einstein's analysis thus brought mathematical symmetry to the electrodynamics of moving bodies. The observations made by all observers in states of relative motion could now be correlated without reference to any uniquely privileged ether frame of reference.[63] In Cunningham's terminology, Einstein's had shown how to bring complete 'geometrical correspondence' between moving frames of reference simply by following up the *mathematical* implications of Larmor's original transformations.

The second important insight read into Einstein's paper by Cunningham was the demonstration that the mathematical symmetry now inherent in the electrodynamics of moving bodies was actually a physical *consequence* of the ETM. For Cunningham, the (now famous) discussion of kinematics and simultaneity with which Einstein began his paper amounted to the following. If the world were ultimately electromagnetic in structure, the only practical way of *defining* simultaneity between two spatially separated events would be by employing electromagnetic signals; and since, according to the ETM, electromagnetic radiation travelled with the same velocity in all directions for all observers, any observer could legitimately employ Einstein's definition of simultaneity. That this was indeed Cunningham's view is fully confirmed by the following remarks with which he began his own major contribution to electrodynamics in 1910:

> The absence, as far as experiment can detect, of any phenomena arising from the earth's motion relative to the electromagnetic aether has been fully accounted for by Lorentz and Einstein, *provided the hypothesis of electromagnetism as the ultimate basis of matter be accepted*, so that the only available means of estimating distances between two points is the time of propagation of effects between the bodies, such propagation taking place in accordance with the equations of electron theory.[64]

In fact, Cunningham required both of the above insights to make sense of Einstein's paper. As a Cambridge wrangler trained in the early years of the twentieth century, Cunningham would have been very impressed by the mathematical symmetry brought to the electrodynamics of moving bodies by

[61] Einstein, 'On the Electrodynamics of Moving Bodies', *op. cit.*, note 4, pp. 153–156. Cunningham was not the only Cambridge electromagnetic theoretician to read Einstein's paper in this way. G. A. Schott appended a note to his paper 'On the Radiation from Moving Systems of Electrons, and on the Spectrum of Canal Rays', in which he pointed out that Einstein had given the 'correct' expression for velocities measured in moving systems if the Lorentz transformations were adopted; see *Philosophical Magazine* **13** (1907), 657–687, 687.

[62] Einstein, 'On the Electrodynamics of Moving Bodies', *op. cit.*, note 4, p. 156.

[63] *Ibid.*, pp. 165–167.

[64] E. Cunningham, 'The Principle of Relativity in Electrodynamics and an Extension Thereof', *Proceedings of the London Mathematical Society* **8** (1910), 77–98, 77 (my italics).

Einstein's work. But the mathematics would have been of little physical significance for him, were it not for the fact that Maxwell's equations transformed exactly under the Lorentz transformations and the fact that the ETM offered a physical explanation (at least in any given frame of reference) of the contraction of moving matter. For Cunningham, Einstein had developed a very powerful mathematical technique in electromagnetic theory (which could, for convenience, be referred to as the principle of relativity) as a consequence of the ETM. Furthermore, the strategy that Cunningham assumed Einstein to have followed in developing the principle of relativity (following up the mathematical implications of the ETM) suggested to him how he might further develop the principle in a similar manner.

7. The New Theorem of Relativity

During the academic year 1906/7, Cunningham was not primarily concerned with defending Lorentz's electron theory. He had, in fact, begun collaborating with another Cambridge mathematician, Harry Bateman, in an attempt to establish, mathematically, what he would eventually call the 'new theorem of relativity': a theorem that would be valid for electrical systems in states of accelerated as well as states of uniform motion.[65] Bateman's early career was almost identical to Cunningham's. A student of Trinity College, Cambridge, he became Senior Wrangler in 1903 and a Smith's Prizeman in 1904.[66] In the autumn of 1906, Bateman joined Cunningham at Liverpool University where he too had been appointed to a junior lectureship in mathematics. During the following academic year (1906/7) the two mathematicians found that they shared common interests in the solution of differential equations and certain aspects of transformation theory.[67] Through these shared interests, and Cunningham's understanding of Einstein's work, they began to work together to establish a new, more general, principle of relativity. The importance of Einstein's work for Cunningham and Bateman was, as we have seen, that it demonstrated how the principle of relativity could lend a pleasing mathematical symmetry to the problem of the electrodynamics of moving bodies. It

[65]Sanchez-Ron notes, correctly, that Bateman's approach to electrodynamics was geometrical, but claims, wrongly, that Cunningham and Bateman worked independently on the covariance of Maxwell's equations; see 'The Reception of Special Relativity in Great Britain', *op. cit.*, note 3, p. 42.

[66]For biographical details of Bateman and a complete list of publications, see E. Erdélyi, 'Harry Bateman' (Obituary Notice), *Royal Society Obituary Notices* 5 (1948), 591–618. Bateman was coached to the Senior Wranglership by the last of the great Cambridge mathematics coaches, R. A. Herman.

[67]Bateman's early research in mathematics was also concerned with the solution of differential and integral equations; for example, 'The Solution of Partial Differential Equations by Means of Definite Integrals', *Proceedings of the London Mathematical Society* 1 (1904), 451–458; 'The Theory of Integral Equations', *Proceedings of the London Mathematical Society* 4 (1906), 90–115.

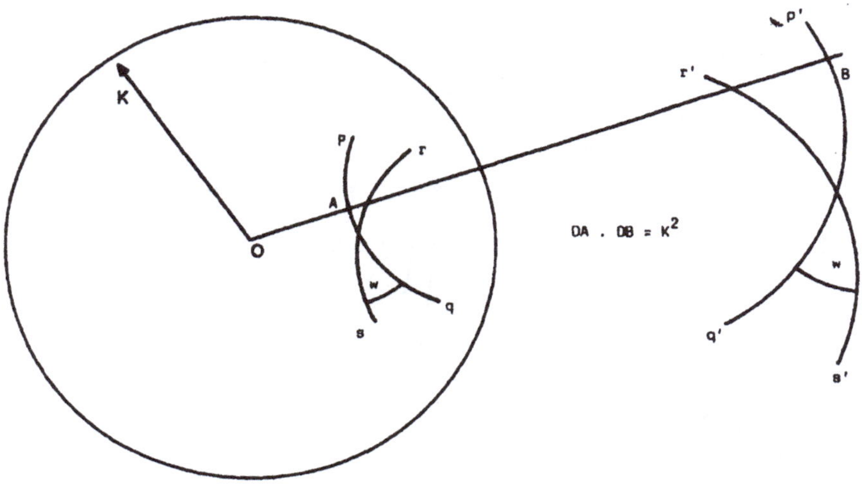

Fig. 1.

suggested, moreover, how a whole new range of problems in electromagnetic theory might be made accessible to mathematical solution by bringing together the principle of relativity and their common expertise in transformation theory.

During 1907, both Cunningham and Bateman became interested in the physical application of a mathematical technique known as *inversion* and its generalization in the form of the *conformal transformation*. Both men had been taught inversion as a geometrical technique that was chiefly of use in electrostatics.[68] As the technique is central to Cunningham's early research in electrodynamics, I shall explain how the technique was employed by Cambridge wranglers. The geometrical interpretation of inversion in two dimensions is illustrated in Fig. 1. Consider a circle of radius K with centre O. For any point A inside the circle, an *inverse* point B can be generated outside of the circle by imposing the condition $OA.OB=K^2$. In this transformation K is known as the *radius of inversion*. If the point A now moves to generate the line pq, the point B will also move generating the *inverse image* of pq which is labelled $p'q'$. A number of geometrical relationships between the inscribed figures and their inverses can be demonstrated, the most important of these for our present purposes being that if two lines inscribed in the circle intersect in an angle ω,

[68]H. R. Hassé (Seventh Wrangler 1905) later recalled that 'to the average student pure geometry at that time [1900–1907] can have meant little more than the use of one or other of the processes of inversion, reciprocation or projection for the solution of problems...'; see 'My Fifty Years of Mathematics', *Mathematical Gazette* **35** (1951/52), 154. On the fundamental status of geometry as the study of space in Cambridge mathematics during the nineteenth century, see J. L. Richards, 'Projective Geometry and Mathematical Progress in Mid-Victorian Britain', *Studies in History and Philosophy of Science* **17** (1986), 297–325.

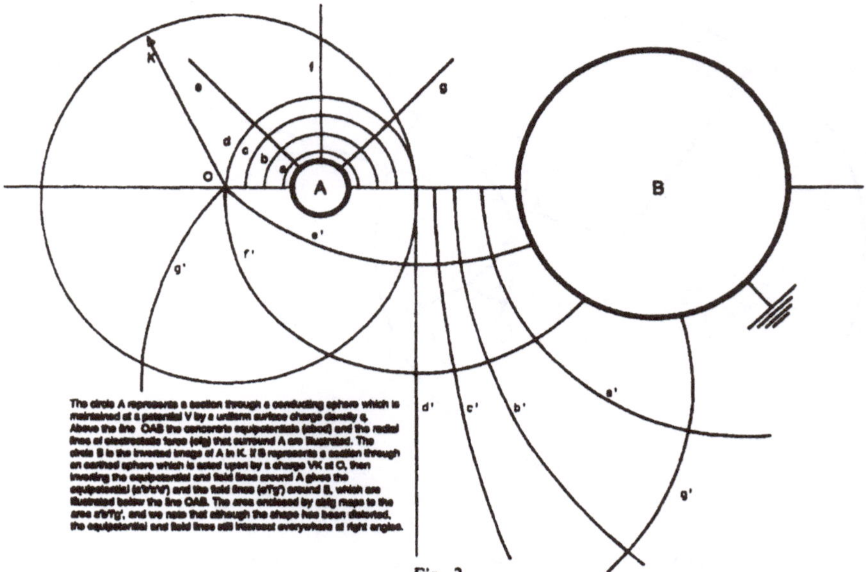

Fig. 2.

then their inverse images will also intersect in angle ω. This property is actually definitive of a much wider group of transformations, known as the *conformal* transformations, of which inversion in a circle may be considered a simple, geometrical, example.

The physical importance of the technique of inversion in electrostatics is illustrated in Fig. 2. If the geometrical structure of a charged conductor and its surrounding field lines is inverted in *K*, a new charged conductor with its surrounding field lines is generated. That the field lines in the inverted image will continue to intersect everywhere at right angles (as they must) is guaranteed by the property, mentioned above, that angles between intersecting lines are preserved under inversion. By inversion, therefore, it is possible to find the form of the electric field surrounding a complicated system of charges and conductors by inverting the known solution to a simpler case. This was, moreover, precisely the kind of physical-mathematical problem that was employed in the Cambridge Mathematical Tripos to introduce students to advanced mathematical techniques.[69]

In February of 1908 Cunningham read a paper before the London Mathematical Society (hereafter LMS) entitled 'Conformal Representation and the Transformation of Laplace's Equation'. Although the paper was never

[69]These particular techniques were first developed systematically in electrostatics by Maxwell in his *Treatise on Electricity and Magnetism* (New York: Dover, reprint of the 3rd edition, 1954), pp. 244–316. For an account by one of Cunningham's contemporaries, see J. H. Jeans, *The Mathematical Theory of Electricity and Magnetism* (Cambridge: Cambridge University Press, (1908), pp. 198–203.

published, it is reasonable to infer that in it Cunningham investigated the possibility of generating solutions to Laplace's equation (a second order partial differential equation) by the transformation of known solutions.[70] In regions where the charge density is zero (outside of a charged sphere, for example), the distribution of electric potential (V) surrounding a charged conductor will always satisfy Laplace's equation:

$$\frac{\partial^2 V}{\partial x^2}+\frac{\partial^2 V}{\partial y^2}+\frac{\partial^2 V}{\partial z^2}=0$$

with appropriate boundary conditions applied. But, as we have just seen, it is possible to move from the solution of one electrostatic problem to another, and hence from one solution of Laplace's equation to another, by the process of inversion. This is the analytical analogue of the geometrical technique described above, and Cunningham was very likely concerned with the generality of the solutions that could be generated by this method.[71]

Bateman was also interested in the physical application of the conformal transformations. In October of 1908 he submitted a paper to the LMS on conformal transformation in a four-dimensional space and its application in geometrical optics.[72] Bateman opened his paper by recounting the history of the application of the method of inversion to problems in electrostatics, explaining that any conformal transformation could be made up out of a series of inversions in suitably chosen spheres. Indeed, he pointed out that the demonstration of this fact had actually been set as a question in the Mathematical Tripos that he had sat in 1903.[73] He then showed that just as new solutions to Laplace's equation could be obtained by the conformal transformation of known solutions (Cunningham's problem), so new solutions of the four-dimensional equations

$$\frac{\partial^2 V}{\partial x^2}+\frac{\partial^2 V}{\partial y^2}+\frac{\partial^2 V}{\partial z^2}-\frac{1}{c^2}\frac{\partial^2 V}{\partial t^2}=0$$

$$\left(\frac{\partial V}{\partial x}\right)+\left(\frac{\partial V}{\partial y}\right)+\left(\frac{\partial V}{\partial z}\right)-\frac{1}{c^2}\left(\frac{\partial V}{\partial t}\right)=0$$

[70]E. Cunningham, 'Conformal Representation and the Transformation of Laplace's Equation', read before the LMS on 13 February 1908; see *Proceedings of the London Mathematical Society* **6** (1908), viii.

[71]Cunningham may not have been aware that a paper on this topic had already been published by the London Mathematical Society (A. R. Forsyth, 'On the Transformations of Coordinates which Lead to New Solutions of Laplace's Equation', *Proceedings of the London Mathematical Society* **29** (1898), 165–206), which perhaps explains why the paper was never published.

[72]H. Bateman, 'The Conformal Transformations of a Space of Four Dimensions and their Applications to Geometrical Optics', *Proceedings of the London Mathematical Society* **7** (1909), 70–89.

[73]*Ibid.*, p. 71 (footnote). See Cambridge Tripos Papers, Mathematics, Part I (1903), Paper 1, Questions 3 and 4.

(which were fundamental to the electromagnetic theory of light) could be obtained by the four-dimensional conformal transformation of known solutions.[74] Finally, Bateman applied his results to some problems in optics and mentioned, in passing, that the fundamental equations of electron theory could also be written in mathematically symmetrical form in four dimensions by employing the new dimension s ($s = ict$).[75] This paper represented Bateman's initial contribution to the collaborative venture that he began with Cunningham during the academic year 1906/7. When combined with Cunningham's interest in the principle of relativity, Bateman's research inspired Cunningham to search for new transformations (other than the Lorentz transformations) that would transform Maxwell's equations exactly.

It was Cunningham's understanding of the principle of relativity that formed his point of departure for the new research project. According to Cunningham, the principle of relativity expressed the fact that the electromagnetic field equations transformed exactly under the Lorentz transformations. The principle was extremely useful in electromagnetic theory because it sometimes enabled one to simplify otherwise awkward calculations in the electrodynamics of moving bodies by transforming from one reference frame to another. This technique could only be applied, however, to systems that were in *uniform* states of relative motion. It could not be applied to accelerated systems.[76] Cunningham now set out to find new transformations that would preserve the form of Maxwell's equations for the case of accelerated motion.

Cunningham's strategy in tackling this problem was to imagine that electromagnetic processes occurred in a four-dimensional space, so that electromagnetic fields that changed with time could be thought of as static geometrical forms, rather like three-dimensional electrostatic fields.[77] Each

[74] *Ibid.*, pp. 72–78.

[75] *Ibid.*, pp. 82–87.

[76] For a technical discussion of why, according to Cambridge electrodynamics, the transformations were valid only for uniformly moving systems, see Warwick, 'On the Role of the FitzGerald-Lorentz Contraction Hypothesis...', *op. cit.*, note 44, pp. 84–88.

[77] Cunningham accredited Minkowski with the observation that in a four-dimensional space (x, y, z, ict) the Lorentz transformations represented a finite rotation of the space around the $y = 0$, $z = 0$ axes, but it must not be thought that Cunningham and Bateman were following Minkowski. Minkowski's 'Das Relativitätsprinzip' lecture of 1907 was not published until 1915 (*Annalen der Physik* **47** (1915), 927–938) and his famous 'Raum und Zeit' did not appear until 1909 (*Physikalische Zeitschrift* **20** (1909), 104–111). Only his 'Die Grundleichungen für die elektromagnetischen Vorgänge in bewegten Körpern' had reached Cambridge (May 1908) when Cunningham and Bateman submitted their papers (*Göttinger Nachrichten* (1908), 53–111). Cunningham and Bateman had, in any case, begun their collaboration two years earlier. Bateman's four-dimensional, integral, representation of the electromagnetic equations had been taken from the work of another Cambridge wrangler (Richard Hargreaves, Fifth Wrangler 1876) also working in Liverpool; see R. Hargreaves, 'Integral Forms and their Connection with Physical Equations', *Transactions of the Cambridge Philosophical Society* **21** (1908), 107–122.

four-dimensional solution could then be inverted in a four-dimensional *hypersphere* of *pseudo-radius* K in order to produce a new solution (see Appendix). Central to Cunningham's paper was the demonstration that Maxwell's equations retained their form under these transformations. This effectively established that if the electromagnetic fields in one four-dimensional frame of reference were solutions of Maxwell's equations, then the inverted images of those fields would also be solutions of Maxwell's equations. Any one of these frames of reference could, therefore, represent a possible physical reality in the sense that an observer working in that frame would find that Maxwell's equations applied to all electromagnetic processes. Since the conformal group of transformations could be built up by repeated inversions (on this point Cunningham acknowledged Bateman's assistance), Cunningham concluded that this must be the most general group of transformations with this property.[78]

In order fully to appreciate the importance of this result, it must be remembered that, for Cunningham, the principle of relativity was valid only insofar as it had been *analytically derived*. Larmor and Lorentz had shown that the Lorentz transformations exactly correlated the fields of electrical systems in states of uniform motion, and that the ETM made these transformations physically plausible. For Cunningham, Einstein had completed this work mathematically, but only for systems in uniform motion. Now Cunningham had shown, analytically, that there were other transformations that could preserve the form of Maxwell's equations between electrical systems in much more complicated states of relative motion. This was the foundation of what Cunningham called the 'new theorem of relativity'.[79]

This whole enterprise was grounded upon a commitment to the ETM. Just as this theory guaranteed the physical plausibility of the Lorentz transformations, so the new transformations could only represent possible physical systems because the world was ultimately electromagnetic in nature. Cunningham pointed out that the kinds of motion of electrical systems represented by the inversion of a simple system were a good deal more complicated than those considered by Einstein. Even in the most simple cases, a stationary system transformed into one that was expanding or contracting radially about a fixed point in a particular way and, as with the electrostatic case, it was often difficult to decide upon the physical problem to which any

[78]Cunningham, 'The Principle of Relativity...', *op. cit.*, note 64, p. 79, pp. 87–89. In discussing the physical interpretation of the four-dimensional representation of electromagnetic processes, J. W. Nicholson remarked, quite explicitly, that it was a 'fiction' which 'in the mathematical sense, was effective after the manner that ordinarily electrical images in electrostatics are effective'; see, S. B. McLaren, *Scientific Papers* (Cambridge: Cambridge University Press, 1925), p. 37.

[79]Cunningham, 'The Principle of Relativity...', *op. cit.*, note 64, p. 87.

given transformation was the solution.[80] Cunningham had, nevertheless, succeeded in finding, mathematically, a new and more general group of transformations that preserved the form of the electromagnetic equations, and this for him was equivalent to extending the field of application of the principle, or 'theorem', of relativity.

8. Re-Defining the Ether

In the discussion so far I have concentrated on the technical practice of Cunningham's physics and have not explicitly considered the status of the electromagnetic ether in his work. The reader will probably have noticed, however, that despite his belief that motion through the ether could not be detected (at any order), Cunningham was not inclined to follow Einstein in abandoning the ether concept as 'superfluous'.[81] In this section I shall describe Cunningham's notion of the ether and suggest how it related to the technical practice of his research. The important point to notice is that the ontological status of Cunningham's ether reflected the changing mathematical practice inherent in his electrodynamics. The ether concept *itself* is therefore of extremely limited utility in explaining Cunningham's reading of Einstein's work.

Cunningham's first published statement on the status of the ether occurred at about the time he was beginning his work on the new theorem of relativity. In 1907, several physicists had objected to the claim, made originally by Larmor, that magnetic effects might be attributable to a real physical flow of the (otherwise stagnant) ether.[82] The objection was made on the grounds that *relative* motion between the rapidly moving Earth and the ether did not appear to give rise to any measurable magnetic effects. In a letter to *Nature*, Cunningham pointed out that this objection was valid only if one assumed that *relative* motion between a moving observer and the ether was *equivalent* to a real flow of ether in the stationary-ether frame of reference. But, he continued, this was not the case because it was 'not permissible to speak of the velocity of

[80] *Ibid.*, p. 39. In Figure 2, for example, it is not immediately obvious how the inverted field should be interpreted physically. For an example of how Cunningham and Bateman's work was used in electromagnetic theory, see H. R. Hassé, 'The Equations of the Theory of Electrons Transformed Relative to a System in Accelerated Motion', *Proceedings of the London Mathematical Society* **12** (1913), 181–206.

[81] Einstein, 'On the Electrodynamics of Moving Bodies', *op. cit.*, note 4, p. 141.

[82] Larmor suggested in the early 1890s that a magnetic field probably represented some kind of real flow of the ether in the stationary ether frame, but this was not to be confused with motion *relative* to the ether which would not generate such an effect. For Lodge's attempt to measure this effect see B. J. Hunt, 'Experimenting on the Ether: Oliver J. Lodge, and the Great Whirling Machine', *Historical Studies in Physical Sciences* **16** (1986), 111–134.

an observer relative to the aether as though the aether were a material medium given in advance'.[83]

Cunningham went on to explain that the ether was *defined* by its electromagnetic properties, and that one of these was that a uniform drift through the ether had no effect upon the observed phenomena. In order to justify this statement he quoted at length from Larmor's *Aether and Matter* and insisted that:

> The aether is in fact, not a medium with an objective reality, but a mental image which is only unique under certain limitations . . . *Two frames of reference imply two aethers*; so long as we restrict ourselves to a single frame, the objection to Larmor's scheme does not arise.[84]

Cunningham clearly saw himself as championing Larmor's theory in arguing that the ether was only a useful concept when referred to any given observer's frame of reference. His notion of a plurality of ethers (one for each inertial frame of reference) was, however, quite new.[85] Following his discovery that Larmor's transformation's were exact (so that the electromagnetic effects of motion through the ether could never be detected) Cunningham must have concluded that the notion of a unique frame of reference (the stationary ether frame) had to be abandoned as meaningless. As he later remarked in discussing Larmor's contribution to electromagnetic theory:

> Every electromagnetic chain of phenomena is in a sense an electromagnetic clock or space-scale . . . . To grant the completeness of Larmor's scheme is therefore to admit the impossibility of determining physically a unique system of variables ($xyzt$) which may be called absolute space and time variables; in other words it is to admit the impossibility of establishing the uniqueness of the aether.[86]

Cunningham nevertheless believed, as did many wranglers, that electromagnetic effects had, ultimately, to be attributed to a dynamical medium if they were to remain comprehensible to the human mind.[87] He solved this problem by re-defining the ether in a way that was compatible with his

[83]E. Cunningham, 'The Structure of the Ether', *Nature* 76 (1907), 222.

[84]*Ibid* (my italics).

[85]Larmor himself immediately rejected Cunningham's claim by asserting that arguments based upon relativity must 'lapse' because absolute motion *could* be measured in relation to the distant 'quiescent' ether. See J. Larmor, 'A Plea for Absolute Motion', *Nature* 76 (1907), 269–270.

[86]E. Cunningham, *The Principle of Relativity* (Cambridge: Cambridge University Press, 1914), 44.

[87]On the role of the ether as a unifying concept in Cambridge physics, see 'The Electrodynamics of Moving Bodies...', *op. cit.*, note 3, pp. 192–196. Cunningham went on to try to develop an ether whose basic physical properties could be associated with Lorentz-invariant quantities (pp. 191–195).

mathematical description of electromagnetic processes. Each observer now associated an electromagnetic ether with his own inertial frame of reference and could, therefore, give a consistent dynamical account of electromagnetic processes *as observed in that frame*. The theorem of relativity (in the form of the Lorentz transformations) could then be employed to reveal how the electric and magnetic vectors of different ethers (observed in different frames of reference) were related. Cunningham's scheme thus retained the ontological purpose of the ether (to render electromagnetic processes dynamically comprehensible) while making it compatible with the new mathematical formalism.

The point I wish to emphasize is that we should not try to explain this conceptual shift in the ether's status as somehow *forced* on Cunningham by his contact with Einstein's work. The shift was not designed to protect the ether from being undermined by the theory of relativity. As the above quotation reveals, Cunningham's notion of a plurality of ethers owed as much to his reading of Larmor's *Aether and Matter* as to his reading of Einstein's 1905 paper. Likewise the shift did not originate in an attempt to bring about a 'reconciliation between the theory of relativity and the ether'.[88] Cunningham neither recognized any theory of relativity in Einstein's work nor believed that Einstein had called for the ether concept to be abandoned.[89]

The shift in status points, rather, to the fact that early twentieth-century British electromagnetic theory was not simply, or even primarily, a theory *of* the ether. Many of the mathematical techniques employed by wranglers originated in continuum mechanics, and the ether concept lent plausibility to the application of these techniques in electromagnetic theory. But, by the early years of the twentieth century, the ether no longer acted as a heuristic guide to the development of electromagnetics in the way that it had from the 1860s to the early 1890s.[90] As we saw in the previous section, the ether played no direct role in Cunningham's research on the conformal transformation of Maxwell's equations. It was, in fact, exactly the other way around; Cunningham's sense of the ontological reality of the ether derived from the theoretical technology that he had been taught in Cambridge. This theoretical technology was, moreover, quite different from that employed by Einstein, so that even if Cunningham had (for whatever reason) been prepared to abandon the notion of the ether, he would still not have been a practitioner of relativity theory.

[88]This remark is from Goldberg, 'In Defense of Ether...', *op. cit.*, note 3, p. 117.

[89]Einstein did not explicitly deny the existence of the ether in his 1905 paper. He stated merely that it was superfluous 'inasmuch as in accordance with the concept to be developed here, no "space at absolute rest" endowed with special properties will be introduced'; see Einstein, 'On the Electrodynamics of Moving Bodies', *op. cit.*, note 4, p. 141.

[90]On the role of the ether in British electromagnetic theory from Maxwell to Larmor, see B. J. Hunt, *The Maxwellians* (Ithaca: Cornell University Press, 1991).

9. Conclusion

I have described Cunningham's work on the transformation of the electromagnetic equations in detail in order to illustrate how misleading it is to think of him as having contributed directly to the development of relativity theory, or else as having struggled to make ether theory compatible with the theory of relativity. As we have seen he did not see Einstein's work of 1905 as a self-contained or self-evident theory, but rather as a modest extension of work due mainly to Larmor and Lorentz.[91] It was Cunningham's interest in the ETM and the mathematics of transformation theory that piqued his interest in Einstein's early work on relativity.

Cunningham's work on the general transformation of the equations of electromagnetism also illustrates the important role played by the local theoretical technology in the work of mathematical physicists. In this case the concept of the ether played no immediate role in the location of the conformal transformations, but Cunningham's training in the Mathematical Tripos was nevertheless crucial. As we have seen, it was his commitment to the ETM that both informed his reading of Einstein's work and conferred immediate value upon the search for the most general transformations of the equations of electromagnetism. Furthermore, the actual techniques that he employed in producing his results were a direct application of the theoretical technology of Cambridge mathematics. Employing the technique of inversion in the search for new solutions to problems in electromagnetic theory was second nature to Cambridge wranglers, and Cunningham's novel reading of Einstein's work enabled him to apply this technique in four dimensions in order to establish the new theorem of relativity.

We have also seen how the ontological status of the ether was generated by the theoretical technology that Cunningham employed. The concept of the ether was sufficiently flexible to allow Cunningham to shift easily from a single to a multiple ether model whilst developing his work on the conformal transformations. This point is very important to grasp. It was not commitment to the ether concept that informed Cunningham's contribution to Cambridge electromagnetic theory. Indeed, it is precisely by inverting the usual hierarchy between the ontologizing ether and technical practice that we can make visible the practical skill-base routinely employed by Cambridge mathematicians and begin to understand how they applied these skills to develop the ETM.

Finally, my discussion of Cunningham's work illustrates one way in which I seek to problematize the received notion of a British *response* to relativity.

[91]Only later did these British mathematicians begin to appreciate that the principle of relativity was part of a more comprehensive *theory of relativity* which some (mainly German) physicists held to deny the existence of the ether. This is discussed in detail in Warwick, 'The Electrodynamics of Moving Bodies...', *op. cit.*, note 3, ch. 3, 4 and 5.

Cunningham certainly 'responded' to Einstein's 1905 paper in the sense that he, unlike the vast majority of his British colleagues, identified it as important. But it was not the essential presence of the theory of relativity in the paper that piqued his interest. To follow that line of argument leads to the absurd conclusion that Cunningham somehow divined the existence and importance of the theory of relativity even though he had not understood it and was committed to a quite different, and incompatible, research programme. I have argued that it is Cunningham's active identification of Einstein's text as relevant to his own research that is to be explained, and that the explanation must engage Cunningham's reading of the text on its own terms. That reading cannot be consistently reconstructed within a relatively-centred historiography that speaks of partial insights and misunderstandings, but only in terms of the wrangler culture to which Cunningham belonged and to which he sought to contribute.

Acknowledgements — I would like to thank Jeff Hughes, Bruce Hunt, Iwan Morus and Simon Schaffer for making a number of helpful comments on an earlier draft of this paper.

APPENDIX

Inversion in Four Dimensions and Maxwell's Equations

In this Appendix, I explain how Cunningham employed the technique of inversion in four dimensions to establish the new theorem of relativity. We saw in Section 7 that inversion on a two-dimensional surface was accessible to straightforward geometric representation (see Fig. 1).[92] If we now imagine the centre of the circle in Fig. 1 to lie on the intersection of Cartesian axes XY then the distance OA becomes $\sqrt{X^2+Y^2}$. If we likewise represent the inverted image of OA on a Cartesian coordinate system xy, then the distance OB becomes $\sqrt{x^2+y^2}$.

Now just as the relationship between OA and OB was represented by $OA.OB = K^2$, so the transformation from one Cartesian coordinate system to the other is represented by:

$$(\sqrt{X^2+Y^2}).(\sqrt{x^2+y^2}) = K^2 \tag{1}$$

where the points XY and xy must lie on a straight line passing through the origin. Cunningham sought to apply the same technique in four dimensions. In the four-dimensional space imagined by Cunningham, the distance from the

[92]The figures referred to in this Appendix are those of the main text.

origin to any point (X,Y,Z,cT) is $\sqrt{X^2+Y^2+Z^2-c^2T^2}$. Using polar coordinates (R,Φ,Θ,cT) this distance becomes $\sqrt{R^2-c^2T^2}$. We can now define the inversion of the point (R,Φ,Θ,cT) in a way that is precisely analogous to the two-dimensional example given above. If we imagine a second coordinate system (r,ϕ,θ,ct), which is the inverted image of (R,Φ,Θ,cT), then transformation from one coordinate system to the other is represented by:

$$(\sqrt{R^2-c^2T^2}).(\sqrt{r^2-c^2t^2}) = K^2 \tag{2}$$

The constant k is now called the *pseudo-radius* of inversion because it defines a *hyper-sphere* in the four-dimensional space.[93] If we also impose the condition that the points (r,ϕ,θ,cT) and (R,Φ,Θ,cT) must both lie on a straight line that passes through the origin, then equation (2) becomes equivalent to the transformations:[94]

$$R = \frac{K^2 r}{(r^2-c^2t^2)} \qquad T = \frac{K^2 t}{(r^2-c^2t^2)} \tag{3}$$

It was these transformations (3) that Cunningham took as the starting point for his work.

Cunningham first investigated the way in which velocities would transform between the two systems. He found that if a point travels with velocity (u_r,u_n) in (r,ϕ,θ,cT) then its velocity (U_R,U_N) in (R,Φ,Θ,cT) will be:[95]

$$U_R = \frac{(r^2+c^2t^2)u_r - 2c^2rt}{2rtu_r-(r^2+c^2t^2)} \tag{4}$$

$$U_N = -\frac{(r^2-c^2t^2)u_n}{2rtu_r-(r^2+c^2t^2)}. \tag{5}$$

A particle that is at rest in (R,Φ,Θ,cT) will appear to move in (r,ϕ,θ,cT) with a radial velocity v:

$$v = \frac{2c^2rt}{(r^2+c^2t^2)} \tag{6}$$

as may easily be seen by setting $U_R = U_N = 0$ in equations (4) and (5).[96] Cunningham next introduced the conventional notation:

[93]The terms *pseudo-radius* and *hyper-sphere* are used because the radius and sphere respectively cannot be represented geometrically in three-dimensional Euclidean space.

[94]In polar coordinates this condition can be imposed on the spatial components R and r simply by requiring that the polar angles Φ,Θ and ϕ,θ remain unaffected by the transformation (that is, $\Theta = \theta$ and $\Phi = \phi$ for all R and r). The condition can also be imposed on the temporal coordinates cT and ct by requiring that $t/T = r/R$.

[95]There is a typographical error (here corrected) in the lower line of expression (4) in Cunningham's paper. The velocity subscripts R (r) and N (n) refer respectively to the velocities parallel and normal to the radius vector.

[96]It follows from equation (4) that the velocity normal to the radius vector cannot be finite in one frame and zero in the other.

$$\beta = \left(1-\frac{v^2}{c^2}\right)^{-\frac{1}{2}} = \frac{r^2+c^2t^2}{r^2-c^2t^2} \tag{7}$$

so that equations (4) and (5) assumed the more familiar looking form:

$$U_R = \frac{u_r - v}{1-u_r v/c^2} \tag{8}$$

$$\beta U_N = \frac{u_n}{1-u_r v/c^2} \tag{9}$$

Having established these kinematical relationships between the coordinate points and velocities in the two systems, Cunningham showed that Maxwell's electromagnetic field equations transformed exactly under transformations (3) provided that the components of the electric (E) and magnetic (H) fields were related in the following way:

$$E_R = \lambda^2 E_r \qquad H_R = -\lambda^2 \beta H_r$$

$$E_\Theta = \lambda^2\beta\left(-E_\theta + \frac{vH_\phi}{c}\right) \qquad H_\Theta = \lambda^2\beta\left(H_\theta + \frac{vE_\phi}{c}\right)$$

$$E_\Phi = \lambda^2\beta\left(E_\phi - \frac{vH_\theta}{c}\right) \qquad H_\Phi = \lambda^2\beta\left(H_\phi - \frac{vE_\theta}{c}\right)$$

where λ has been written for $(r^2-c^2t^2)/K^2$. The electric charge density P will also transform exactly under inversion provided that:

$$P = -\lambda^3\beta(1-u_r v/c^2)\rho$$

where *ur* is the velocity of convection and ρ the charge density at a point in the frame (r,ϕ,θ,cT).

Since the conformal group of transformations could be built up from repeated inversions (see Section 8 of the main text), Cunningham claimed that he had now established the 'theorem of relativity' for that group of transformations.[97] Notice that Cunningham's procedure in establishing the new 'theorem' is quite different from Einstein's. Einstein *derived* the Lorentz transformations from two physical principles (the *principle of relativity* and the *principle of the constancy of the light velocity*).[98] Cunningham, by contrast, established what he called the new 'theorem' of relativity by showing that Maxwell's equations could be made to transform exactly under the conformal transformations. In this sense Cunningham was following Larmor in establishing the new theorem of relativity. For Cunningham, as for Larmor, it was the ETM that guaranteed the *physical* plausibility of the new transformations.

[97]E. Cunningham, 'The Principle of Relativity in Electrodynamics and an Extension Thereof', *Proceedings of the London Mathematical Scoiety* **8** (1910), 77–98, 89.

[98]Einstein 'On the Electrodynamics of Moving Bodies', *op. cit.*, note 4, pp. 140–171, pp. 145–151.

Acknowledgments

Einstein, Albert. "On the Electrodynamics of Moving Bodies." In *Albert Einstein's Special Theory of Relativity: Emergence (1905) and Early Interpretation (1905–1911)*, translated by Arthur I. Miller (Reading: Addison-Wesley Publishing Company, 1981): 392–415. Reprinted with the permission of Addison-Wesley Pubhshing Company.

Holton, Gerald. "Mach, Einstein, and the Search for Reality." In *Thematic Origins of Scientific Thought: Kepler to Einstein,* rev. ed., edited by Gerald Holton (Cambridge: Harvard University Press, 1988): 237–277. Reprinted by permission of *Daedalus*, Journal of the American Academy of Arts and Sciences.

Holton, Gerald. "Einstein and the Cultural Roots of Modern Science." *Daedalus* 127 (Winter 1998): 1–44. Reprinted with the permission of Russell Sage Foundation.

Darniol, Olivier. "Henri Poincaré's Criticism of Fin De Siècle Electrodynamics." *Studies in History and Philosophy of Modern Physics* 26 (1995): 1–44. Reprinted with the permission of Elsevier Science Ltd.

Janssen, Michel. "Reconsidering a Scientific Revolution: The Case of Einstein Versus Lorentz."*Physics in Perspective* 4 (2002): 421–446. Reprinted with the permission of Birkhäuser Verlag.

Miller, Arthur I. "The Special Relativity Theory: Einstein's Response to the Physics of 1905." In *Albert Einstein: Historical and Cultural Perspectives,* edited by Gerald Holton and Yehudah Elkana (Princeton: Princeton University Press, 1982): 3–26. Copyright © 1982 by Princeton University Press. Reprinted by permission of Princeton University Press.

Galison, Peter. "Einstein's Clocks: The Place of Time." *Critical Inquiry* 26 (Winter 2000): 355–389. Reprinted with the permission of the University of Chicago Press, publisher.

Cassidy, David. "Understanding the History of Special Relativity." *Historical Studies in the Physical Sciences* 16 (1986): 177–195. © 1986 by The Regents of the University of California.

Pyenson, Lewis. "The Relativity Revolution in Germany." In *The Comparative Reception of Relativity,* edited by Thomas Glick (Dordrecht: Reidel, 1987): 59–111. Reprinted with the permission of

Reidel Publishers.

Glick, Thomas. "Cultural Issues in the Reception of Relativity." In *The Comparative Reception of Relativity,* edited by Thomas Glick (Dordrecht: Reidel, 1987): 381–400. Reprinted with the permission of Reidel Publishers.

Goldberg, Stanley. "In Defense of the Ether: The British Response to Einstein's Special Theory of Relativity, 1905–1911" *Historical Studies in the Physical Sciences* 2 (1970): 89–125. © 1970 by The Regents of the University of *California.*

Warwick, Andrew. "Cambridge Mathematics and Cavendish Physics: Cunningham, Campbell and Einstein's Relativity, 1905–1911. Part 1: The Uses of Theory." *Studies in the History and Philosophy of Science* 23 (1992): 625–656. Reprinted with the permission of Pergamon Press Ltd.